Urban Energy Systems

Cities are home to over half of the world's population, with millions attracted by the economic and social opportunities of urban life. But urban lifestyles demand reliable and sustainable energy services – to heat and cool our buildings, to power our homes and offices, and to transport ourselves and our goods. *Urban Energy Systems* looks at the technical and social systems that satisfy these service demands and asks how they might be provided more sustainably.

Adopting a holistic approach, this book describes urban energy systems as complex interconnected systems. It shows how cities have evolved from consuming locally available fuels to centres of global energy networks with significant environmental impacts. New building and transportation technologies offer significant potential for improvement but only if technologies, demands and institutions are analysed in an integrated manner. Drawing on analytical tools and case studies developed at Imperial College London, the book presents state-of-the-art techniques for examining urban energy systems as integrated systems of technologies, resources, and people.

Urban Energy Systems offers a comprehensive introduction to the field. It will be valuable for students and researchers looking to understand how these important infrastructures will contribute to urban sustainability, as well as planners, engineers, policymakers and other experts who wish to learn more about innovative analytical techniques.

James Keirstead is a researcher at Imperial College London working on the BP Urban Energy Systems project.

Professor Nilay Shah is Director of the Centre for Process Systems Engineering and co-director of BP Urban Energy Systems project, Imperial College London.

Professor David Fisk is Director of the Laing O'Rourke Centre for Systems Engineering and Innovation and co-director of BP Urban Energy Systems project, Imperial College London.

Urban Energy Systems is a must read book for all professionals working with or interested in energy and energy-related environmental issues in urban areas. The editors and authors have skillfully presented the complexities of urban energy system holistically, as well as teased out the key elements such as technologies, urban sectors, modeling and the transition pathways.

– *Shobhakar Dhakal, Ph.D., Associate Professor, Energy Field of Study, Asian Institute of Technology*

This is an important book by researchers at the forefront of a rapidly growing discipline. Using an integrated approach, the authors skilfully knit together cutting edge topics from ecologically based design to activity based models, helping us to understand the energy systems that will underlie cities of the future. It will help to inspire engineers, planners and policy makers in the transition to greener, healthier cities.

– *Christopher Kennedy, Professor of Civil Engineering, University of Toronto; and author of The Evolution of Great World Cities*

Urban Energy Systems

An integrated approach

Edited by James Keirstead and Nilay Shah

Routledge
Taylor & Francis Group

LONDON AND NEW YORK

First published 2013
by Routledge
2 Park Square, Milton Park, Abingdon, Oxon OX14 4RN

Simultaneously published in the USA and Canada
by Routledge
711 Third Avenue, New York, NY 10017

Routledge is an imprint of the Taylor & Francis Group, an informa business

© 2013

British Library Cataloguing in Publication Data
A catalogue record for this book is available from the British Library

Library of Congress Cataloging in Publication Data
Urban energy systems : an integrated approach / edited by James Keirstead and Nilay Shah.
   p. cm.
Includes bibliographical references and index.
ISBN 978-0-415-52901-3 (hbk.: alk. paper) – ISBN 978-0-415-52902-0 (pbk.: alk. paper) – ISBN 978-0-203-06678-2 (ebook) 1. Cities and towns–Energy consumption. 2. Power resources. 3. Energy development–Environmenal aspects. 4. Sustainable urban development.
I. Keirstead, James, 1979– II. Shah, Nilay.
HD9502.A2U735 2013
333.7909173'2–dc23   2012032650

ISBN: 978-0-415-52901-3 (hbk)
ISBN: 978-0-415-52902-0 (pbk)
ISBN: 978-0-203-06678-2 (ebk)

Typeset in Sabon
by Cenveo Publisher Services

Printed and bound by CPI Group (UK) Ltd, Croydon, CR0 4YY

Contents

Figures

Tables

Contributors

Dr James Keirstead is a Lecturer in Department of Civil and Environmental Engineering, Imperial College London. His research focuses on the integrated modelling of urban energy systems and the links between urban form, consumer behaviour, public policy, technical systems and resource consumption. His work has been published in leading international journals and major reports, including the 2012 *Global Energy Assessment* and the 2011 *First Assessment Report* of the Urban Climate Change Research Network. He is a Chartered Engineer, Member of the Energy Institute and a board member of the International Society for Industrial Ecology's Sustainable Urban Systems section. He received a DPhil from the University of Oxford in 2006, where he was a Commonwealth Scholar.

Professor Nilay Shah is a Professor of Process Systems Engineering at Imperial College London, where he is also the Director of the Centre for Process Systems Engineering. His research interests include the application of multi-scale process modelling and mathematical/systems engineering techniques to analyse and optimize complex, spatially- and temporally-explicit low-carbon energy systems, including bioenergy/biorenewable systems, hydrogen infrastructures, carbon capture and storage systems and urban energy systems. He is also interested in devising process systems engineering methods for the design and operation of complex systems such as large-scale supply chains and bio-based processes (especially combining the life sciences with physical science and engineering), and in the application of model-based methods for plant safety assessment and risk analysis. He has published widely in these areas and is particularly interested in the transfer of technology from academia to industry. He has MEng and PhD degrees in Chemical Engineering and is a Fellow of the Institution of Chemical Engineers.

Dr Salvador Acha received a BSc(Eng) degree in Electronics and Communications Engineering from Monterrey Tech, in Mexico, in 2003. After working in the private sector, he joined Imperial College London in 2006 where he obtained a PhD degree in Electrical Engineering in 2010 for his thesis on optimally integrating distributed energy resources into infrastructures. Currently, he is a Research Associate at Imperial with interests including energy efficiency and management programmes, roll out of plug-in hybrid electric vehicles and the optimization of distributed energy sources.

Professor David Fisk is Director of the Laing O'Rourke Centre for Systems Engineering and Innovation at Imperial College London, where he previously held the Royal Academy of Engineering Chair in Engineering for Sustainable Development. He was Chief Scientific Adviser to the Office of the Deputy Prime Minister from 2002. From Head of the Mechanical and Electrical Engineering Division Building Research Establishment in 1978, he went on to take on a number of policy roles at the Department of the Environment, including the negotiation of the UN climate change treaties. He became Department of the Environment Chief Scientist in 1988. He has been a Member of the Gas & Electricity Markets Authority since 2009, and became President of the Chartered Institution of Building Services Engineers in 2012.

Mark Gerard Jennings has a BE(Hons) from the National University of Ireland, Galway, an MSc from Stanford University and is currently a PhD candidate in Imperial College London. His research interests include developing strategies towards the reduction of existing fuel and power demand in the building sector.

Dr Pierluigi Mancarella received MSc and PhD degrees in Power Systems Engineering from the Politecnico di Torino, Italy, in 2002 and 2006, respectively. After working as a Research Associate at Imperial College London, he is currently a Lecturer at the University of Manchester, where he teaches 'Power Systems Economics' and 'Smart Grids and Sustainable Electricity Systems'. His research interests and project activities include modelling and analysis of multi-energy systems, integrated energy infrastructure development and smart cities, energy systems environomics, business models for new energy technologies, and generation and network investment under uncertainty. He is the author of two books, five book chapters and more than 60 peer-reviewed and highly-cited papers on technical, environmental and economic analysis of distributed multi-energy systems.

Antonio M. Pantaleo is an electrical engineer. He joined the University of Bari in 2001 as a research associate in the field of energy efficiency and renewable energy sources for agricultural and agro-industrial applications. He also joined Edison Energie Speciali in 2000, working on business development of biomass power plants and wind energy, and the Italian Transmission System Operator in 2003, working on the integration of intermittent generation in the power system. In 2006, he became a permanent researcher at the DISAAT Department, University of Bari, in the field of renewable energy and sustainable use of energy in agro-industrial sector. He has been aggregate professor since 2009. He joined the Centre for Environmental Policy, Imperial College London in 2006 as visiting research associate and he is now working on a PhD with the Centre for Process Systems Engineering, Imperial College London. His research interests include the optimization of bioenergy conversion routes for stationary applications, biomass potential assessment and related sustainability issues, the integration of renewable and embedded generation into existing energy systems and energy efficiency in the agricultural and agro-industrial sectors.

Nicole C. Papaioannou received an MEng degree in Chemical Engineering from Imperial College London, in 2006. Her interest in the energy and environment sector motivated her to study for an MSc degree in Environmental Technology and Energy

Policy at the same university, and she successfully completed it, in 2007. After a year in the private oil and gas sector, she returned to Imperial to pursue a PhD degree in Chemical Engineering. She is currently in her final year and the title of her thesis is: 'Environmental Impact Assessment and Optimization of Urban Energy Systems'. She has had the opportunity to present her findings locally and at international conferences.

Dr Paul Rutter is a Visiting Professor at Imperial College London. He has spent over 30 years as a research scientist and manager in the medical, chemical and petroleum industries. He has published technical papers on a wide variety of subjects and has served as an advisor for several academic and international organizations.

Dr Nouri Samsatli holds MEng and PhD degrees in Chemical Engineering from Imperial College. His research has included biochemical processes, fine chemicals as part of the Britest Project, supply-chain management and most recently, urban energy systems. He also enjoys teaching at Imperial College and holds a part-time teaching fellow position at University College London. He is currently working for Process Systems Enterprise Ltd. on a Systems Modelling Toolkit for Carbon Capture and Sequestration.

Dr Aruna Sivakumar is a Lecturer in travel behaviour and demand modelling at the Centre for Transport Studies, Imperial College London. She also leads the transport work streams in multi-disciplinary research projects such as the BP-sponsored Urban Energy Systems project and the EPSRC-sponsored Digital City Exchange project. Aruna is a young member of the Transportation Research Board (TRB) Committee on Travel Behaviour and Values, an elected member of the executive board of the International Association for Travel Behaviour Research (IATBR) and a member of the Methodological Innovations committee for the European Transport Conference. Aruna holds a PhD and MSc in Transportation Engineering from the University of Texas at Austin, and her research interests include econometrics, travel behaviour and the role of ICTs, integrated urban system models and transport policy.

Preface

Reliable supplies of energy are the lifeblood of our cities. Whether we live in a small town of just a few hundred or a megalopolis of more than 10 million, we depend on energy to deliver vital services every day. Energy heats and cools our buildings, illuminates the dark, powers our offices and factories and fuels our vehicles. Even in cities without access to modern commercial fuels like natural gas and electricity, supplies of animal waste and biomass are essential for meeting routine needs like heating and cooking. But these energy services must not be taken for granted, and urban energy systems face many challenges in the coming decades.

Urban Energy Systems: An Integrated Approach is an attempt to capture the changing nature of these complex systems. It presents a set of case studies and cutting-edge analytical tools to inform and stimulate our understanding of these often over-looked infrastructures. We have subtitled it *An Integrated Approach* to reflect our belief that the efficiency of urban energy systems can be greatly improved by better integration of individual technologies and patterns of demands, and by careful consideration of the unique social and economic context of each city.

This book has arisen out of the BP Urban Energy Systems project at Imperial College London. This project was established in late 2005 and also helped to inaugurate Imperial's Energy Futures Lab, a focal point for energy research across the College. The project's main aim was to employ a multidisciplinary, systems-based approach to identify the potential benefits of an integrated analysis of the design and operation of urban energy systems, with a view to identifying large reductions in the energy intensity of and greenhouse gas emissions caused by cities. It was conceived out of the growing recognition that cities and their associated energy systems and resource-consuming infrastructure will be critical in the drive to establish more sustainable ways of living.

The project took as its inspiration the great success of modelling, optimization, systems engineering and process integration in complex engineered systems such as petroleum refineries. It was concerned with establishing a modelling framework for urban energy systems and identifying in broad terms, the potential of such a framework to establish novel ways of designing and operating urban energy systems, estimating the associated benefits and helping to identify the means and business models required to achieve such benefits.

The project was predicated on the assumption that, while individual urban energy systems may have been optimized, there has been no attempt to cross-optimize different resource systems and over cities' total energy consumption. Such system-wide optimization has produced efficiency gains of several tens of per cent in other systems such as refineries and petrochemical complexes. Another assumption was that modelling frameworks and computational tools have evolved very quickly, as has computing power. Nevertheless, the research challenges associated with cities are arguably an order of magnitude more complex, involving diverse energy vectors and being confounded by the daily decisions of millions of agents.

Somewhat over 6 years on, the project is drawing to a close, and this book represents a major milestone in its progress. Much has been achieved in terms of understanding urban energy systems, developing a modelling framework and effective models, and using these tools to generate unique insights into the design and operation of urban energy systems. However, new challenges were unearthed along the way and the field remains a fascinating one for researchers from almost every discipline.

We hope that the book will appeal to a wide audience. For practising engineers, the modelling techniques presented should offer new insights and the case studies will illustrate how innovative solutions might be achieved in practice. For policy-makers and non-engineers, the book aims to provide an introduction to the major urban energy technologies so that you may appreciate the key features of these technologies and how they relate to other systems. And for students, we hope that the book will be a one-stop-shop for urban energy systems providing a good overview of the major issues in this challenging interdisciplinary field.

James Keirstead and Nilay Shah
London
August 2012

Acknowledgements

The editors and contributors would like to acknowledge the support of many individuals and organizations, without whom this book would not have been possible. First and foremost, the underlying research has been funded with the generous support of BP, and encouraged by BP's Chief Scientist at the time, Dr Steve Koonin. Advice from BP staff, and its Imperial College liaison Graham Elkes in particular, has been valuable throughout the project. They have encouraged us to explore the world of urban energy systems fully and we believe that the resulting research is stronger for that freedom. We have also learned from collaborations and conversations with a number of industrial and public sector partners including Arup, Foster and Partners, the World Bank and others. Academic colleagues, both in the UK and abroad, provided encouragement for our work throughout the project and it is rewarding to be part of an emerging global community studying urban energy systems. We are also indebted to the research project's advisory board members: Arnulf Grübler (International Institute for Applied Systems Analysis, Austria), Peter Head (Arup, UK) and the late Bill Mitchell (Massachusetts Institute of Technology, USA).

Specific sections of the book arose from collaborative projects and case studies. The input, guidance and advice of Newcastle City Council climate change and energy master planning teams was vital for the work presented in section 12.4 and was conducted in collaboration with Carlos Calderon at Newcastle University. The advice on technological options and obstacles for commercial building retrofits (Chapter 4) by Tyler Haak of The Trane Company, and Deirdre McShane and her colleagues at Taylor Engineering is greatly appreciated. The Nakuru section (13.4) was based largely on the MSc thesis of Chancel (2010). Many thanks to these individuals and groups for sharing their data, expertise and experience with us.

Thanks to Kat Holloway and Michael Fell at Earthscan for their enthusiasm for this project and their patience as the manuscript was prepared. This book contains copyrighted images used with permission from the London Transport Museum collection, Eureka Entertainment Ltd., skyTran LLC, and the Science Museum/Science & Society Picture Library. Open-source clip-art from http://openclipart.org includes: 'block house' by rg1024, 'simple vector tree' by bobocal, 'reference desk' by SteveLambert, and 'Shopping Basket' by gnokii, as well as the Crystal 'user-home' icon from http://openiconlibrary.sourceforge.net

Finally, we would like to thank our families and colleagues for their ongoing support.

Abbreviations

With some exceptions, standard SI units are used throughout, e.g. W = watts for power, J = joules for energy with appropriate order of magnitude prefixes. The following abbreviations are also used.

ABM	activity-based model or agent-based model
AD	anaerobic digestion
AMMUA	agent-based micro-simulation model of urban activities
ASHP	air-source heat pump
BIPV	building integrated photovoltaics
BMW	biodegradable municipal waste
CA	cellular automata
CCHP	combined cooling, heat and power
CHP	combined heat and power
CoP	coefficient of performance
DER	distributed energy resources
DG	distributed generation
DH	district heating
DMG	distributed multi-generation
EF	ecological footprint
EHP	electric heat pump
EIA	environmental impact assessment
ESCo	energy service company
EU	European Union
EV	electric vehicle
GA	genetic algorithm
GDP	gross domestic product
GIS	geographic information system
GHG	greenhouse gas
GPS	global positioning system
GSHP	ground-source heat pump
GT	gas turbine
HVAC	heating, ventilation and air conditioning

ICE	internal combustion engine
ICT	information and communication technologies
IEA	International Energy Agency
ITS	intelligent transport systems
LCA	life-cycle assessment
LDR	less developed regions
LHV	lower heating value
LP	linear programming
LPG	liquefied petroleum gas
LUT	land-use transport (model)
MDR	more developed regions
MFA	material flow analysis
MILP	mixed-integer linear programming
MINLP	mixed-integer non-linear programming
MSW	municipal solid waste
ORC	organic Rankine cycle
PHEV	plug-in hybrid electric vehicle
PV	photovoltaics
ROC	renewable obligation certificate
RTN	resource technology network
SP	separate production
STN	state task network
TCO_2ER	trigeneration CO_2 emission reduction
TURN	technology and urban resource network
UES	urban energy systems
UK	United Kingdom
US/USA	United States of America

Part I
Introduction

1 The growing importance of urban energy systems

Nilay Shah and James Keirstead

Imperial College London

Urban energy systems are a somewhat neglected and unrecognized part of our cities. Yet without reliable and abundant supplies of electricity, heating, transport fuel, and other services, all functioning smoothly behind the scenes, the more visible elements of a city would not be possible. The bright lights of central shopping districts would grow dim, the flow of urban traffic would grind to a halt, and our homes would not be as comfortable. This book is an exploration of these systems and the links they have with our lives, economies and the environment. In this first chapter, we introduce the subject by defining urban energy systems and highlighting major global trends in urbanization and resource consumption, as these basic statistics and working hypotheses have been the motivation for the research presented here.

1.1 Motivation

Almost all the recent news stories about energy and climate change focus on the national and international aspects of energy systems. For example, at the time of writing, the manifesto of incoming French President François Hollande calls for a reduction of France's nuclear power capacity to 50 per cent of consumption by 2025. Behind the scenes, this headline policy will have been informed largely by an analysis of the energy system at a national scale. Energy analysts will draw their conclusions based on computer models[1] that describe energy demand and supply using national annual aggregate figures. Results may be broken down by end-use sectors and there may be limited spatial and temporal disaggregation, for example, in terms of national or international regions. However, cities and urban energy systems rarely appear.

There are four reasons for revisiting this situation:

Urbanization trends Over 50 per cent of the world's population now lives in urban areas, and much of the future growth of the world's population will take place in cities.

Spatial concentration The spatial concentration of activities within urban areas provides a major driver for the growth of urban populations, as people move to cities for economic opportunities. The related concentration of energy service demands also offers untapped potential for system integration and optimization.

Climate change and other risks Pressures associated with climate change, energy security and resource scarcity are increasing the need for change in energy systems. Urban governments are increasingly recognizing these threats, and fortunately cities also offer access to the capital and markets needed to support systemic changes.

Local institutions Cities are organizational structures, governed by institutions with relatively short decision-making cycles (in contrast with national governments). They are primary innovation locations with ready access to capital, skills, technologies and markets.

1.1.1 Urbanization

UN projections indicate that in the twenty-first century, around 90 per cent of population growth will be in urban areas; these are expected to account for 60 per cent of the population, 80 per cent of the wealth, and according to IEA estimates, 73 per cent of direct energy use by 2030 (UN 2011, IEA 2008b). Hence, the pattern of future energy demand will be increasingly characterized by the networks of the city, and solutions to the grave challenges faced by cities in a world of over 9 billion all have implications for urban energy systems.

These projections and trends are evident in Figure 1.1. There are four key points of interest. First, the more developed regions (MDR) have a broadly stable population and a society which is already highly urbanized and gradually becoming more so. Second, the less developed regions (LDR) are expected to experience population growth where almost all of this growth is driven by growth in the urban population (this trend holds for the LDR regions excluding China as well). Third, the growth in total population comes from the LDR countries which will need to be supported by large investments in new energy systems infrastructure. Finally, these trends show a marked increase in global urbanization and a marked tailing off of the absolute global rural population after 2015. By 2050, the share of urban population is expected to be around two-thirds.

Figure 1.2 illustrates that established cities, like London and Chicago, are in relative decline when compared with the surging growth of cities in the less developed regions, particularly in Asia (e.g. Delhi, Mumbai, Dhaka, Karachi). Only a handful of established internationally-connected metropolises, such as New York–Newark, Shanghai, and Tokyo, will continue to grow and maintain their relative rank. Other researchers have noted the volatility of these rank series (Batty 2006), and this will have significant implications for urban energy systems. In Chapter 3, we will take a closer look at how urban growth affected the energy systems of London.

Although a great deal is made of these so-called mega-cities, today only 10.7 per cent of the global urban population lives in cities of over 10 million inhabitants. A large proportion of the urban population (31.4 per cent) live in mid-sized cities of sizes between 500,000 and 5,000,000 inhabitants, and 50.9 per cent live in small cities of fewer than 500,000 residents (UN 2011). For urban energy systems, this means that scale-appropriate solutions must be sought.

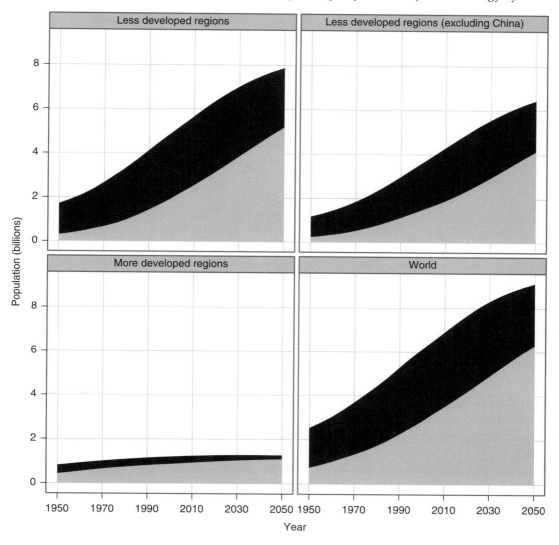

Figure 1.1 Global and regional urbanization trends. Light grey represents the urban population, black the rural population. MDR = more developed regions, LDR = less developed regions. *Data source*: UN (2011).

1.1.2 *Spatial concentration*

The urban economics literature provides many clues as to why this urban growth is happening. While traditional issues like physical security and the cultural attractions of a big city area are important (Kotkin 2005), Fujita *et al.* (1999) and others quantify the rationale behind continual urban aggregation using increasing returns to scale which come about through self-reinforcing geographical concentrations. The reason for the ongoing process of urbanization is explained through the formation

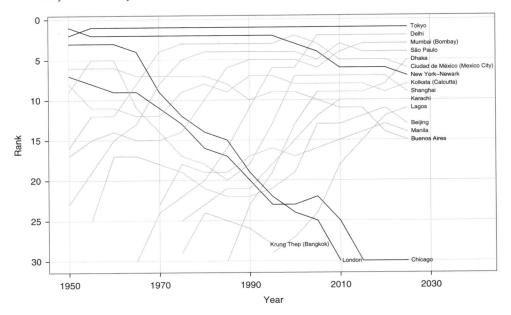

Figure 1.2 Relative ranking of different cities by population; past and projection. Cities from 'more developed regions' are in dark grey; 'less developed region' cities in light grey.
Data source: UN (2011).

and growth of agglomeration economies, 'in which spatial concentration itself creates the favorable economic environment that supports further or continued concentration' (p. 4), including those of labour markets, skills and knowledge, capital and of course demand.

A recent publication from the World Bank (2008) concurs and observes that half the world's production fits into only 1.5 per cent of its land, in large part due to the enormous concentration of economic activity in the major global cities and the increasing trend towards further concentration. This is attributed to agglomeration, migration, specialization and trade. Density itself is recognized as having a self-reinforcing characteristic by encouraging development and growth through these effects, which increases migration to urban centres. There is strong empirical evidence that spatial concentration rises as average incomes increase, as evidenced by the high level of urbanization in the more developed regions of Figure 1.1.

Hall's (2002) history of urban planning describes how different types of cities have evolved both through organic processes of growth and development and proactive planning (see also Morris 1994). He documents the phenomenal growth and sheer complexity of the projects associated with the development of the Pudong district in Shanghai and the Shenzhen region. Here a conscious decision to urbanize resulted in the creation *ab initio* of concentration and agglomeration effects, followed by the rapid migration of capital and labour – shoehorning decades' worth of organic growth and system feedbacks into a single decade. This process is not complete; the next phase is the complex of networked cities in the Pearl River delta which is expected to evolve into a 'polycentric megacity'. Once merged, the megalopolis is expected to have a population of 42 million (Moore and Foster 2011).

These observations have led to cities being conceptualized as 'complex systems'. In early work, Forrester (1969) explored the concept of cities as complex dynamic systems with feedback loops and time correlations, where the structure and policies of the system are important governing factors and the ones which can be used to ensure stability and effectiveness. More recently, Batty (2007) applies contemporary complexity theory to the realm of the city and uses cellular automata and agent-based models to explain phenomena such as self-organization and emergent properties. This systems-based view rather than silo-based functional view of cities and their inhabitants and infrastructure is a rapidly growing academic discipline; we shall utilize it in the context of urban energy systems (Chapter 2).

Of course the requirements of cities vary depending on their level of development and the World Bank report makes some recommendations to ensure an equitable geographical economy:

- In countries with a primarily rural population, proactive establishments of institutional foundations for urbanization combined with good land policies and the ability to provide basic services to everyone are important. Costa Rica is used as an example of good practice.
- In rapidly urbanizing countries, governments must establish healthy institutions, and 'connective infrastructure' to ensure that benefits of rising economic density are shared. Chongqing is used as an example of good practice.
- In countries where urbanization has advanced beyond the recommendations for rapidly urbanizing countries, effective interventions, land and basic service policies and good transport infrastructure are needed to avoid the growth of peri-urban settlements. Bogotá is used as an example of good practice.

Note that for all of these cases, urban energy systems play a central role as part of the basic services and infrastructure necessary for a successful city. It should be clear that analysing and modifying the global and national systems for energy collection, conversion and consumption cannot be achieved without an understanding of the city and its activities.

1.1.3 Climate change and other risks

One of the major motivations for research on urban energy systems is climate change. The IEA estimates that cities account for more than 70 per cent of global CO_2 emissions (IEA 2008b) and therefore by improving the efficiency of urban energy use and using low-carbon energy sources, cities can play an active role in climate change mitigation.

Cities will also feel the effects of climate change, both in terms of infrequent, high-impact events such as floods which are exacerbated by limited natural drainage capacity and gradual sea-level rises (note that several of the cities in Figure 1.2 are low-lying coastal cities). Lloyd's 360 Risk Project Report (Lloyd's 2006) states that a 4 m sea-level rise would inundate almost every coastal city in the world. They categorize urban climate change-related risks into: floods, subsidence, heatwaves and water quality and shortage problems. Climate change is also likely to have an impact on urban health and major infrastructures including water, wastewater, transportation networks and energy systems (Rosenzweig *et al.* 2011).

A particular sub-theme of academic and practical interest is the relationship between urbanization and climate both at the microscale (urban heat island effect) and macroscale (changes in global albedo). According to the US glossary of meteorology, the urban heat island effect (which arises out the replacement of vegetation with built environment) will on average result in a temperature elevation of 1–2°C in a city of a million inhabitants, rising to up to 12°C on calm clear nights. On regional and even a global scale, changes in land use, including urbanization, have been shown to affect the global albedo (positively or negatively, depending on the materials used) and therefore, the surface energy balance. For example, Zhang *et al.* (2011), attribute some proportion of observed regional climate variations in China to urbanization, both from albedo effects and pollution (aerosols and particulates).

1.1.4 Local institutions

Many regions have stringent climate change mitigation targets. For example, the UK's Climate Change Bill requires emissions reductions of 80 per cent, with respect to 1990, by 2050. It has become evident that cities will be at the forefront of the battles against climate change. Almost every major global city has published some form of climate action plan. For example:

- Paris's climate protection plan targets greenhouse gas (GHG) emission reductions of 75 per cent by reducing local GHG emissions by 25 per cent, using 25 per cent less energy and targeting a 25 per cent contribution of renewable energy (Mairie de Paris 2012).
- London has separate climate change mitigation and adaptation strategies (Mayor of London 2011, 2012). The mitigation strategy sets the target of a 60 per cent reduction in CO_2 emissions by 2050 through retrofitting programmes for London's homes and public buildings, as well as aiming to provide 25 per cent of London's energy from local low carbon sources.
- New York City's PlaNYC has a two-part climate change plan: to reduce GHG emissions by more than 30 per cent between 2005 and 2030, and to increase the city's resilience to climate change risks (The City of New York 2011).
- Hong Kong also has an explicit Climate Change Action Plan, which targets a reduction in energy intensity of 25 per cent between 2005 and 2030 (EPD Hong Kong 2010).
- Mumbai is expected to have a climate change action plan in 2012.

In countries like the United States, where political action on energy and climate issues at a national scale can be difficult, such local initiatives are increasingly important (Lutsey and Sperling 2008). This arrangement provides flexibility, allowing cities to address those challenges of greatest concern to local residents and making use of local resources at their disposal. As noted in Bulkeley and Betsill (2003), these local policies often offer important 'co-benefits' with other urban priorities as well.

However, cities typically have limited scope for action, particularly in energy systems where regulatory and financial powers lie with other levels of government (Keirstead and Schulz 2010). For example, consulting engineers Arup, recently reported that only 53 per cent of surveyed municipalities directly own or operate energy procurement services (Arup 2011). As a result, many cities have come together in major networks, such as the C40 cities partnership (C40 Cities 2011), to share expertise in this area. Chapter 13 discusses some of these governance issues in more detail.

1.2 Working hypotheses and issues to consider

These trends describe the motivation for research on urban energy systems. Clearly an important aspect of urbanization is the energy system that provides the energy services (i.e. power, heating, cooling and transport) that supports the daily operation of the city. The interdependence of the growth and development of the energy system and the city as a whole is self-evident. An inefficient energy system could easily act as a brake on urban development or the growth of the urban area. By 'urban energy system' we mean the integrated whole system of energy (typically electricity and fuels, but sometimes also including heat) supply into the city boundaries, the interconversion of these energy vectors, the delivery of final energy services and the flows and disposal of wastes. This system comprises resources (e.g. natural gas, petroleum), technologies (e.g. combined heat and power engines) and networks (e.g. heat and power networks). A more formal definition is offered in Chapter 2.

We believe that it is an opportune time to study this vital element of city infrastructure. Accumulating urban datasets confirm that, behind the individualism of each city, there are scaling laws for urban services that span across the spectrum of cities (Bettencourt *et al.* 2007), giving the prospect of estimating the impact of new technologies that reflect city economies of scale. Although the resource flows inherited from twentieth century cities are characteristically independent and un-integrated, in part because neither the data nor systems technology was available to realize the economies from process integration and optimization, advances in research and data handling promise to remove these barriers. Given the pressures cities will face, there are some radical changes that could take place within cities over the next 30 years from integration of services in response to a changing energy supply and environmental context. This coincides with the need for infrastructure renewal in existing cities and infrastructure development in rapidly developing cities.

Finally, if these changes take place, the city unit is likely to be the dominating organizational structure behind the innovation. Integration played a key role in the improvements in process efficiency during the 1970s and 1980s. It is plausible that, given a framework to create the appropriate business opportunities, similar changes in resource use could come about. As a consequence, there is an unrecognized potential to deliver equivalent or better services in cities at substantially reduced resource flows, with improvements in resource intensities per unit of service certainly better than 20 per cent, but possibly as much as 50 per cent or more.

Early work in this area was concerned with mapping out the flows of urban energy on Sankey diagrams (ICLEI *et al.* 1996). These powerful representations show the flow of energy in a system, from primary energy sources through conversion system to final energy demands. They indicate the quantity of wasted energy along the way and are therefore useful in identifying sources of waste and quantifying overall system efficiency. The diagrams for Helsinki and Toronto are shown in Figure 1.3.

The figures clearly show that Helsinki has a higher efficiency than Toronto on the basis of the first law of thermodynamics (68 per cent versus 50 per cent); this is because the waste heat from proximate power generation is used for district heating. These figures also show an important aspect of urban energy systems: the interaction with the hinterland. As with most other resources (e.g. food, minerals), the scope for production of these resources within the urban boundary is limited. Cities have become concentrated

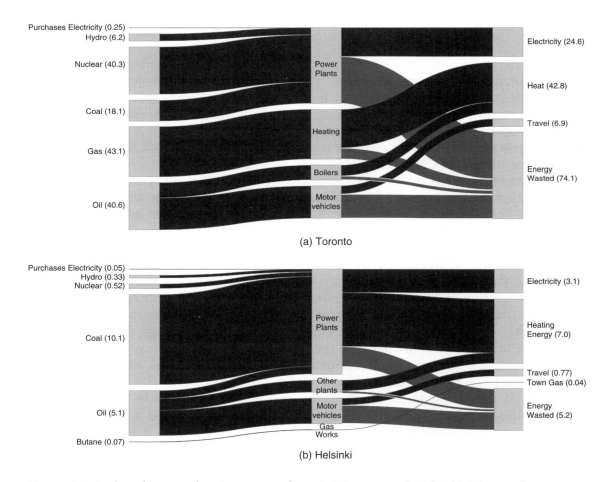

Figure 1.3 Sankey diagram showing energy flows in Toronto and Helsinki. Figures show annual energy demand in TWh.
Data source: ICLEI *et al.* (1996).

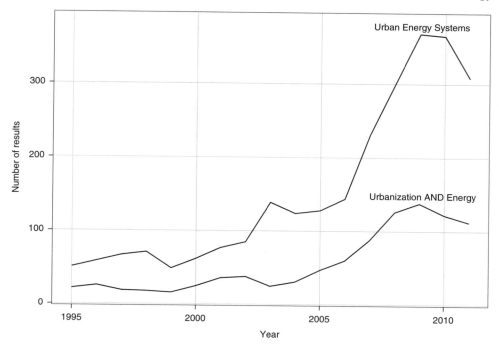

Figure 1.4 Published articles in the urban energy field by year.
Data source: Web of Knowledge (http://webofknowledge.com/).

and specialized converters and consumers of resources, often requiring supply chains with global reach.

The academic study of urban energy systems and the implications of urbanization for energy has followed a trend of rapidly increasing interest in recent times, as shown by Figure 1.4, which illustrates the number of results for literature searches using Web of Knowledge for the topic keywords 'Urban Energy Systems' and 'Urbanization AND Energy'.

In both cases, there is a large increase in interest in this cross-cutting, interdisciplinary topic from about 2005 onwards. The research is varied, and includes studies of how to supply cities with their energy needs in more sustainable ways. But clearly, urban energy systems as a concept bring with them some interesting issues for consideration, including:

• How is the urban area of interest defined? Is it a geographical, administrative or functional (i.e. associated with the engineering infrastructure) boundary?
• The challenge and opportunity of high energy demand density and load diversity (i.e. different loads at different times and/or locations) and the relatively low amount of primary energy harvesting with the city boundaries.
• The need to ensure installations that account for the changing nature of cities, e.g. changes in social demographics leading to increased household formation and

the changing nature of cities' economic activities from industry to services with concomitant changes in energy demand patterns (e.g. high-grade heat to cooling).

- The very open nature of cities with large material and energy flows across their boundaries. They are dependent, directly or indirectly on fossil fuel which supports long global supply chains, and have become less dependent on the near hinterland; this contrasts with the original, simple rural–urban economic geography model of von Thünen (1826).
- The contribution of energy conversion (including transport) systems to local air quality problems.

To summarize, urban energy systems provide a rich subject for research and development. This book arises out of a project which explored the hypotheses that:

- There is potential for massive systems integration across all urban energy systems including power, heat, transport, water, and waste
- One can draw on the experience of process industry where full systems integration was introduced some 20 years ago
- Today, we have much more powerful optimization and modelling algorithms than were available in industrial research of the 1970s, and new ways of modelling multi-scale systems are being reported
- Cities have much more powerful ways of collecting spatial and real-time data than they have ever had before
- We have a much clearer idea how global optimizations can emerge through self-organizing properties of the system.

These challenges are explored in the chapters that follow.

1.3 Plan of the book

As the subtitle suggests, we believe in an 'integrated approach' to the study of urban energy systems. This is reflected in the book's layout which, over four parts, addresses the technical, economic, social, political, environmental and methodological aspects of urban energy systems analysis.

Part I sets the stage with an introduction to the topic. Chapter 2 explores some of the ideas mentioned here in greater detail, examining alternative analytical concepts for urban energy systems and leading to a formal definition of 'urban energy systems'. Chapter 3 then provides a narrative history of the development of energy systems in London. The idea here is to give the reader a feel for the complexity of the subject and the range of technological, social and economic factors that shape the performance and structure of our current (and future) urban energy systems.

Part II examines urban energy use and technologies. In Chapter 4 the use of energy in buildings is explored with particular reference to how existing buildings can be retrofitted to improve their energy efficiency and climate resilience. Chapter 5 looks at important energy supply technology: distributed multi-generation and district energy systems. As shown in Figure 1.3 above, the density of urban areas enables us to capture more of the useful energy contained within raw fuels, thus providing a valuable method of improving urban energy systems. A second supply technology is urban renewables (Chapter 6). While we focus primarily on urban biomass resources as a powerful illustration of the links between cities and their hinterlands, we also briefly describe other technologies such as solar thermal, solar photovoltaics, micro-wind, and heat pumps. Chapter 7 then turns from the supply of heat and power to consider the transport sector. What options are available for improving the efficiency of energy use in urban transport systems?

Part III considers modelling technologies for urban energy systems. Computer models are important tools for analysing urban energy systems, but there are a range of different approaches available. Chapter 8 provides a general introduction, considering basic questions such as 'what is a model?' as well as reviewing recent literature in the field. Chapter 9 looks at optimization models as a particular modelling approach with relevance to the design of efficient energy supply systems. These models need to be driven by detailed descriptions of energy services demands and Chapter 11 discusses activity-based models, the state-of-the-art approach for generating such demand profiles within a wider system of models for studying urban land use and transport. Chapter 10 examines optimization approaches to urban energy systems design, but taking the perspective that objective functions should reflect wider ecological impacts and not just minimize cost. Finally, Chapter 12 discusses uncertainty and sensitivity analysis. The output of complex models can be difficult to understand and these techniques are essential to the production of high-quality and robust model inferences and decision-support. Each chapter also contains detailed examples and case studies.

Finally, Part IV concludes by looking at some of the practical issues surrounding the implementation of the highly efficient urban energy system designs. Chapter 13 considers how we can manage the transitions of highly complex socio-technical systems. The examples of Copenhagen, Denmark and Nakuru, Kenya are used to illustrate these ideas. In Chapter 14, the idea of 'urban energy futures' is discussed. In other words, how can visions of future cities be used to guide planning processes and build support for transformative change? Chapter 15 provides a brief conclusion, revisiting the hypotheses proposed here.

Note

The most common energy systems models at a national or international scale include MARKAL, POLES, MESSAGE, TIMES and OSEMOSYS. See http://www.iea-etsap.org/web/Markal.asp for example.

2 Conceptualizing urban energy systems

James Keirstead

Imperial College London

The word 'city' can mean different things to different people. For some, it may evoke images of towering spires and neon lights, seemingly endless opportunity, and the buzz of millions of people going about their business; for others however, a darker image of dirty streets, overcrowded trains, noise and crime may dominate. These views reflect the complexity of modern cities and suggest that multiple interpretations of the physical and social urban environment are possible.

Of course, this book is not about cities in general but about the energy systems which power them. How can they be interpreted? Should they be defined as technical systems networks of infrastructure providing vital services with minimal human influence beyor design and operation? Or should the scope be wider, considering the myriad ways which urban energy systems reflect our social and economic lives?

Perhaps one of the difficulties is that energy is such an essential part of modern that we often don't think about it in an explicit way. We simply turn on the light sw drive our cars, or crank up the thermostat without giving thought to the underlying and infrastructure systems that deliver these services. Ultimately, this is because e consumption represents what economists call a 'derived demand'. We don't co barrels of oil or kilowatt-hours of electricity for their own sake, but only incu demands in the provision of services such as mobility, space comfort, lighting on. And this means that, as consumers, we may not recognize the vastness of th supporting our actions.

This chapter aims to provide a theoretical introduction to the analysis of urb systems. As in any field of study, having a robust theoretical foundation (to build analytical tools and interpret new data. It also helps us to unde similarities and differences between cities and their energy systems, and to potential of the many opportunities for improving urban energy efficiency. V is, therefore, a brief review of both the technical and non-technical means urban energy systems.

2.1 Physical models of urban energy systems

As engineers, our interest in urban energy systems is tied largely to the p that constitute these systems. We want to understand the mechanics of

Figure 2.1 A schematic showing a city as a thermodynamic system. Energy coming into the city, E_{in}, is transformed by urban processes into useful work, W, and waste energy, E_{waste}. There may also be a change within the internal energy of the city, e.g. through the urban heat island effect.

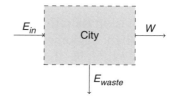

involved, and to use the laws of physics, chemistry and mathematics to predict the behaviour of these systems in a variety of situations. Unfortunately, there is no single 'physical' model of an urban energy system and in this section, we present three complementary views: cities as thermodynamic systems, metabolic systems and complex systems.

2.1.1 Cities as thermodynamic systems

Thermodynamics is the study of the relationships between heat and work in a system. Cities can be thought of as thermodynamic systems in this sense, as a sort of 'black box' which consumes energy and materials to produce work (Figure 2.1).

A key question in the analysis of any system is to decide on its boundaries. As an example, suppose we are interested in the properties of an internal combustion engine. In this case, our choice of system boundaries is fairly obvious. Fuel goes into the engine, something happens inside, and the engine emits hot exhaust gases and performs some useful work by turning a shaft. An analysis of the system's performance therefore is a relatively straightforward matter of comparing what goes into the engine with what comes out of it. In contrast, defining the boundaries of a city is not a trivial question. Of course a city may have a clearly defined administrative boundary, but in terms of the physical processes going on inside the city, this delineation may be relatively meaningless. For more information on the difficulties of how to define a city's boundary, see Box 2.1.

Box 2.1 Defining an urban boundary

Imagine yourself taking a train trip through a country. Most people would feel relatively confident in being able to decide whether the view passing by the window was of an urban or rural area. Lots of open space, forests and fields? It's probably rural. High flats, heavy traffic and neon lights? Probably a city. However in practice, the division is not so clear-cut and this presents significant challenges to those wishing to study urban systems of all kinds.

Take the simple example of trying to calculate the population of a city. In some countries, a city's boundary will be defined by law (e.g. an administrative area) whereas in others, the definition may be based on measurable data, such as population density or economic activity or not defined at all. For demographers trying to determine the world's urban population, this means that some sort of standard definition is needed to compare between countries and the UN Population

Division therefore maps local definitions onto one of three categories: 'city proper', 'urban agglomeration' or 'metropolitan area' (Buettner 2007, UNSD 2011). As an example, if we measure the population of London within the administrative Greater London Authority boundary, the 2000 population was approximately 7.1 million. However, if we take a more functional view, based on the larger urban agglomeration, the population is closer to 13 million (SEDAC-CIESIN 2012).

As a recent World Bank report notes, 'which spatial scale to use, or how best to define a subnational area, depends on the issue and the information available'. (World Bank 2008: 78). In the case of urban energy systems, researchers have been limited by the poor availability and quality of urban energy consumption data and have had to rely on a pragmatic approach, assessing each city's data on a case-by-case basis. Fortunately, efforts are underway to standardize these data collection processes (IEA 2008b: 181). But researchers should be aware that the choice of boundary is important, particularly when making comparisons between cities, as the resulting conclusions can change significantly (World Bank 2008).

Urban boundaries also shift over time and administrative definitions are often to slow to catch up to the reality on the ground. A notable example is the concept of the megalopolis, areas where neighbouring cities effectively grow together into one super-city or city-region. Notable examples include the Boston-Washington (BosWash) corridor in the north-eastern United States and the Pearl River Delta area in China. Faced with these amorphous urban areas, many researchers are turning to alternative problem-specific urban definitions, for example, based on water-sheds, commuting areas or mobile phone service areas (Berg 2011).

For the moment, our concern is not about where this boundary actually lies, but whether or not it is *open* or *closed*. The key distinction here is whether materials and energy traverse the boundary of the system. Early in the development of systems theory, Bertalanffy (1950) noted that many important fields of scientific study, such as physics and chemistry, can be studied from a closed system perspective but that a richer understanding is gained by considering them as open systems. In the case of cities however, there can be no ambiguity. Without flows of materials and energy across its boundaries, a city simply could not exist (Nicolis and Prigogine 1977, Filchakova *et al.* 2008).

We can therefore analyse a city as an open thermodynamic system, since it is not in equilibrium with its surrounding environment. The city itself represents a highly-ordered, i.e. low entropy, system and maintaining this state requires an exchange of materials across the boundary. Coming into the city are highly-ordered materials and forms of energy, such as concrete, steel, electricity, natural gas, and leaving the city are highly-disordered materials and energy, such as waste products including waste heat. This view of cities as entropy producing systems can also be seen in analysis of social and economic activity (Batty 2008).

Perhaps the most vivid demonstration of this distinction between ordered and unordered spaces is the Ukrainian city of Prypiat. Located only 3 km from the Chernobyl Nuclear Power Plant, the city was originally founded in 1970 to house workers from the power station, and grew to a peak population of nearly 50,000 residents. However, following the meltdown in 1986, the city was abandoned and it has now fallen into disrepair with leaky buildings, trees growing on roofs and through floorboards, and wildlife reclaiming streets and playgrounds. Alan Wiseman's exploration of what the Earth would look like if everyone suddenly disappeared, paints a similar picture, noting that some residential neighbourhoods could become forests in just 500 years (Weisman 2007).

An advantage of the thermodynamic view is that it clearly captures the unsustainability of urban life, not in the sense that urban life is undesirable but that cities exist in tension with their surrounding environments. To maintain themselves, they can be seen as processes with a fairly linear metabolism taking in materials, consuming or converting them, and then exporting wastes to their hinterlands (e.g. Robinson 2009). By closing this loop and recycling materials, many researchers see significant promise for increased urban sustainability (e.g. Filchakova *et al.* 2008). However, the thermodynamic view of cities tells us that there is no such thing as a free lunch, and converting high entropy waste materials back into low entropy usable goods will necessitate additional inputs of energy. Therefore, while we may be able to make progress in reducing the material metabolism of cities, the energy requirements are likely to grow.

To analyse a city's energy use from a thermodynamic perspective, there are two laws of primary importance. The first law of thermodynamics, which states that energy must be conserved, means that energy flows entering the city must be balanced with flows leaving the city (and any changes in in-city stocks, through storage). A notable example of this is the urban heat island effect, whereby cities become warmer through changes in albedo (e.g. paving surfaces, so that solar radiation is trapped in local materials) and heat rejection from processes such as air conditioning (e.g. Hamilton *et al.* 2009, Dhakal and Hanaki 2002, Mirzaei and Haghighat 2010). The second law of thermodynamics states that thermodynamic systems tend to an equilibrium with their environment over time. This statement offers insight into the boundary issue noted above, but also indicates that many energy system processes with a city are irreversible. Burning natural gas in a domestic heating system, for example, will produce hot water and waste heat with the same total energy as the input fuel, but the temperatures of these relative flows mean that the amount of *useful* energy available (that is, its *exergy*), in terms of providing space heating, for example, is not conserved. As a result, all energy conversions are subject to irreversible losses and can never be 100 per cent efficient.

It is worth noting that the ecology community also has developed a concept called *emergy*, which is similar to energy and exergy and has been applied to the energy analysis of cities (Huang and Chen 2005, Odum 1988, Tilley 2004). However, a recent comparison noted that the approaches have significantly different paradigms driving them and that exergy analysis represents 'the best engineering method for system optimization' (Sciubba and Ulgiati 2005: 1953).

2.1.2 Cities as metabolic systems

One of the most powerful analogies for urban systems is that of a living organism. Like a body, cities ingest resources, expel wastes, and circulate materials through distribution networks. Indeed, recent work has shown that many of the same principles apply to the sizing of urban infrastructures as for analogous systems in human bodies; for example, there are economies of scale in the size of urban road networks in much the same way as seen in the cardiovascular systems of elephants or mice (Bettencourt *et al.* 2007, West *et al.* 1997, Kühnert *et al.* 2006).

The idea of an urban metabolism has been formalized within the field of industrial ecology. Beginning with Wolman's 1965 article, researchers have looked at the flows of energy, food, water, waste and other materials to understand the physical impacts of a city. Such analyses are typically conducted at the scale of the whole-city for a single year (e.g Kennedy *et al.* 2008), which means that some of the minute-to-minute variations in resource demand and infrastructure performance get lost. Overall however, urban metabolism is a powerful tool for studying urban energy systems.

As with the thermodynamic analysis of a city, an important issue in urban metabolism studies is the definition of the system boundaries. For example, suppose that we want to the measure the greenhouse gas emissions from a city. One common protocol recommends that emissions be divided into those that occur within the boundaries of the city (Scope 1), those that are directly linked to the city by infrastructures (e.g. electricity consumption, Scope 2) and other emissions driven by urban consumption (Scope 3) (WRI/WBCSD 2011). But other authors have suggested that a broader perspective, looking at the life-cycle emissions (Kennedy *et al.* 2009) or more relevantly here, those emissions linked to 'essential infrastructures' such as commuter and airline transport, energy supply, waste and water systems (Ramaswami *et al.* 2011). In this perspective, urban infrastructure provides us with a motivation for moving across the physical city boundary, arguably opening up a more accurate representation of urban activities and their impact than if we simply stayed within the official city limits.

The so-called geographic-plus, or infrastructure, view of urban emissions begins to reflect the social dimension of urban energy systems. Infrastructures are essentially the 'glue' that binds together both cities and modern society together, both socially and physically (Graham and Marvin 1994). Flow management, for example, is a concept which examines the resource flows that support economic and social systems (e.g. Moss *et al.* 2000). In other words, the physical flows are only meaningful in the context of the services creating the associated demand. Furthermore, owing to their size and complexity, urban infrastructures are expensive and difficult to plan and build; they reflect the wider social and economic relationships within our cities. Moss *et al.* (2000) provide examples of urban water, waste and energy systems to illustrate the political economy of these systems. In the case of energy networks in the UK, for example, local environmental policies and infrastructure aspirations have to work with the privatized structure of the UK energy market and the authors suggest that authorities trying to improve their urban energy systems should devote more of their efforts to understanding the complex logic driving energy use and networks, rather than simply mapping the resource flows. Again, we are into the territory of complex systems: 'Chains of related innovations bind infrastructure networks closely to broader technological systems; these, in turn, are seamlessly woven into the fabric of social and economic life … Only very rarely do single infrastructure networks develop in isolation from changes in others'. (Graham 2000: 114).

2.1.3 *Cities as complex systems*

As Michael Batty notes in the introduction to a recent working paper, '[c]ities have been treated as systems for fifty years but only in the last two decades has the focus changed from aggregate equilibrium systems to more evolving systems whose structure merges from the bottom up' (Batty 2008: 1). This, in essence, is the complex systems perspective of the city.

The idea has a long history and was closely linked to the evolution of systems thinking in general. In an influential early paper, Simon (1962) offers a basic definition:

> Roughly, by a complex system I mean one made up of a large number of parts that interact in a nonsimple way. In such systems, the whole is more than the sum of the parts, not in an ultimate, metaphysical sense, but in the important pragmatic sense that, given the properties of the parts and the laws of their interaction, it is not a trivial matter to infer the properties of the whole.

(Simon 1962: 468)

Cities and their energy systems clearly fit this description. Urban energy systems contain many individual components that interact in a multitude of ways. In some cases, these interactions will be fairly straightforward. For example, if we consider a home heating system, we may have a gas boiler providing hot water which is connected to a radiator delivering the heat and a thermostat that controls the operation of pumps and valves within the system. If the room is cold, we can predict fairly easily how the boiler will perform and the room will heat up to meet our level of comfort. But even in this simple case, complexity hoves into view. In a block of flats, for example, the usage of a boiler in any single flat will depend on the use of the boilers in neighbouring flats, as their warmth will affect the heat demand in surrounding flats.

The complexity of urban energy systems truly comes into its own when considering the whole city scale. For example, in August 2003, a major black-out struck New York City. This event started with a relatively small single trigger, but cascaded across most of the north-eastern United States leading to widespread blackouts (US-Canada Power System Outage Task Force 2004). In this case, the system was comprised of multiple cities, each with a complex pattern of energy demands that add up to a single load curve, supported by a networked infrastructure spanning thousands of kilometres. Understanding how the change in any one component in the system would affect the whole chain of events is nearly impossible.

Another good example of complexity in urban energy systems is the feedback loop that exists between land use planning and transportation demands (Mackett 1985, Wegener 2004). Consider a green field location next to a major city. An enterprising developer may feel that there is a market for a new shopping centre. He or she purchases the land, builds the new shopping centre, and then waits for customers to arrive. But those customers must travel to the location first. Therefore part of the development process will involve analysing the available transportation options and perhaps building a new infrastructure to reach the site. This new transportation infrastructure will shift patterns of travel and, with all these new visitors, a second developer may decide to renovate some nearby buildings to provide new housing. This cycle of land use change and transportation altering the accessibility and attractiveness of different locations is a fundamental driver of the shape of our cities (Figure 2.2). And in turn, the shape of our cities has a significant impact on its energy consumption.

These three perspectives – cities as thermodynamic, metabolic and complex systems – highlight important features of urban energy systems. First, the structure of urban energy systems emerges from the bottom-up. The activities of individual citizens create demands for energy services, which then require complex infrastructures in order to supply them. This sets up the dynamic of urban metabolism, the city importing energy resources and expelling wastes in order to maintain order and function. The laws of thermodynamics

Figure 2.2 A four-car 1923 Standard stock train approaches Burnt Oak Underground station on the Hampstead and Highgate line (now Northern line). New suburban houses occupy the land on both sides of the track. On the right, houses of the LCC Watling Estate are under construction. © TfL from the London Transport Museum collection. Reprinted with permission.

can be seen to govern these processes, for example, highlighting the open nature of the urban boundary and the inexorable loss of exergy, or energy quality, along the steps of the energy system.

2.2 Socio-technical models of urban energy systems

It would be hard to deny that these physical theories are relevant to the understanding of urban energy systems. After all, the laws of physics can't be cheated. However, a purely physical view of urban energy systems obscures many of the vital social and economic processes that determine the actual design and operation of these systems. Cities are primarily social and economic structures. They offer people opportunities for security, culture and commerce (Kotkin 2005) and their infrastructures in turn suppport their position as nodes in local and global networks of economic and social activity: 'cities ... become, in a sense, staging posts in the perpetual flux of infrastructurally mediated flow, movement and exchange' (Graham 2000: 114).

In this section, we will review some of the key theories used to interpret urban energy systems as socio-technical systems. From the level of small-scale domestic energy technologies, all the way up to the development and evolution of large networked infrastructures like district heat or electricity systems, theories of socio-technical systems enable us to understand the limits of physical models and to be aware of where the messy realities of politics, culture and economics lie.

2.2.1 Domestic energy technologies

Energy consumption can begin with a simple action, like flicking a switch. A good place to start thinking about these wider questions of energy technologies is therefore in the home. How did domestic energy consumption technologies get there, and how do we use them in ways that might not be obvious at first glance?

Suppose someone invented a new widget that would radically change the cost or environmental impact of domestic energy consumption. It could be an energy display monitor, an improved heating technology or even a backyard nuclear reactor (Wilson 2007). For this technology to have an effect, it clearly has to be adopted by individuals; that is, they have to buy the technology and start to use it in a certain way within their daily life.

At a societal level, this spread of new technologies through a market is known as the diffusion of innovation, 'the process by which an innovation is communicated through certain channels over time among the members of a social system' (Rogers 2003: 5). In such an analysis, the characteristics of the technology are important. Potential users might first evaluate a technology's relative advantage (Is the innovation perceived as better than the idea it supersedes?); compatibility (Is the innovation seen as being consistent with the existing values, past experiences and needs of potential adopters?); complexity (Is the innovation perceived as difficult to understand and use?); trialability (Is there an opportunity to experiment with the innovation on a limited basis?) and observability (Are the results of an innovation visible to others?).

Diffusion of innovation studies note that there are different types of adopters, each with different preferences and risk tolerances. Early adopters, for example, can be characterized by high levels of knowledge with which to understand new technologies and crucially, a desire to experiment with them (and afford the potential costs of an unsuccessful adoption). For example, in the case of solar hot water, researchers found that these early adopters evaluated solar systems to be 'less financially risky, less socially risky, less complex, more compatible with their personal values, and less observable by others' (Labay and Kinnear 1981: 275). However, once these individuals have successfully adopted a technology, they begin to tell others about the experience and this helps to spread the technology further. In other words, a good technology doesn't spread purely on its own merits; it needs to be evaluated, trialled and then spread through social networks.

Once a technology is in the home, the next question is, how will it be used? Sociologist Max Weber provides a useful typology of 'social action' which can be applied to differentiate domestic energy consumption behaviours (Weber *et al.* 1968: 24). The first mode, arguably the economist's favourite, is *instrumentally rational* behaviour 'for the attainment of the actor's own rationally pursued and calculated end', for example, turning off the lights in order to save money. Alternatively, behaviour might be *value rational*, that is, turning off the lights in order to reduce greenhouse gas emissions and protect the environment. However, there are also two important modes of non-rational behaviour: *affectual*, driven by an individual's feeling at any given moment (e.g. a petulant teenager turning off the lights to sulk in the dark); and *traditional* or basic habit (e.g. turning off the lights when leaving the house).

Technology can also be used in unexpected ways, both physically and in a social context. A book, for example, might have been designed to be read, but it also functions well as a doorstop, paperweight or fly swatter. Personal stereos play music, but they also enable listeners to shape their daily experience, playing upbeat or downbeat songs as appropriate, on a commute (Bull 2005). 'Green' buildings are a classic example of this, where the expected technical potential fails to be realized owing to unexpected modes of user behaviour, such as leaving windows open in a building that relies upon mechanical ventilation with heat recovery (Chappells and Shove 2005, Rohracher and Ornetzeder 2002). Similarly, improvements in thermal energy efficiency might be 'taken back' as increased comfort (Milne and Boardman 2000, Hong *et al.* 2009), a phenomenon known generally as the 'rebound effect' (Sorrell 2007).

In a commercial setting, these personally and socially-motivated behavioural factors are arguably suppressed and decisions on adoption and use become more instrumentally rational. As we will see later, this motivates the use of optimization techniques in the design of large-scale energy systems. But as this brief discussion of domestic energy consumption behaviour has hopefully shown, one cannot expect that an energy technology will be passively adopted and used by consumers as its designer originally intended. The motto of the 1933 Chicago World Expo – 'Science finds, industry applies and man conforms' (Smits 2002: 863) – reflects the technological determinism of the time, but is now widely recognized to be false. Users bring their own meanings to technology and use them within a complex web of social norms, personal values and decision-making contexts.

2.2.2 *Large technical systems*

Urban energy systems also contain what might be described as *large technical systems*, i.e. 'spatially extended and functionally integrated socio-technical networks such as electrical power, railroad, and telephone systems' (Mayntz and Hughes 1988: 5). A city does not consist of houses, sitting demurely apart from one another, each with their own boilers and private fuel supply; these buildings are connected to one another by electricity, gas, heat and transportation networks. As these systems are large and complex, it takes time for them to be built and it can be hard to quickly change them. Therefore we must understand the evolution of such large technical systems if we are to assess the realistic potential for change in a given city.

The classic study in this field is *Networks of Power: Electrification in Western Society, 1880–1930* (Hughes 1983). The book showed how, although physical changes in an electrical system could be represented by a network diagram, the *explanation* of those changes necessitated a consideration of 'many fields of human activity, including the technical, the scientific, the economic, the political, and the organization'; in other words, 'power systems are cultural artifacts' (Hughes 1983: 2). Hughes describes the development of electrical systems in cities around the world, including Chicago, Berlin and London noting in particular the role of 'system builders', such as Thomas Edison, who were able to develop not only the technology but also a supporting system of institutions that necessitated the use of the innovation.

This analytical approach has been very influential, highlighting that successful technologies are the result of people and technology working together. For example, a later study showed how the social and cultural contexts of American and Danish life (e.g. individualism versus collectivism) had a major impact on how the electricity systems of these nations were perceived and designed (Nye 1990). However, this earlier literature created a divide between the system and the environment (i.e. things external to the system but influencing it) and, as this split has an influence on any subsequent analysis, one is again faced with the problem of trying to locate the boundary between system and its environment (Joerges 1988).

A more modern take on this problem can be found in the literature on multi-level technological transitions (for a review, see Geels 2002). In this literature, the analysis is slightly different and doesn't try to explain past developments as much as it tries to offer insight on possible future transitions. The question of transitions in urban energy systems will be considered further in Chapter 13, but for the moment, the key point is that transitions in technological systems – including urban energy systems – are not the result of a single determined 'system builder', talented though they might be. The technological transitions literature instead demonstrates how factors at three levels – the wider socio-technical *landscape* (e.g. societal trends, political forces, economic cycles); specific suites of technologies, institutions and use patterns known as *regimes*; and in small-scale experimental *niches* – must co-evolve and align for any particular technology to thrive.

These socio-technical perspectives on urban energy systems, at both the level of individual technologies and when considering larger infrastructures, demonstrate two key lessons. First, the technological potential of an artefact can only be realized in concert with its users. In other words, creating more efficient urban energy systems will not be possible by simply inventing a technological magic bullet. Even if such a technology existed, it would have to be adopted by consumers and used in a way that supported their personal desires and needs, alongside the constraints of culture, institutions and economics. Second, large infrastructures – such as energy networks – evolve slowly and require the coordination of multiple factors. Inventors, business people or politicians may propose grand solutions, but like a network's constituent technologies, these solutions must fit within the wider context of social, cultural and economic factors. Understanding the performance of urban energy systems, past, present and future, requires an appreciation of these factors.

2.3 Defining urban energy systems

This chapter has shown that there are several different ways of viewing cities and their energy systems. Each perspective has its own set of analytical tools and techniques and its own motivating questions, but none of them seem to speak exclusively to our research question: what is the potential improvement in the efficiency of urban energy systems? Answering this will require a mix of technical understanding, of cities as thermodynamic, metabolic and complex systems, and an appreciation for the wider social and economic forces at work in our adoption and use of energy systems and their component technologies.

   The challenge therefore is to find a definition of urban energy systems that recognizes these viewpoints. For the purpose of this book, we will adopt a modified version of Jaccard's 2005 definition of an energy system as 'the combined processes of acquiring and using energy in a given society or economy' (p. 6). This definition highlights three features of urban energy systems:

combined processes Delivering energy services requires many different steps including resource extraction, refining, transportation, storage and conversion to end service. While the urban environment may be physically separate from these processes, they should be considered in an overall analysis if they are ultimately being used to service urban demands. This references the thermodynamic and metabolic views of the city as an open system. For example, urban greenhouse gas emissions inventories typically include emissions from ex-urban electricity generation as a minimum and can be expanded to include a range of life-cycle or upstream emissions (Kennedy *et al.* 2009).

acquiring and using Energy systems represent a balance between supply and demand. Historically, cities might be seen as centres of passive demand which must be supplied from an ex-urban source, but recent work suggests that there are now significant opportunities for in-city energy generation (Foresight 2008). Given these possibilities, urban energy systems should be conceived of as including both sides of the supply and demand equation.

given society or economy An energy system is a socio-technical system, comprised of more than just pipelines, fuels and engineering equipment as discussed above. Markets, institutions, culture, consumer behaviour and other factors affect the way technical infrastructures are constructed and operated. Urban energy systems therefore need to be viewed more widely and account for local context.

The most difficult part about adapting this generic definition of an energy system to one that is specifically 'urban' arguably lies in defining the boundaries of a city. As discussed in Box 2.1, this is largely a question of data availability and we therefore take a pragmatic view of this problem. Recognizing the open nature of cities, we attempt to take a life-cycle view of urban energy systems moving beyond the physical boundaries of the city where necessary to understand relevant impacts and drivers. This will not always align with available datasets, but by focusing on urban activities as the driver for these wider systems, a tangible system definition can be reached.

These considerations lead to the following definition of an urban energy system:

the combined processes of acquiring and using energy to satisfy the energy service demands of a given urban area.

3 A brief history of urban energy systems as exemplified by London

Paul Rutter

Imperial College London

3.1 Introduction

On average, each one of us now consumes nearly 25 times as much energy per year compared with our hunter–gatherer ancestors (Smil 2010). This consumption is powered by a range of modern convenient energy sources and we use electricity and fossil fuels where once we relied primarily upon fodder for draft animals, biomass for heat, animal and vegetable oils for light, with a very small but significant contribution from waterwheels and windmills. The history of urbanization is therefore as much a history of the search for ways to provide sufficient food, shelter and energy services for a growing population, as it is one of monarchs and empires. Indeed in his book, *Energy in World History*, Vaclav Smil (1994) quotes the anthropologist Hoyt Alverson who said

> the most salient aspect of the ecologic dimension of culture, looked at over millennia of cultural revolution, is the correspondence between the size and density of culture bearing populations on the one hand and the amount of potential energy per capita that must be captured from the environment and transformed into material and energy forms on the other.

This chapter considers how modern urban energy systems have evolved to provide us with the heating, cooling, lighting, mobility and communications on which we now depend.

There is great uncertainty surrounding the reasons that caused the hunter–gatherer way of life to be abandoned around 10,000 years ago in favour of settlements based on proto-farming supplemented by hunting. Inevitably, some settlements grew to dominate others and in small towns, a system of agriculture developed which was dependent upon the cooperative efforts of the inhabitants. In time, trade grew between towns and their surrounding hamlets, which led to sufficient surplus for some division of labour and specialization. The first cities appeared around 3000 BC in Mesopotamia, and subsequently in the Indian subcontinent, North Africa and China. These rapidly achieved many of the city characteristics that we recognize today: palaces, grand public buildings, artisan districts, markets and busy activity. Food production was carefully managed, often using extensive irrigation and storage systems. Trees were a critical part of the agricultural system, since they provided the essential commodities of timber for buildings and tools as well as fuel wood for cooking. In some cases, the need for increasingly scarce land around the growing cities and towns for the production of food resulted in forest clearance and fuel shortage. Jared Diamond, in his 2005 book, *Collapse*, describes a number of instances where forest clearance either for fuel or to create more arable land, led to cities being abandoned.

The struggle to support increasingly elaborate civilizations, empires and populations from relatively short-cycle biomass (i.e. food crops and trees) continued for about 3,500 years. But it was not sustainable and in about AD 1600 urban energy systems began to change. Using the example of London, we describe a transition from traditional wood fuel to coal, a shift that was an important factor in the development of the Industrial Revolution since it not only led to the invention of the steam engine, which enormously increased the power available for industrial growth, but also had the effect of releasing large tracts of woodland for food production. As Hobsbawm (2008: 43) points out, at some time in England during the 1780s, 'the shackles were taken off the productive power of human societies, which henceforth became capable of the constant, rapid and up to the present limitless multiplication of men, goods and services'. The impact of this change on London was dramatic and, although much of Britain's productive capacity was located in the industrial cities of the Midlands and the North West, London was Britain's financial centre and, by 1830, it was the only western city of more than 1 million inhabitants. Its importance as a showcase for new energy technologies led to early demonstrations of gas lighting in 1804 and electric lighting in 1881. These technologies, together with Bazalgette's new water and sewage system begun in 1869, radically changed life in London and set the scene for life in the city today.

Urban energy systems represent the 'combined processes of acquiring and using energy' to meet the energy service demands of an urban population (after Jaccard 2005). However, unlike a century ago when most of London's energy was derived from coal brought into the city by barge and rail, virtually the only fuel that is now transported into the city in bulk today is petroleum either as diesel or petrol. The majority of the city's energy now enters via pipes or wires, which knit together urban areas as part of national grids. But even this is beginning to change; solar photovoltaics, integrated wind power and run-of-river hydro, together with combined heat and power technologies are once again enabling the city to generate its own energy.

These centuries of changes are detailed below and summarized in Table 3.1.

3.2 Transition from hunter–gatherer to settler

As the ice receded around 14,000 years ago, conditions in many mid-latitude areas appear to have been about as warm as they are today; forests began to spread back and food supplies are likely to have improved. There was then a sudden return to cold accompanied in the north by a brief resurgence of the ice sheets. This period is referred to as the Younger Dryas and it lasted for about 1,300 years until 9500 BC when the Earth entered several thousand years of warmer and moister conditions. This sequence of climatic events led to a number of ideas for the reasons why evidence of the first large permanent settlements date from soon after the end of the Younger Dryas. The earliest settlement described in detail so far is at Jericho in the Jordan valley and dates from 8500 BC. It consists of round houses made with sun-dried mud bricks enclosed by a substantial stonewall. The presence of an apparent fortification supports the ideas of Keeley (1997) and others that frequent skirmishing between family groups

Table 3.1 The development of energy services in London, 1500–2000

Year	Population	Heating	Stationary power	Transport	Lighting	Water
Before 1500	80,000	Fuel wood demand around 1.8 dry tons per capita per year plus some coal	Largely animal and human power supplemented with tide mills, water mills and windmills from the 11th century	Human carriers and hand pulled carts, horse wagons, ships and river craft	Rush lights, oil lamps and tallow candles	River Thames, shallow wells, piped water from springs to cisterns, water carriers
1500–1525	55,000	100,000 tons of dry wood				
1525–1550						
1550–1575		~16,000 tons of coal imported into London in 1550				Water mills to supply drinking water constructed under London Bridge
1575–1600	220,000	Price of fuel wood double the price of coal per BTU	Iron masters prevented from using charcoal made within 22 miles of London			
1600–1625		182,000 tons of coal imported into London in 1614	Parliament banned the use of wood in glass making	River Thames 'improved' to facilitate river transport		Construction of New River Canal
1625–1650		283,000 tons of coal imported in 1637				
1650–1675	450,000	Great Fire of London 1666				Failure to implement Wren's improved sewage system
1675–1700		314,000 tons of coal into London imported in 1676				

Period						
1700–1725	500,000		Newcomen's 'fire' engine Derby's coal fired blast furnace		35,000 oil lamps light the streets of London	
1725–1750		Over half Britain's net coal used for domestic heating		London Building Act widens London's streets London supplied largely by horse-drawn wagons		
1750–1775	675,000	677,000 tons of coal imported into London in 1750			Royal Society offer prizes for improved lighting. Westminster Paving Act improves lighting	
1775–1800				Grand Junction Canal opens		
1800–1825	1 million	1.5 million tons of coal imported into London in 1815		Stephenson's rocket	Winzer demonstrates gas lighting at Lyceum theatre	Effluent discharge directly into Thames made illegal
1825–1850			Faraday's work on Electromagnetism	11,000 horses including horse-drawn omnibuses; Euston railway station opened	40,000 gas street lamps light 213 streets of London	Rapid increase in use of water closets often connecting with sands supplying local wells. 1st cholera epidemic 1831
1850–1875	2.36 million	A 16 ton block of coal displayed at London's Great Exhibition	Commercial electric motors	76,000 licenced horses; First underground opened	Metropolis Act allows local monopolies; Gas main laid in Piccadilly	Work begins on new London sewage system

(Continued)

Table 3.1 Cont'd

Year	Population	Heating	Stationary power	Transport	Lighting	Water
1875–1900			Edison opens the first electric power station in 1881; Ferranti's plans for giant power station at Deptford turned down in 1890	300,000 horses. Electric trams, railways and first deep electric underground	Electric Lighting Act; Welsbach incandescent gas mantle. Electric arc followed by incandescent electric lighting	Discharge of sewage in Thames stops
1900–1925	6.6 million		65 Electrical Utility Companies supply London			
1925–1950		Gas nationalization	Electricity Supply Act; National Grid; Battersea Power station opened	London Passenger Transport Board unifies London's bus, tram and underground services		
1950–1975		National gas grid and supply switched to Natural Gas	2nd Clean Air Act ends London's smogs; 14 Power stations in London in 1970	Heathrow airport opens		
1975–2000	7.5 million		Last coal and oil fired power station in London closed down in 1990	Docklands light railway		Construction begins on 80 km London Water Ring Main
2000–2025			London wind turbine array	Lot's Road power station which supplied 60% of power for the underground closed in 2002	LEDs replace incandescent bulbs in London's traffic lights. All incandescent bulbs phased out by 2011	London's new overflow sewer to be completed in 2020

and tribes, possibly competing for range, eventually led to nomadic groups coming together in permanent or semi-permanent settlements for mutual protection. Others, including Renfrew (2008) and Mithin (2003), believe that settlement was a gradual process that evolved from summer and winter camps and an increasing reverence for the dead coupled with growing skills as proto-farmers and herders. Hodder's (2006) excavations at Çatalhöyük, a settlement in central Turkey, occupied by 3,000–8,000 people around 9,000 years ago (7400–6000 BC), suggest a reverence on the part of the inhabitants for their ancestors who they buried beneath the floors of their houses. Çatalhöyük was in no way an urban centre characterized by the division of labour and provision of common services we see in modern towns and cities. It was more a collection of family units which acted independently yet lived clustered tightly together in houses that were accessed through the roof.

The earliest evidence of a settlement in England is at Star Carr in Yorkshire dating from 9,000 years ago, probably about 5,000 years after humans appeared in Britain. By 6,000 years ago Britain's population had probably risen to around 100,000 with permanent and semi-permanent settlements throughout the islands, including the Thames valley.

3.3 Early urbanization

Successful settlements require sustainable sources of water, food and fuel for cooking and warmth. Management of the amount of land allocated to food and fuel production to satisfy a growing population was a challenging task. Gradually, however, the weight of numbers would have had an impact on the surrounding environment, making wood and combustible biomass difficult to find close to the settlement. Timber for building huts would also have become scarce and would have needed to be transported over increasing distances. Successful settlements managed these problems by improved farming techniques and trade. However, archaeologists and historians list many examples of once powerful settlements and even cities that fell into ruin because of the denuded land's inability to sustain the population (e.g. Diamond 2005).

Around 4,000–5,000 years ago (c. 3000–2000 BC) large towns and cities such as Ur, Uruk and Lagash appeared in Mesopotamia, the strip of land between the rivers Indus and Euphrates in modern Iraq. These drew upon the surrounding villages for food and fuel, which were either traded for manufactured goods made in the cities or paid as tribute. Importantly, most of these cities were located on navigable rivers that provided a low-cost method of bringing bulk goods, such as building material and fuel wood, from the surrounding countryside. A village field plan proposed for Ur by Postgate (1994) suggests that the raised dykes along the river Euphrates were used to grow sustainable fuel wood possibly through coppicing, whilst the easily irrigated land below the dykes was devoted to crops. Harvested wood and possibly other biomass, e.g. reeds or straw, would have been loaded into boats and transported by river to be unloaded at the wharves in Ur.

Although Europe and Britain were both settled extensively around 5,000 years ago, in Britain (unlike in mainland Europe) there is little or no evidence of towns before the Roman colonization of AD 43. Pre-Roman settlement in Britain took the form of over 3,000 hill forts dating from around 500 BC, each with its dependent hinterland of smaller enclosures and farmsteads (Darvill 1987). The Romans understood the strategic importance of rivers and built a bridge at the highest fordable point on the Thames (near present-day Westminster). They founded Londinium about 10 years later, a short distance down-river on the two low hills of Ludgate and Cornhill, either side of the Wallbrook stream where it entered the Thames. The city was destroyed by fire barely 10 years after building began and then was sacked by Boudicca in AD 60. The city was rebuilt and, in AD 200, a defensive wall was built round the city, enclosing an area of nearly 163 hectares and probably housing around 25,000 people, although it may have been as little as 6,000 (Ackroyd 2000).

In common with Ur, Rome and contemporary settlements large or small, Londinium required a wood supply system to provide construction timber for buildings and ships, as well as fuel. Wood fuel was not only required for domestic cooking but also for brewing, glass manufacture, metal-working and the bath houses. To illustrate the amount of wood consumed by the Roman way of life, Williams (2002) quotes a figure of 23 acres of coppiced woodland being required to sustainably fuel the Roman baths at Welwyn in England. In fourth century Rome, traders known as *lignari* supplied the city with its annual demand for 1–1.5 m^3 per capita of *lignum* (fuelwood). Sands (2005) states that whole forests close to Rome were cleared to provide fuel, and towards the end of the empire, there were clear indications of the shortage of wood for fuel. Charcoal was substituted for wood, fired bricks became thicker and used less mortar, glass was recycled and there was a return to mud-brick construction. Londinium's population was somewhat smaller than that of Rome during the latter part of the Roman Empire so it is likely that sustainable supplies of timber and wood fuel within easy reach of the Thames would have been more than sufficient. Political instability led to a decline in London's population from about AD 300. By AD 1100 however, the population was perhaps 15,000 people. Wood fuel was supplemented by materials such as reeds, straw and stubble by London bakers and brewers but their use was forbidden soon after AD 1212 because of the danger of uncontrolled fires.

By AD 1300, London had grown to be a city of 80,000 people. The demand for fuel was increased by industrial metal working, textile finishing, pottery and glass making, baking, brewing, lime burning and firing tiles for roofing. Domestic cooking and heating added to the demand. Although coal was already being imported into the city, predominantly for lime burning in the western area, wood was the preferred fuel for domestic use, food preparation and glass making. *Buscari* or *Wodemongere* traded in wood grown in the well-wooded counties surrounding London. It was brought into the city by barge and landed at Woodwharf upstream of London Bridge, Queenhithe and Timberhithe. Fuel wood logs, faggots (bundles of rods tied up with bands of twisted hazel) and charcoal were produced from underwood (the residue from trades using timber) and the branches of timber trees (Rackham 2010). Charcoal was produced both as an urban fuel and for iron smelting, for which there was no substitute. Galloway *et al.* (1996) estimate that the annual per capita demand for fuel was 1.76 tons of dry wood (equivalent to 0.75 tons of coal), creating a demand for London of 140,000 tons of wood per annum in AD 1300 which would have required the entire underwood output of around 70,000 acres (28,000 ha) of intensively managed woodland. Around 16 per cent of the counties surrounding London probably lay under woodland in AD 1350 (Galloway *et al.* 1996) suggesting that London had an adequate local fuel supply. However, the cost of land transport would have made transporting faggots more than about 14 miles by cart prohibitive. River transport, on the other hand, was much cheaper and this had the effect of creating a supply zone for faggots that extended along the Thames from the Kent coast as far as Henley upstream but no more than about 10 miles from the river. This supply system continued into the fifteenth century, although the increasing scarcity of wood around some towns required overland transport from as far as 20 miles away (Lee 2003).

3.3.1 Early renewable energy

Whilst the majority of mechanical power was provided by draught animals and humans, wind had been harnessed as early as 4000 BC by the Egyptians who used sails extensively on the Nile. The power of flowing water was captured by water-wheels somewhat later in the western Mediterranean around 500 BC, and was rapidly utilized for grinding cereals, thus eliminating many thousands of painful repetitive hours of human toil. Water-power was soon developed to drive a variety of machines used, e.g. in blacksmithing, tanning, fulling and wood turning.

A Roman tide mill relying on harnessing the water column between high and low tides may have been built as early as before AD 200 on the river Fleet, close to London's walled city. There were numerous tide mills along the Thames-foreshore in the early medieval period. An Anglo-Saxon water mill was built in Carshalton close to the source of the river Wandle. There were also water mills located in the arches of London Bridge which used the tidal flow to drive pumps supplying water to the city. The first was built around 1580 by a Dutchman named Peter Morice. The original wheels were destroyed in the Great Fire but were replaced and, by 1731, three water wheels operated 52 pumps delivering 132,120 gallons of water an hour to a height of 120 feet. The pump continued to provide water to the city until 1822–7 years after house waste was permitted to be discharged to the Thames via the sewers (Halliday 2003).

London also had a useful wind resource. The first wind turbines were probably invented around AD 900 in the Middle East but it was not until the fourteenth century that windmills began to be widely used in mainland Europe. In Britain, seven mills had been erected in Stepney by the eleventh century (Ackroyd 2000) and in subsequent years, over 200 windmills were built throughout London to provide flour for the growing population. Many survived only a few years but several remain today.

Although water wheels and eventually windmills were capable of providing about 25 times more power than a horse, or 50 times more than a man, in Britain as a whole they provided less than 10 per cent of the total power capacity available from humans and animal muscles. This situation continued until about 1800, when steam began to dominate inanimate sources of power (Smil 1994, Wrigley 2010).

And so, whilst many aspects of thirteenth century life in London would have seemed extraordinary to a visitor from 2000 BC Ur, the energy system would have been familiar. For about 3,500 years very little had changed in the urban energy system. Life in these cities moved slowly and depended upon what could be harvested the previous year and what could be carried on a person's back, by horse and cart or by boat. Improvements in technology and the increased use of draught animals and industrial heat resulted in a doubling of the energy consumed per capita from 15 GJ per capita per year in 1500 BC Egypt (Smil 2010) to about 30 GJ per capita per year in seventeenth century Europe (Malanima 2006).

3.4 London's expansion and the move to coal

The next major transition was from resources harvested instantaneously or within a limited number of years to the use of the vast reserves of fossil fuels that represent hundreds or thousands of years of equivalent energy stored in a compact form. Coal was one of the first fossil fuels to penetrate the urban energy system. The amount of coal burned in late thirteenth-century London was relatively small, but the smell was sufficient to incur the wrath of Edward I who banned its use, threatening to confiscate the forges where it was burned. The use of coal in London persisted however, causing a second monarch, Elizabeth I, to complain of the smell of coal smoke nearly 300 years later.

Between 1520 and 1550, London's population grew from 55,000 to 120,000. It was also getting colder. During the winter of 1564–5, Queen Elizabeth is said to have taken a daily stroll on the frozen river Thames (Freese 2006). The growth in demand for fuel put such a strain on the previously sustainable fuel wood supply that it began to fail and the price of wood at the London wharves increased sharply. The Domesday Book estimated that about 15 per cent of Britain was covered in forest during the eleventh century, but by the end of sixteenth century this had diminished to about 6 per cent (Allen 2010) due largely to woodland clearance around large towns and cities to create agricultural land for food production. This forest shrinkage was particularly significant for London which relied on a ring of woodland now 20–40 miles away, resulting in a doubling of the price of wood by 1550 and a trebling to 12 grams of silver per GJ 100 years later (Allen 2010).

The price of coal, however, remained low at between 2 and 4 grams of silver for the same heating value. This discrepancy was almost entirely due to the cost of transport. The weight of wood that could be cut by a woodsman in 1 year was similar to the amount of coal that could be dug by a miner (Rackham 2010). The difference in heating value, coupled with the fact that coal was only transported a short distance from the mines on the Tyne to ships that took it to London, meant that, by the time both fuels reached the city wharves, the cost of fuel wood per GJ was much higher than that of coal.

London's ability to draw on a supply of coal brought by sea and river from Newcastle more cheaply than wood transported 20 miles or so overland illustrates the importance of the position of London and other successful commercial cities on a navigable waterway. Not all European capitals were so fortunate. In 1561, Philip II decreed that Madrid should be the capital of Spain as it was located at the geographical centre of the Iberian Peninsular. Madrid was at least 350 km from the sea and accessible only by land, making it the largest city without access to water transport in European history. In addition to this logistical disadvantage, Madrid's hinterland was thinly populated, with wool being the only product that could profitably bear the cost of transport to the coast.

As a result, by the early 1780s, one-third of the Crown's net revenues was spent on supporting the city (Reader 2005).

Although many people were reluctant to use coal and few houses were able to burn it successfully, substantial quantities of coal were imported into London during the first quarter of the seventeenth century (Hatcher 1993). Most houses were heated by a central fire which allowed the smoke to gather above the hearth and thence to an open flue, as it had since medieval times. As the century progressed, Allen (2010) suggests that some enterprising builders added chimneys and fire grates to take advantage of the cheaper coal fuel without filling the house with noxious fumes. After experimentation and 'collective invention', functional coal-burning houses emerged and it is probable that this trend of building houses with chimneys accelerated during the rebuilding after the Great Fire of London (discussed below). The commercial use of coal also increased as bakers and brewers developed technology to prevent their products being tainted by coal smoke. The 'iron masters' had come under considerable pressure to limit the amount of wood they consumed for charcoal from a statute of 1580 which prohibited ironworks using charcoal made within 22 miles of outer London. However, it took a further century for iron manufacture to turn to coal, following the development by Abraham Derby of a coke smelting process, and another 50 years or so of improvements before coke iron production started to replace charcoal. Similar pressure was brought to bear on the glass industry, which was reluctant to use coal because the fumes impaired the quality of their product. Nevertheless, Parliament banned the use of wood for glass making in 1615. This compelled the glass manufacturers to adopt new technology that enabled them to burn coal in modified furnaces, which resulted in relocating the industry away from London towards the northern coalfields.

The move away from relatively short rotation biomass fuels to coal took even longer in continental Europe. Whilst there had been relatively little government intervention in controlling fuel prices in London, the situation in seventeenth century Berlin was different, where wood prices were controlled by state edict (Sieferle 2001). These proved to be ineffective and a central firewood administration was established in 1694 to regulate the private wood trade. In 1702 in Konigsburg, firewood was rationed according to the rank of the household. Wood conserving stoves were known as early as 1325 but in 1763 an official contest took place in Prussia to design a domestic stove that consumed minimum quantities of wood. Iron production had benefited from the use of dedicated wood supplies in parts of Germany but this privilege was removed in 1783 with disastrous results on their ability to export cheap iron. By the end of the eighteenth century, the state authority that maintained the cheap price of fuel wood in Berlin allowed the price to rise and also began to sell coal at a loss to promote its use. In Silesia, the use of coal was also increasing rapidly in spite of the widespread propaganda against it during the first half of the eighteenth century. Authoritative articles and pamphlets had been published in Germany and France describing the health hazards of burning coal. One claimed that one-third of all the inhabitants of London died of wasting diseases and lung ailments caused by the corrosive effects of smoke from coal fires. Another obstacle to the widespread use of coal in continental Europe was the inland location of the coal pits and commercial centres, and the lack of good navigable waterways. This meant that the price of coal was relatively high and as late as 1886 in Hamburg, Ruhr coal could not compete with imported coal from England.

By 1800, the consumption of coal in Britain had risen to 15 million tons per annum providing both domestic and industrial heat (Allen 2010). The potential to convert heat into mechanical power, which would substantially increase the demand for coal, had yet to be realized, although the Royal Society had focused their attention on the problem as early as the late 1600s. They had been intrigued by Denis Papin's demonstration of a brass cylinder containing a piston that was made to move by heating water in one end of the cylinder. Newcomen, who probably did not know of Papin's invention, built a much larger piston with a separate steam boiler and convinced the mine owners that steam power was a practical proposition. His first 'fire' engine was installed in a mine in 1712 and was much cheaper to run than the 50 horses it replaced, even though it was extraordinarily inefficient. In spite of the Royal Society's indifference to Newcomen's demonstration of the potential for steam power, the eighteenth century saw a flurry of related inventions. These both widened the application of steam and improved its efficiency to the point where, in 1801, Trevithick was able to demonstrate the Cambourne road engine. Stephenson's Rocket appeared in 1825 and was operated on the Stockton to Darlington railway, heralding a new age for transport which was to have a resounding impact on city life. From this time on, people would be able to travel further for the same time commitment (or rather physical energy expenditure, see Kölbl and Helbing 2003).

3.5 Networks of roads, rails and pipes

In the mid-seventeenth century, London was England's economic powerhouse with an estimated population of almost half a million. The city's architecture had changed little from the Middle Ages. Narrow, cobble-stoned, foul-smelling streets doubled as the city's sewers. Many of the streets were lined with homes made of wood and pitch up to four storeys high. The upper storeys of these homes overhung the lower ones and projected into the street, effectively blocking out the sun and decreasing the distance between the buildings. This typical construction and London's uncontrolled growth had created ideal conditions for the spread of disease and a fireman's nightmare: a city dominated by old, dry, wooden structures, tightly packed into a confined space just waiting for a spark to ignite disaster. In 1665, plague struck the city and claimed 100,000 lives. A year later, on 2 September 1666, a fire started in Pudding Lane and within 4 days, the Great Fire of London had destroyed 13,000 houses within the old city walls and left about 80,000 people homeless.

Charles II was sympathetic to the concept of a complete redesign of London and a number of plans were drawn up by notable architects and scientists of the time, including Robert Hooke, John Evelyn and Christopher Wren. All the plans featured wide streets and grand squares with Wren and Hooke favouring a grid system, but the formidable task of raising revenue for land acquisition and compensation for houses and businesses that could not be rebuilt meant that the city was eventually reconstructed along existing street lines and property boundaries. However, the first Rebuilding Act passed by Parliament in 1667, imposed regulations for wider roads and strict conditions for the types of stone or brick houses that could be built. Timber buildings were forbidden.

By 1667, most of the city had been rebuilt, although the new buildings were again divided by lanes and small alleys, giving 'the general impression [of] once more dense and constricted life' (Ackroyd 2000: 240). Any attempts at improving the means by which goods and people were distributed about the city appear to have failed. The urban energy system remained based on the need to supply food daily to the city markets, either on the hoof or by boat and cart. Wood fuel was still in use although its high cost, together with the new grates and chimneys that were being incorporated in the stone and brick houses, served to increase the demand for coal.

As London grew in size and commercial importance, improved road transport became critical. The Roman roads had long since deteriorated and until the Highways Act of 1555, which made local parishes responsible for their maintenance, they were in a poor state and barely passable in winter. Matters improved slightly by 1706 when Turnpike Trusts were allowed to levy tolls to invest in new roads. Eventually there were about 1,100 trusts in Britain responsible for 36,800 km of engineered roads. London and other large cities were now well connected by stagecoach reducing the time of a journey from London to Edinburgh from almost a fortnight to less than 2 days until the arrival of the railways and riots about the cost of tolls contributed to the abandonment of the Turnpike system in the mid-nineteenth century.

London was additionally connected to its hinterland by water – not only by the river Thames, which had undergone improvements since 1624 – but also by a series of canals. The Grand Junction canal provided London's principal link with the rest of the Britain's canal system. Work began in 1793, following an Act of Parliament authorizing its construction, and was completed 12 years later connecting the Oxford canal to the Thames at Richmond. Part of the plan was to create an arm from the canal in West London to Paddington nearer the centre. Opened in 1801, the Paddington Arm was of great benefit with a large basin acting as a distribution centre for goods to be taken to other parts of London by cart. A successful passenger boat service also ran for a number of years on the canal from Paddington to Uxbridge; a town about 15 miles away. Meanwhile, the river Thames continued to serve as an important means of transporting people as well as goods into London, and in 1850 at least 15,000 people travelled to work by paddle steamer along the river (Taylor 2002).

Although the canals brought coal and non-perishable goods into London, by the middle of the nineteenth century, passenger transport was dominated by the thousands of wagons pulled by six or eight horses entering or leaving London each week. Horses were also the chief method of moving people around within the city. The first omnibus service appeared in 1829, pulled by three horses as in Paris, but after a few years most were drawn by two. By the end of the nineteenth century 200,000 horses were working in the streets of London (Turvey 2005) and traffic congestion was severe (see Figure 3.1). Rivalry between omnibus companies was great and it was not uncommon for pedestrians to be injured or even killed. In one instance, a driver was charged with manslaughter for running over and killing a man whilst racing another omnibus at a speed of 12 miles an hour in a crowded thoroughfare (Knowledge of London 2011). Despite these disadvantages, passenger and goods traffic within the centre of London

Figure 3.1 Illustration of Ludgate Hill by Gustav Dore, showing the railway bridge with a train crossing billowing smoke. Both Ludgate Hill itself and Fleet Street are completely packed with people, horse-drawn buses and other vehicles. The dome of St Paul's cathedral can be seen in the background. © TfL from the London Transport Museum collection. Reprinted with permission.

still depended upon horses for many years as, aided by the new railways, London grew and more people commuted into London from the growing suburbs.

The railways revolutionized land transport and enabled fresh food, fuel and people to be brought into the city centres on a daily basis at a relatively low cost. London's Euston station was opened in 1837. Mainline stations operating from London were prohibited from the centre of the city because Parliament was concerned about the disruption their construction would cause to the city but, even so, at least 100,000 Londoners had their homes destroyed. London became so congested that Charles Pearson, a City of London solicitor, set out plans in 1845 for an underground railway to alleviate the problem. After a number of false starts and reluctant investors, construction began in 1860. Sadly, Pearson died in 1862, a year before the first underground railway was opened between Farringdon and Paddington (Wolmar 2004). This relied on steam engines for motive power with chimneys to enable smoke and steam to escape from the tunnels.

Changes in transportation networks were paralleled by the growth of new energy services. Parliament had already responded to the growth of urban populations and the resultant need for additional health and safety measures. During the eighteenth and early nineteenth centuries it had passed a number of Town Improvement Acts that compelled towns to do such things as clean, pave and light the streets, and provide a clean water supply. Initially, street lighting was by means of 'parish lamps' consisting of a small tin vessel, half-filled with fish oil containing a piece of cotton twist as a wick (Ackroyd 2000). Lamplighters were employed to light, trim and fill the lamps. By the end of the seventeenth century, oil lamps were the dominant form of street lighting. This improved matters but the Royal Society saw the need for further improvements and offered prizes from 1770 for new lighting technology. By the early nineteenth century there were 35,000 lamps lighting the streets of London. Various illuminants were used in domestic lamps, including animal fats, vegetable oils and beeswax. One of the cheapest forms of lighting was the rushlight, made by dipping reeds repeatedly in tallow to build up a candle. One and a half pounds of rushlight would last a family 1 year. Whale oil, particularly from sperm whales, produced a bright light and spermaceti wax from the whale's brain cavity gave a better light than tallow candles. Technical improvements to oil lamps improved their efficiency and the cost of oil lighting fell by two-thirds between 1750 and 1820 (Fouquet and Pearson 2006). The price of lighting continued to fall but, by the middle of the nineteenth century, whales were becoming increasingly difficult to find and a new alternative, 'kerosene oil', derived from coal or crude oil entered the lighting market.

A key innovation at this stage was the development of physical networked infrastructures for supporting urban energy services. At the end of the eighteenth century, William Murdoch in Britain and Philip Lebon in France were independently experimenting with the gases that were produced by heating coal or wood under controlled conditions. In 1798, Murdoch used coal gas to light a room in a house in Cornwall and in 1801, Lebon staged a demonstration of gas lighting in Paris (Williams 1981). Albrecht Winzer, a German professor of commerce, was quick to realize that this provided the potential for a new method of lighting and began public demonstrations of gas lighting in London in 1804. Increasing public interest encouraged steam engine pioneers Matthew Boulton and James Watt, together with Murdoch, to install six cast-iron retorts to provide the gas for lighting a cotton mill in Manchester in 1806. In the same year, Winzer, who realized the commercial advantages of centrally manufactured gas that could be piped to many customers, anglicized his name to Frederick Winsor and attempted to gain a government charter allowing him to form a commercial gas lighting company. His first attempt failed but, in 1812, he started the Gas Light and Coke company that would eventually control most of the London gas market. The new technology caught on quickly and by 1849, gas lighting provided by local gas works had been installed in 700 large towns in the UK.

Both the growing domestic and commercial markets for lighting, however, were about to undergo further change. The industry was extremely competitive and rival firms supplying the same area caused great disruption in the streets by laying separate gas mains. In 1860, Parliament took action to regulate what was regarded as wasteful competition and the Metropolis Gas Act granted the London gas companies an effective monopoly in their respective areas of supply.

3.5.1 London's sewer

By the time of the Great Exhibition in Hyde Park in 1851, London's energy system provided gas lighting for those who could afford it. It had railways that brought commuters in from the growing suburbs and food from the hinterland, ships and canal boats that brought in commodities and coal, and streets that were liberally covered with the ordure from thousands of horses pulling trams and hansom cabs. Although not normally thought of as part of the urban energy system, the problem of how to dispose of London's sewage had a major impact on the management and construction of the city's infrastructure. The majority of the waste from the activities of nearly two and a half million people ended up in the river Thames. Since the Middle Ages, London's sewage was discharged into cesspools where it was collected for sale and carted to the surrounding farms as agricultural manure. During the first half of the nineteenth century, the number of cesspools had grown to 200,000 and the cost charged by the 'nightsoil men' to empty them and transport sewage to the increasingly distant farms was becoming prohibitive. The growing popularity of the flushing water closet exacerbated the problem by increasing the volume of untreated sewage to the point where, in 1850, about 150 million tons was flushed into the Thames.

The problem of London's sewers had been recognized by Wren in 1678 but nothing was done. The result was inevitable and 1831 saw the first cholera outbreak in London, with 6,536 deaths. The second outbreak in 1848 killed 14,137 and the third in 1853 resulted in the deaths of 10,738. The smell from the river was so bad by 1858 that it was heralded as 'The Great Stink' and after six separate Royal Commissions had attempted to arrive at a satisfactory scheme for sewage collection in the metropolitan area, Joseph William Bazalgette, the chief engineer of the Metropolitan Board of Works, began work on a comprehensive new sewage system (Halliday 2003).

Bazalgette's plan proposed a network of main sewers, running parallel to the river, which would intercept both surface water and waste, conducting them to outfalls at Barking on the northern side of the Thames and Crossness, near Plumstead, on the southern side. Work began in 1858. The press followed the building of the sewers, embankments, pumping stations and reservoirs with considerable interest and generally praised the work in glowing terms. This, and the increase in the rateable value of properties in the metropolitan area made the task of raising a further £1,200,000 in 1863 relatively easy. In June, John Thwaites, the chairman of the Metropolitan Board of Works wrote to the then Chancellor, William Gladstone, requesting authority to borrow this sum against the security of the additional rates and, 1 month later, Parliament granted the Board the power to borrow the sum required.

The Prince of Wales officially opened the new sewage system on 4 April 1865, although it was not completed for another decade. The system had been designed for a population of 3,450,000, about 25 per cent more than that existing in 1850 but by the time of Bazalgette's death in 1891, the number of people had risen to 4,225,000 with per capita water consumption at 90 gallons per day, treble that allowed for in the original plan.

By 1865 large numbers of fish had returned to the Thames but below the outfalls, the river was still polluted. Arguments about the condition of the river below the outfalls continued for 17 years, resulting in a Royal Commission in 1882. The Commissioners produced two reports in 1884, the second of which declared that 'it is neither necessary nor justifiable to discharge the sewage of the Metropolis in its crude state into any part of the Thames'. Once again, the press with scathing articles in *The Times* and the *Pall Mall Gazette* applied pressure forcing the Board to engage a contractor to build 13 precipitating channels at Barking at a cost of £406,000 and a similar scheme at Crossness at a cost of £259,816. In addition, six sludge vessels were ordered to take the sludge and dump it at sea. The first vessel, arriving in 1887 from Barrow at a cost of £16,353 was named the 'Bazalgette' and remained in service until 1998, when it was replaced by an incinerator at Barking. Bazalgette's scheme was largely completed in 15 years, including construction of the Victoria, Albert and Chelsea embankments. The Metropolitan Board of Works, created to oversee the project was replaced by the London County Council in 1889 and the pumps powering the water supply and sewage disposal network probably consumed less than 2 per cent of London's annual total coal imports (Caller 2012).

London's sewers are therefore an important part of the overall transition to the creation of rationalized network infrastructures for vital urban services. Figure 3.2 shows an artist's impression of the Victoria embankment drawn in 1867, illustrating the sewer, the underground railway, pipes for water, gas and later electricity, and a projected pneumatic railway under the Thames, which was never built.

3.6 The electric city

Lighting produced by either a gas or oil flame had improved considerably since the days of candles and rushlights but suffered from a number of disadvantages. It was a potential fire hazard, the flames produced soot, and light levels were low, especially outside. Gas was expensive and slow to penetrate the domestic market. An experiment with gas lighting in the newly built Houses of Parliament in 1838 was abandoned since the cost was nearly four times that of traditional wax candles (Barty-King 1984). Growing urban populations, not only of London and Britain needed a cheap, safe, easily distributed form of energy. Electricity promised to be the answer.

Figure 3.2 Impression of Victoria Embankment drawn in 1867 showing pipes for water, gas and later electricity (1), the sewer (2), the underground railway (3), and a projected pneumatic railway under the Thames, which was never built (4). © Science Museum/Science & Society Picture Library. Reprinted with permission.

Since Faraday demonstrated that an electric current could be produced by moving a conductor through a magnetic field, several devices had been developed utilizing the 'new electricity'. The first commercial application was Henry's electric telegraph in 1844, followed by Bell's telephone in 1876 and Edison's phonograph a year later. But it wasn't until 1878 that the first practical electric lighting system was demonstrated by Jablochkoff, in Paris and London. This consisted of an electric arc struck between two carbon electrodes. The year 1878 also saw the first floodlit football match when Sheffield Football Association played the first evening game under 8,000 candle power arc lights provided by two Siemens generators. While the system was difficult to operate and produced noxious fumes, it was clearly a threat to the gas industry. A concerned gas company appointed a committee who reassured shareholders by reporting 'we are quite satisfied that the electric light can never be applied indoors without the production of an offensive smell which undoubtedly causes headaches and in its naked state can never be used in rooms of even a large size without damage to sight'. However, arc lighting was adopted extensively in the USA and by 1890, there were over 200,000 arc lights providing illumination to American city streets. By contrast Britain had only 700 (Shiman 1993).

Incandescent light bulbs suitable for indoor lighting were invented independently by Swan in Britain and Edison in the USA in 1879. By 1881, 1,200 of Swan's bulbs were used to light the Savoy theatre in London, in front of an astonished audience. Edison quickly realized the advantages of Winzer's centralized gas distribution system and built a generating system which provided electricity to a network of customers through copper wires. He demonstrated his lighting system at the Paris exhibition in 1881 and at Crystal Palace in London a year later. In the same year, he opened steam-powered central electricity generating stations in New York and London. By May 1882, 16 new companies offering similar electricity supplies appeared, creating considerable speculation in electricity stocks.

The UK government both encouraged and hindered the development of the fledgling gas and electricity industries. In the same way as the 1860 Metropolis Gas Act had allowed district monopolies, the 1882 Electric Lighting Act gave similar monopolies to electricity companies, only this time a so-called 'scrap-iron' clause was included. This allowed the companies to have the monopoly supply for a specific district for 21 years, after which the company would be subjected to compulsory purchase at a value based on its material assets. The Electric Lighting Act additionally imposed severe conditions over the way in which electricity could be supplied and distributed including the restriction of overhead cables. Coupled with the lower cost of gas brought about by gas companies consolidating to provide larger more efficient production facilities, this proved disastrous for investment in the UK electricity industry. By 1884 Edison's power station in London had closed.

Welsbach's invention in 1886 of the incandescent gas mantle, which provided a brighter light, also impacted the fortunes of the fledgling electricity industry. In an attempt to revitalize it, in 1888 Parliament abandoned the 'scrap iron' clause. London's relatively slow adoption of electric lighting seemed to be based primarily on cost, together with fears about the implications of granting a monopoly to an electric company. One local authority announced it was taking bids for electric lighting in order to force gas prices down. Further delays to the electrification of London were caused by the Board of Trade's inquiry into the London Electric Supply Company's (LESCo) plan to invest £1 million in what would have been the world's largest central generating station at Deptford. Designed by Sebastian Z. Ferranti, it was to consist of four 500-ton alternators generating sufficient electricity to supply 2 million lamps. Work began in April 1889 and LESCo applied for the necessary statutory authority to lay cables under streets within the boundaries of 24 local authorities. The application was on an unprecedented scale and coincided with a number of smaller applications from other companies. Sensing an opportunity to regulate the operations of the electricity supply undertakings by laying down the principles under which the power to break up the streets of London would be granted, the Board of Trade ordered a local inquiry. This also gave them a chance to examine the whole scope of the Deptford station. As a result of fears expressed by the inquiry that the centralization of generating capability at Deptford would risk supply reliability to London, the LESCo directors agreed to put two of the four 10,000 horsepower generating units in a station to be built elsewhere. By the end of 1889, a smaller generating unit was brought into service at Deptford but, as a result of technical failures and investors' loss of confidence, Ferranti's dream was never realized. Deptford struggled on and, after a major refurbishment in 1900, progressed to become one of the major London power stations until it eventually closed in 1983 (Cochrane 1986).

At the beginning of the twentieth century, there were over 800 gas businesses supplying the UK market and by 1935, the gas supply industry provided employment for about 230,000 people, supplied about a quarter of the population and had capital assets worth about £200 million. By 1920, the UK electricity system consisted of over 600 suppliers owned both by local authorities and private companies acting independently, resulting in about 75 per cent more generating plant throughout the country than was required to supply peak demand. As an example of the diversity that had grown in the industry, by 1918 in London alone there were 70 authorities, 50 different types of system, 10 different frequencies and 24 different supply voltages (Hughes 1983). It was clear that both systems needed to be rationalized and improved.

In 1925, a government report produced by Lord Weir recommended that electricity generation should be restricted to a limited number of power stations connected to a national grid. This resulted in the 1926 Electricity Supply Act and the formation of the Central Electricity Board, which was nationalized under the Electricity Act of 1947. It took much longer to nationalize the gas industry. The Gas Regulation Act of 1920 was the first to allow the exchange of gas from one undertaking to another but this was complicated by the demand pattern for gas, such that by 1938, only 0.4 per cent of all gas sold was via this type of exchange. However, the implied threat of nationalization accelerated the amalgamation of undertakings into holding companies. The 1939–45 War led to valve systems being installed between adjacent undertakings in London, Liverpool and Manchester so that supply could be continued and sections of main isolated for repair. The UK gas industry was eventually nationalized in 1948. The two nationalized utility companies laid the foundations of today's UK energy supply by connecting rural and urban consumers through an electricity grid completed in 1933 and a natural gas grid finally completed in 1978.

These changes in both electricity and gas networks marked a significant shift in the structure of urban energy systems. Whereas, before energy resources were imported to each individual city from its hinterland or wider markets on an ad hoc basis, cities were now connected as parts of national energy systems. This had the advantage of physically removing many of the externalities of the energy system from the city (e.g. pollutants from combustion in electricity generation). Significantly, it also meant that consumers were increasingly separated from the impact of their consumption and were largely ignorant of the complexity of the underlying system.

3.7 Conclusion

In the last 2 millennia, London has grown from a small, fortified town to a city of 7.5 million people, covering an area of about 600 square miles. Prior to the sixteenth century, London was relatively sustainable with a small population, food grown on surrounding farms, and wood fuel produced from managed woodlands along the Thames; the annual per capita energy demand was about 26 GJ with about a quarter being consumed by baking and brewing (Galloway *et al.* 1996). However, in the first decade of the twenty-first century, London's population had an annual per capita energy demand around five times that of their medieval counterparts and the city's 'ecological footprint' has grown from the counties along the river Thames to an area double the size of Britain with supply chains that extend around the world (GLA 2003b).

The history outlined here shows how London's urban energy system has changed over time. From its early beginnings with a biomass-based system, the city switched to coal imported by rail and ship to provide heat, town gas and electricity. These services were produced within Greater London for many years, until the advent of national grids in the twentieth century. All the gasworks in London disappeared when Britain switched to natural gas and only one electricity power station still operates in Inner London at Willesden, with two more in Greater London. In transport, although most of our daily journeys are still taken by car (37 per cent) about one-fifth of us still walk to work compared with the vast majority in the medieval city.

The technologies of steam and electricity and the new fuels, gas and oil, were incorporated into London's urban energy system largely due to the efforts of individual entrepreneurs initially backed by private investors. In the past, attempts to improve London's infrastructure and energy services were confronted by problems over land ownership, loss of investor confidence, inefficient competition between suppliers and delayed, and often poor, legislation. Overall however, these innovation processes have been successful and technology has reduced the cost of heat, light and power (Fouquet 2008), and mechanized transport has reduced the cost of moving goods and people round the city so effectively that there has been little incentive to improve the efficiency of the systems that provide London's urban energy supply in recent decades.

The future, however, holds several challenges. London's mayors have committed to reducing the city's carbon footprint, and energy generation and efficiency have been the subject of numerous reports. In addition, the subjects of energy security and cost have added to increasingly urgent discussions in city institutions regarding the future energy supply. Fortunately, the pace of global urbanization means that city planners not only have a wealth of information describing past energy transitions to consider but are also able to share process and systems design ideas from the many new cities and urban developments currently under construction throughout the world. Perhaps all this attention will lead to a coherent approach to the design of urban energy systems that consider the mistakes as well as the successes of the past.

Part II
Urban energy use and technologies

4 Building energy service demands
The potential of retrofits

Mark G. Jennings

Imperial College London

4.1 Introduction

From Hausmann's boulevards in Paris to the informal settlements of Kibera and the skyscrapers of Asian megacities, much of a city's character is determined by its buildings. These structures are the location for the majority of urban activities, such as work and leisure, and they correspondingly represent the origins and destinations of most urban journeys. It is estimated that 60 per cent of global energy demand originates from buildings (IEA 2010a). Improving the energy performance of buildings is therefore an important part of any overall plan to enhance the performance of an urban energy system.

The focus of this chapter is on retrofitting existing buildings. Since buildings are long-lasting infrastructures, much of the existing building stock will need to be improved if short and medium-term energy efficiency and greenhouse gas reduction targets are to be met. The core query of this chapter is therefore: *What are the best strategies of planning and organizing building stock retrofits such that the efficiency of an urban energy system may be improved?*

Efficiency is considered in this chapter in terms of operations, excluding life-cycle analyses, and the chapter offers an introduction to building energy service demands in the context of retrofits. The narrative presents an overview of key considerations of technology, temporal dynamics of heat and power demand, social constructs of buildings, and benefits of and obstacles to effective retrofits. Mathematical programming strategies are offered, partly as an introduction to the analysis techniques of later chapters. The key conclusions are clear: fuel and power demand is well understood, although the drivers of retrofit choices are not. Nonetheless, mathematical programming offers a robust decision-making tool for improving the decisions made when financing, planning, and organizing building retrofits.

4.2 Fuel and power demand in buildings

Buildings can be divided into those used for residential purposes and those for commercial purposes. Commercial and residential buildings require different services, though both share requirements for fuel or electrical power input. In the last 200 years the global fuel and power demand from buildings has risen dramatically, modelled as

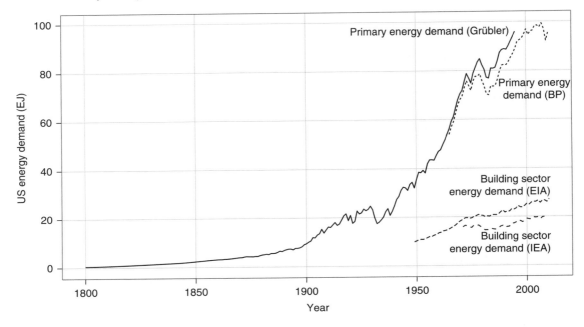

Figure 4.1 Historical estimates of US primary and building sector energy demand.
Data sources: Grübler (1998), BP (2011), EIA (2011), IEA (2011a).

rising from 57.6 EJ in 1971 to 115 EJ in 2009 (IEA 2011a). Figure 4.1 illustrates historical demand growth in the United States. The total primary energy supply is seen to have increased since the beginning of the nineteenth century, and this increase appears to be also true for the building sector demand (at least for when data is available).

While the USA is not necessarily a proxy to be used for other countries, Figure 4.1 does nonetheless reflect the magnitude of changes to global building sector demand in recent years. Whether existing building sector demands in urban energy systems can be reduced is one of the key questions posed by this chapter.

Considering these rising demands for fuel and power, what building services do fuel and power supply? The services depend on a number of factors, such as the purpose of the building and the country and culture in which the building is located. Figure 4.2 illustrates the demand shares of residential and commercial buildings by end-use in China, the USA and Great Britain. This is a static snapshot however, and service demand shares continue to evolve in many countries. Nevertheless, it is clear that space heating, water heating and lighting and other electrics tend to dominate the majority of building sector demand.

4.2.1 *Fuel demand in buildings*

Fuel is required to provide certain building services. The primary building services that typically require fuel input (rather than electrical power input) are those of space and water conditioning. There are a large number of fuels and associated technologies that

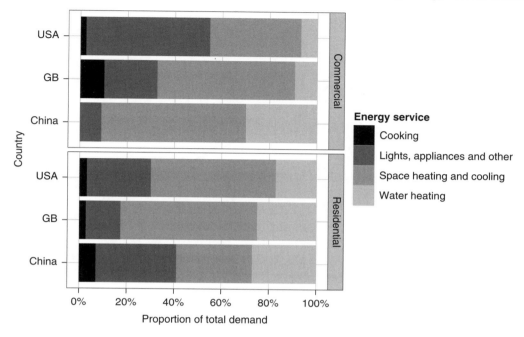

Figure 4.2 Annually aggregated end-use demand shares for buildings in the USA (2010), Great Britain (2000 for commercial, 2006 for residential) and China (2000).
Data sources: DOE (2010), BRE (2002, 2008), LBNL (2007).

can be used to provide these services. In the context of retrofits, it is important to consider the efficiency with which these services are being met.

Building services efficiency can be measured in a number of ways (e.g. see Randolph and Masters 2008: 168–81), but is usually considered in terms of energy efficiency (useful energy out/input energy). Exergy efficiency (Fermi 1956) or energy use indicators (e.g. kWh/m^2 per year) are also used and are more appropriate indicators in many situations. For the rest of this chapter, energy efficiency is the prime indicator of the effectiveness of a retrofit. The energy efficiency of a technology is a function of key variables including the degree of fuel combustion, the source and sink temperatures and the requirements of a particular heat transfer with respect to the designed rate of heat output.

Take an example of fuel demand in residential buildings. Typically, the service being provided is low temperature heating (i.e. less than 100 degrees Celsius). Figure 4.3 shows the estimated improvement in 'exergy' efficiency for the low temperature heat supply in the USA during the twentieth century. Exergy efficiency improved slightly in this time period, rising from about 2 per cent to just over 3.4 per cent. Also shown in Figure 4.3 is the share of useful work output in the US economy from low temperature heat processes. Increases in the useful work output of the USA from rising shares of mechanical drives and electricity have driven down the contribution of low temperature heat. Despite the falling share of low temperature useful work in total US useful work output, small improvements to the exergy efficiency of low temperature heat supply have been made.

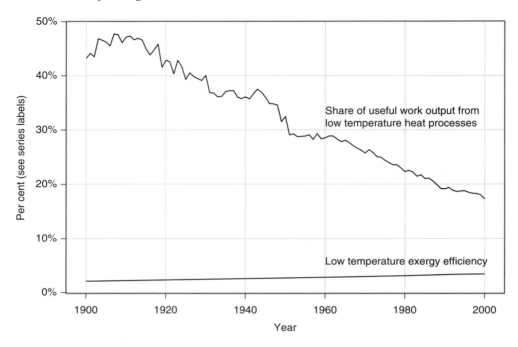

Figure 4.3 The changing role of low temperature heat in the US economy, 1900–2000. Low temperature heat = thermal energy at less than 100°C.
Data source: Warr *et al.* (2010).

Keeping in mind the importance of temperatures at which heat is provided, it is common for analysts to use temporally averaged efficiencies for comparison, such as an annual coefficient of performance (CoP) for heat pumps. The CoP may be defined as 'the amount of useful heat, in kWh, that the pump can generate for each kWh of electricity it consumes while operating' (Ofgem 2008: 31). Certain technologies are more efficient at providing useful heat at certain temperatures.

Heat always flows from a hotter temperature in the direction of a colder temperature, described by the second law of thermodynamics (for a background to thermodynamics, see Fermi 1956). Heat can be stated as the kinetic energy of atoms and molecules within a system, and is transferred by means of conduction, convection or radiation. Heat is typically transferred through a building by the use of water, steam or air. Details of the exact mechanics of heat transfer can be found in any standard textbook, such as Holman (1997) or Incropera and DeWitt (2002).

Roughly stated, conduction describes the heat transfer from moving molecules to one another. Convection is the process whereby a moving fluid transfers heat by means of the stored internal energy of the fluid. In a building, convection will occur when moving air passes over the surface of a material. Whereas conduction and convection require a temperature gradient for heat to flow from molecule to molecule, thermal radiation will occur from any system regardless of the surrounding environment – even in a vacuum. Indeed humans emit heat at a rate of between 80 W and 380 W, varying largely on the activity of the body and body's surface area.

However, the radiant heat transfer of humans to their external environment does not always satisfy the thermal comfort requirements of building occupants. The thermal comfort requirements of an occupant are a function of environmental variables, including but not limited to: the clothing and metabolic heat rate of the occupant; the thermal transmittance of the building (usually measured by U-values, in W/m^2K, e.g. see BRE 2006); the climate external to the building, and any internal heat transfer within the building. Internal heat transfer in a building is primarily controlled by means of combustion of a fuel and then transfer of the heat to a fluid media for distribution to another location in a building.

Passive and mechanical heating systems of buildings are designed to control the internal heat transfer of a building to a suitable degree. The configuration of heating systems varies by building type and country. Chapter 3 gave a brief history of the evolution of demand in London, stemming from the initial use of traditional biomass, such as wood, towards organic materials with higher heat content, such as natural gas. This evolution has been reflected in the modern technologies used in buildings. Natural gas fired water boilers with distribution pipes and panel radiators are now the standard central heating configuration in the UK. Table 4.1 gives an indicative (and non-exhaustive) overview of the evolution of heating systems in buildings by fuel, but does not include the distribution system or controls. A more detailed description of these technologies will be found from merchants and manufacturers or from guides (e.g. Wulfinghoff 1999, CIBSE 2004, ASHRAE 2006).

The technologies used for heating media and occupants in a building vary in a number of ways. The decisions of which technology to install may be determined by a number of factors unrelated to the technologies' efficiency at converting fuels into useful heat. For instance, a typical HVAC design should take account of psychometrics, system

Table 4.1 Characteristics of heating systems for individual buildings by fuel input, UK focus

Fuel input	Initial technology	Modern technology	Premium technology
Wood	Firepit/fireplace	Boiler/range/Aga	Pellet boiler/gasifier
Charcoal/briquettes	Stove	Furnace/range/Aga	Boiler
Coal	Stove	Furnace/boiler/range	Batch fed/gravity feed boilers
Oil	Oil burner	Furnace/boiler	Condensing boiler
Natural gas/LPG/ town gas	Stove	Furnace/condensing boiler/fuel cells	Heat recovery boiler/micro CHP
Electricity*	Convective and radiant heaters	Arc furnace/ HVAC/immersion	Single zone VAV HVAC
Hydrogen*	Heating torches	Fuel cells	Fuel cell stacks
Air	Hypocausts/windows	ASHP	Split-unit heat pumps
Geothermal	Pit/geyser	GSHP	ATES
Solar	Thermal mass/windows	Solar thermal collector	Solar shields

Notes: CHP = combined heat and power, LPG = liquefied petroleum gas, HVAC = heating, ventilation and air conditioning, VAV = variable air volume, ASHP and GSHP = air source and ground source heat pumps, respectively, ATES = advanced thermal energy storage. General waste, wasted heat and biogas are not included as fuel inputs.
*Electricity and hydrogen are not strictly considered here a fuel input, as they must be converted from another state before use as a fuel.

design, controls, air/hydronic distribution systems, codes and standards and indoor air quality. Yet cost and sociological aspects may be equally strong drivers for the building owner as the design guides are for the engineer. Contractors too will have their own set of constraints. Taking an integrated view of decisions to be made in building energy systems requires at least an understanding of the requirements of technologies, and the agents that choose them. The technologies for supplying electrical services and profiles of electrical power demand in buildings are now briefly discussed.

4.2.2 Power demand in buildings

The main building services that require electrical power input are lighting, cooling and powering of appliances. Heat for cooking may also be provided by power input. Power demand in buildings, in contrast to heat, is usually supplied by means of centralized power plants. Decentralized power supply is a relatively new concept.

Electrical power is produced by the phenomenon of electromagnetism. The reader is referred to any basic physics textbook for an introduction. Simply put, a gathering of electrons in a conductor will cause an imbalance in the local electric charges. A potential of electric charge may be built up in a conductor, similar to water behind a dam, and this electrical potential energy will seek a state of equilibrium with its environment. The charge will cause electrons to push their neighbouring electrons by means of the repulsive force between two negative charges. This 'push' conducts the electric power from a centralized power plant through conducting transmission and distribution systems to the electrical power demand in a building. In residential buildings, potential is supplied at a constant magnitude (e.g. about 230 V in the UK). The particulars of circuit analysis, distributed generation, and electric power systems in general are provided elsewhere (e.g. Von Meier 2006, Jenkins *et al.* 2010).

The configuration of electrical power systems and appliances in buildings vary in terms of power demand but may be deemed similar in terms of the basic circuit arrangement. A power source (normally a battery, photovoltaic cell, generator or wall socket) is connected to electrical loads (such as a microwave) by means of conducting wires through which electric current flows. Of interest here are the *resistive loads* in a building; the elements of building services that draw power and convert it to mechanical or thermal energy.

The technologies that supply electrical services in a building are well-known, and may include the following:

- lighting fixtures
- air conditioning units, mechanical ventilation and distribution systems
- electric motors and drives
- office equipment
- refrigerators and freezers
- ovens and other cooking appliances
- televisions, computers and mobile phone chargers.

The size of each power demand (measured in kW) is dependent on a number of variables. For instance the purpose of the building may determine how many, what type, and

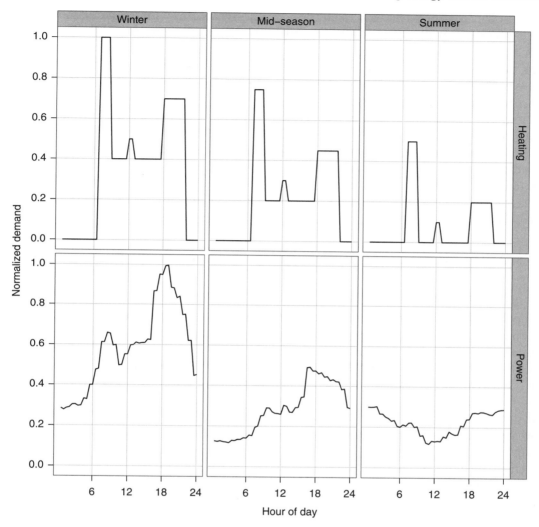

Figure 4.4 Typical load profiles of large samples of residential buildings (DTI 2006, Weber and Shah 2011).

when the lighting fixtures are drawing power to produce light. The temporal demand for aggregated electrical power in a building, similar to heating demand, may vary by time of day, type of day and season. Figure 4.4 gives indicative demand profiles for a residential building by season and type of demand. These standardized profiles display typical temporal patterns of demands in residential buildings in the UK. Residential and commercial buildings in other countries and climates will demonstrate quite different profiles. As seen in Figure 4.4, residential demand tends to peak for heat during the early morning and for electricity during the early evening. Note, however, that these smoothed estimates of demand profiles represent a large sample of residential buildings, with the 'spikiness' of individual buildings smoothed due to diversity between buildings

(Strbac 2008). In reality, the representation of one type of day and three seasons may not be sufficient for allocating electricity bill settlements. Elexon, the administrator of the UK balancing and settlement code, uses weekdays, Saturdays and Sundays for five seasons per year to create estimates of load profiles. The use of these 15 representative days are perhaps the best tradeoff between fidelity and accuracy for modelling building demand profiles during the course of a year.

An understanding of, and familiarity with, the basic concepts, technologies and temporal dynamics of fuel and power demands in buildings is required for effective integrated consideration of retrofits to existing energy demands. However, without a consideration of *why* these demands exist, and *how* they may change over time, an integrated framework may lose some of its intended purpose, particularly if the intended purpose of such a framework seeks to optimize decisions on governmental energy policies. The next section gives a brief overview of the social role that buildings play.

4.2.3 Sociological aspects of fuel and power demand

Buildings represent more than just bricks, mortar and any other combination of materials. Buildings have the functions of comfort, durability and esteem (Osbourn and Greenho 2007). It is esteem that many buildings are built to symbolize – think of the Taj Mahal, the White House or simply a neighbour seeking an extension. The argument presented here, which is by no means a physical law but neither is it thought unrealistic, is that buildings form part of the social strata and associations of success. Current social strata often consist of classes into which society may be divided. Class structure may be less restrictive than previous incarnations of stratification within a society (think feudalism!), but many sociologists argue that current urban areas can still be mapped out by separate areas with unique social characteristics (Giddens 2009). Occupying a building in a certain urban area may well symbolize an association with the prevailing social class of that area. Assuming symbolism still exists, then what drives it?

Humans have a desire for approval, which may be argued to be one of the strongest driving forces of mankind (Smith 2002). This desire for approval takes shape in many forms, but perhaps no forms are more visually obvious than the buildings we choose to live in and the technological innovations we surround ourselves with. Theoretically, it can be argued that displays of wealth are normally provided through the act of consumption – *conspicuous consumption* as Veblen (1899) called it. Consumption is usually represented through symbolism, such as a large house in a nice neighbourhood. The size, appearance and location of a building are presented here as a function of the wealth of the occupants. Wealth correlates to the esteem of the building owner and that wealth is often used to buy bigger buildings with concomitant larger fuel and power demands. It is the appearance of wealth (as a proxy for success) that is important. People in general seek larger houses and more obvious consuming patterns such that others may notice. This notice can help confirm their higher social status and subsequent approval. One doesn't buy an office on Wall Street, a shop on the Champs-Elysées or a flat in One Hyde Park with the intention of being inconspicuous!

The accompanying counteraction to the theory of conspicuous consumption is that of *conspicuous conservation* (Griskevicius *et al.* 2010). Obvious displays of conservation may include environmentalism or self-sacrifice due to, e.g. climate change. Such conservation is deemed here to be caused by the same underlying psychological desire for approval. In some social strata, conspicuous conservation may be the dominant force, but it is the author's view that conspicuous consumption will continue to be the main driver of building fuel and power demand in the short-term at least. Both conspicuous consumption and conservation are convenient theories for the discussion above, and cannot be disproved with the same rigour as a purported physical law. Nonetheless, ignoring psychological forces and sociological trends in building stock is to treat them as irrelevant and this is believed by the author to be an error worse than including them, at least in the discussion of integrated building energy systems models.

The human desire for approval is, thus presented here to be a main driving force for changing fuel and power demands in buildings. How then have these demands changed over time? Surely humans are rational consumers that pick the least expensive and most efficient energy technology available to them? This may be true for some but what if the least-cost efficient option does not fit with an individual's aspirations and finances? Innovations are taken up by 'first-movers', which in the building sector is often the building owner who is seeking to display their wealth. A counter-example is where innovations are tried out purely because of their functionality, e.g. in hospitals for better internal space conditioning.

Regardless of the cause, it has been argued that technological innovation in the building sector is initially taken up by the sub-sector of buildings that can most afford it, i.e. industrial and premium buildings – where the profits are largest (Fisk 2008). It is common for innovative technologies to be initially taken up by industry, then transfer into the premium building's sub-sector; move from these niches into commercial buildings; migrate from commercial buildings into premium housing; and then lastly spread from premium housing into general housing. Examples in each building sector include factories, hotels, offices, mansions and row/terraced housing, respectively.

The theory describes the diffusion of innovations, as detailed by Rogers (2003), as a social process whereby different levels of information about a new idea are communicated from individual to individual, or from firm to firm. Usually, new users find out about a new innovation and are then persuaded to use it through a highly respected agent in a particular social system. The science is by no means settled, but innovation theory presents a strong case for the explanation of technology trajectories. The typical innovation narrative describes the evolution of a technology from its initial invention to that of a fully diffused innovation. Box 4.1 describes the trajectory of air conditioning units in the US building sector as one such example.

Box 4.1 The diffusion of air conditioning units in the USA

First patented by John Gorrie in 1851, refrigeration machines were used for cooling of medical patients in Florida. Whilst developments in refrigeration continued, the first electrical air conditioning units was invented over 50 years later by

William Haviland Carrier in 1902, for a lithography and publishing company seeking to control temperature and humidity. Air conditioners were demonstrated, adopted and developed in factories and other high profit-margin buildings from the early 1900s onwards. Textile factories installed them to control the moisture content of the fibres (Cooper 2002). Following initial industrial acceptance, demand for air conditioning grew in the 1920s and 1930s and requests came primarily from owners of premium buildings, such as from soundproof broadcasting studios and hotel dining rooms, and from Rivoli theatre in New York (Gann 2000). Shortly thereafter, commercial buildings began to employ air conditioning and refrigeration as standard. Around this time, General Motors and DuPont synthesized the chlorofluorocarbon, Freon, while Carrier replaced piston-driven compressors with centrifugal compressors. These innovations allowed for the production of smaller more efficient air conditioning units (Ormrod 1990). Compact air conditioning units entered the premium housing market, and in most cases, general housing owners could not afford them. By the beginning of the Second World War, only a small number of houses had installed air conditioning, almost all of which had wealthy owners (Arsenault 1984). After the Second World War, air conditioning units diffused into the general housing sector as their prices reduced. By the 1970s, it was considered normal to have air conditioning units installed in most buildings in the USA.

The three main insights of the air conditioning example for the building sector are thought to be as follows:

1 Diffusion of technologies is a social process, and is tied to the esteem (or lack thereof) associated with a building. Technologies are generally not chosen because of their energy efficiency, rather to fulfill a particular function or improve upon an existing service.
2 Successful product (Schumpeterian) innovation (Unruh and Einstein 2000) usually occurs in the building sub-sector where the investors can most afford to try out new innovations (Fisk 2003). Thereafter, most improvements to a dominant design are incremental (Usherian) improvements (Ayers 1991). Technological trajectories flow in one direction only; innovations introduced to general housing do not migrate upwards into premium or industrial buildings.
3 Radical changes to existing energy technologies have taken long periods of time in the past – general examples exhibit periods anywhere between under 20 years to replace black and white televisions with colour televisions, and over 50 years for automobile infrastructure to be developed (Grübler 1998).

With the main conclusions of this section in mind, the next section explores the benefits of retrofitting buildings.

4.3 Potential benefits of and practical obstacles to retrofitting buildings

Retrofits on both the individual building and city-wide scales offer the potential of great improvements to existing building stock. The UK's Department of Energy and Climate Change presents the example that residential building sector energy demand in Great Britain would have been nearly twice as large today if insulation and efficiency measures had not been installed in buildings (DECC 2011b). Levine *et al.* (2007) and LBNL (2009) provide other estimates of historical demand reductions due to retrofits.[1] Similarly, some retrofit policies appear to have caused reductions in demand (e.g. Nordqvist 2006, Korytarova 2006). Given the potential benefits of retrofitting building stock, a number of real-world obstacles remain in the way of effective retrofits. 'Effectiveness' is defined here in terms of producing the desired effect of a building retrofit, i.e. a more efficient energy system. Both the benefits of and obstacles to effective retrofits are discussed in this section.

First a note on terminology. In this chapter, the term 'retrofit' is used when referring to planned improvements to existing buildings by means of altering, replacing or removing an existing technology or technologies. 'Retro-commissioning' is also used in this context. Other terms such as 'refurbishment', 'refit', or 'renovation' may include structural changes to existing buildings. When a building is being demolished to use the land for another purpose, suitable terms include 'gentrification', 'regeneration', 'transformation' and 'renewal'. Regardless of the name employed, buildings are usually retrofitted as a matter of maintenance or upgrading, e.g. 'The boiler is broken. We should replace it'. Or 'The grandchildren might be here during holidays. Let's replace the radiators with underfloor heating'. Retrofits occur regardless of the actions of the analyst or policy-maker.

It is this 'natural' rate of retrofit that is of interest to analysts of urban energy systems. Later in this chapter, strategies for increasing these natural rates, by means of mathematical representation and analysis of logistics, are discussed. Table 4.2 gives an indication of typical annual rates, although with the caveat that differences in metric definitions mean that data cannot be directly compared across nations and also that refurbishment activity may be used by administrations as a proxy for retrofit.

Table 4.2 Typical rates of retrofit in the building stock

Country	Annual rate of retrofit (% of building stock)		
	Residential	*Commercial*	*All buildings*
Germany	–	–	0.8–1
UK	1–1.9	–	1–1.5
USA	–	2–2.2	–

Note: '–' denotes data not available, as identified by the author.
Data sources: Olgyay and Seruto (2010), Pike Research (2010), Zhai *et al.* (2011), Climate Policy Initiative (2011), Modes (2010), DCLG (2011), Better Buildings Partnership (2010).

4.3.1 *Potential benefits*

The retrofit of existing buildings offers great benefits on an energy basis, which can be divided into two camps: those for reducing existing demands and those for improving the efficiency of supply. The level to which demand is addressed ranges from putting a roll of insulation over the hot water tank, to passive redesign[2] of the whole shell and core (CIBSE 2004). The level to which supply is addressed ranges similarly, from installing a solar thermal collector to retrofitting neighbourhoods with district heating pipelines and a combined heat and power plant. On the small scale, retrofits to individual buildings offer the benefits of single building technologies, particularly on the demand side (e.g. improved thermal envelopes to reduce the heating requirements). Retrofitting groups of buildings in an urban area can provide the benefits of individual demand-side measures, but more importantly can also provide the benefits of centralized supply-side technologies (e.g. the use of waste heat from industrial buildings to supply nearby low temperature demands). Retrofitting at the large-scale also offers improvements to the actual process of retrofitting. Within the above framework, benefits attached to improving individual buildings are discussed first, and then the advantages of large-scale integrated retrofit approaches are described.

Retrofitting individual buildings

The typical ordering of improvements to an existing building centres on lowering the fuel and power demands first (i.e. energy conservation and use efficiency) and improving efficiency of supply second (Mayor of London 2004, CIBSE 2004). The strategy of lowering existing fuel and power demands focuses on improving the thermal envelope of a building and using appropriate controls to minimize superfluous heat or power supply.[3] Note that the demands which can be reduced are usually confined to space heating, cooling, lighting and other electrical demands; cooking and water heating are not included. Table 4.3 offers a non-exhaustive list of current demand-side options for improving technological elements of existing buildings. The given costs are indicative only, and suppliers will offer a range of products and prices for each option. It should also be noted that in many countries, there are planning guidelines, building regulations/codes, and standards that have to be met for particular interventions. The impact of these on the choice of technologies available for retrofits is beyond the scope of this book. The final point of note is that the options available to, and suitable for, commercial *vis-à-vis* residential buildings are dramatically different, although the objective is often the same.

   Demand-side options for individual buildings are premised on the empirical evidence that retrofitting buildings has been effective in many cases. For instance, thermal measures primarily work by altering the relative ease of transferring heat through a building, e.g. by installing cavity or solid wall insulation, reinsulating damaged piping and ductwork insulation. The objective is to reduce the wall's thermal transmittance. Savings from retrofits are not always as expected however, and there is healthy debate surrounding the reductions in demand (e.g. regarding the rebound effect of energy efficiency, see Greening *et al.* 2000, Sorrell 2007). Others would argue that occupants

Table 4.3 Characteristics of selected demand-side retrofit options for individual buildings

Technology	Current GB stock	Benefits (Reduces ...)	Indicative cost (£ per building)
Cavity wall insulation	$U = 1.10$ W per m^{2}K	Heat demand	1,600
Chiller-tower optimization	–	Power demand	–
Critical valve reset	–	Power demand	–
Double glazing	$U = 2.80$ W per m^{2}K	Heat demand	100–500 per m^2
Draught proofing	–	Air changes per hour	120–240
Exterior wall insulation	$U = 1.10$ W per m^{2}K	Heat demand	4,800
Fan pressure optimization	–	Power demand	–
Full insulation	37% of residential stock	Heat demand	–
Higher efficacy lighting	–	Power demand	25–100 per lamp
Loft insulation	$U = 0.32$ W per m^{2}K	Heat demand	280
Low flow showerhead	–	Power demand	15
Natural ventilation	Rare in commercial	Power demand	–
On demand water heater	–	Heat demand	1,500
Passive redesign	< 50 fully certified	Demand < 15 kWh m^2 per year	140,000–250,000
Self-optimizing thermostats	Rare	Heat demand	100
Single zone variable air volume	–	Power demand	6,000
Smart appliances	–	Power demand	15–30 extra per appliance
Smart meter	< 200,000	Power demand	140
Supply air temperature reset	–	Power demand	–
Thermostatic radiator valves	50% have TRVs	Heat demand	50–250
Ventilation optimization	–	Power demand	–
Voltage control hardware	–	Power demand	16,000–80,000
Water tank insulation	55 mm insulation	Heat demand	15

Note: '–' denotes data not available, as identified by the author.
Data sources: Ofgem (2008), DECC (2011b), industrial and author estimates.

are simply doing reasonable things (Shove 1998). Regardless, it remains true that the combination of modern technologies, building codes, design guides and mathematical modelling can allow for better control of the internal environment of a building than many existing building configurations currently permit.

For ease of explanation, demand-side retrofit options in residential and commercial buildings are divided here into building envelope options, appliance improvements, control system options, and design improvements. Building envelope options are quite similar in commercial and residential buildings, with insulation and glazing alterations the primary means of changing the thermal transmittance. 'Smarter' appliances are becoming more popular in residential buildings, symbolized by smart meters and washing machines, although commercial buildings are beginning to take such options, particularly towards reducing the bills for their information technology hardware. Appliance improvements are perhaps the easiest demand-side option for policy-makers to regulate, however appliances may offer the lowest demand reductions if comparing their average half-life to the half-lives of building envelopes, control systems and overall designs.

Control systems are commonly found in commercial buildings, with the focus on efficient control of the heating, ventilation and air conditioning (HVAC) units. Lower cost options include checking, changing and/or fixing the following elements: operating schedules, economizers, humidity sensors and controls, CO_2 sensors, occupancy sensors, actuators, damper linkages and filters. Other typical controls designed in a commercial building retrofit include adding variable flow drives where possible, changing fixed heating setpoints to demand-based setpoints, and allowing resets. More costly controls (as measured on a '£ per square metre' basis) are characteristically innovative and disruptive to the current set-up. Direct digital control (DDC) can be applied to the 'wet-side' or 'dry-side' of HVAC distribution, such as ventilation controls or valve/chiller tower controls as well as upgrades to existing pneumatic-based controls. The premise here is that DDC can optimize HVAC system energy demand. In this regard, a standard control sequence would be applied to variable air volume boxes to improve efficiency. A rarely seen demand-side measure is that of voltage control hardware, which has been tested for improving the power factor[4] of power delivered to commercial buildings that have suitable induction motors (e.g. Mohan 1980).

In residences, space heating is the main application of control systems (for a detailed perspective, see Fisk 1981). Thermal control can be instituted by means of wireless thermostatic valves on radiators, or via self-optimizing thermostats. However, residential control systems are not as sophisticated (nor do they need to be in many cases), where programmable thermostats are often the highest order of programming. Moving from controls to (systems) design improvements, commercial and residential applications differ again, largely on account of the magnitude of demands.

In commercial cases, there are various adjustments that can be made to improve existing pipe and ductwork layouts and technological capacity (e.g. thermal energy storage). As an example, chilled water systems can be improved by means of primary–secondary pumping (for constant volume HVAC), variable primary flow, series counterflow or ice thermal storage. Beyond improving elements of an existing HVAC system, changing the main equipment may be required. Towers, chillers and boilers are often replaced towards the end of their design life. Converting a constant volume to a variable air volume system is also a commonly exercised option. Similar principles of improved design can be found in residential buildings, albeit at different scales.

As improved design of interdependent technologies offers benefits on the demand side, so too does improved design on the supply side. Systems improvements to more than one building are defined in part by their relevance to supply-side improvements, e.g. a single building will receive the benefits of insulation, while many buildings may be supplied from the same centralized heat source. Supply-side technologies are chosen to improve the efficiency of heat and power supply to buildings, preferably *after* demand-side interventions have reduced demands as far as economically viable (to the building owner's preferences). The next chapter deals much more thoroughly with distributed generation and district heating so they are not described in detail here, but Table 4.4 offers a non-exhaustive list of supply-side options for both individual and groups of buildings.

Mainstream options for improved individual residential heat supply include heat recovery boilers, micro-CHP, ground source and air source heat pumps (GSHP and

Table 4.4 Characteristics of selected supply-side retrofit options for buildings

Technology	Current UK stock	Benefits	Indicative cost (£ per building)
ASHP	169 (in 2008)	Renewable heat source	4,500
ATES	0	Renewable heat source	1,000,000
Biomass boiler	1,400 (in 2008)	Improve heat supply efficiency	30,000
(Condensing) boiler	78% seasonal efficiency 28% are condensing	Improve heat supply efficiency	3,000
District heating	<2% total heat demand	Improve heat supply efficiency	500 per m
Fuel cell CHP	5 (in 2005)	Co-generation of heat and power	–
GSHP	3,400 (in 2008)	Renewable heat source	15,000
Heat recovery condensing boiler	9,000 per annum	Reuse heat from flue gases	3,500
Micro CHP	600 (in 2006)	Co-generation of heat and power	6,500
Solar thermal	100,000 (in 2007)	Renewable heat source	3,000
Solar PV	3,000 (in 2007)	Renewable power source	9,000
Use of waste heat	186 GWh (all uses)	Use waste heat for low-temperature heating	–
VAV HVAC	–	Improve HVAC efficiency	–

Notes: CHP = combined heat and power, HVAC = heating, ventilation and air conditioning, VAV = variable air volume, ASHP and GSHP = air source and ground source heat pumps, respectively, ATES = advanced thermal energy storage, PV = photovoltaics. '–' denotes data not available, as identified by the author.
Data sources: Ofgem (2008), DECC (2011b), and author estimates.

ASHP, respectively) and district heating. Each has its own benefits. An ASHP is relatively cheap. Solar thermal collectors can provide much of certain residences's hot water demand, even where diffuse radiation provides most of the incoming solar energy. District heating includes the benefits of fuel flexibility and use of low-grade heat so that higher grade heat sources can be set aside for more appropriate demands. In Denmark, 60 per cent of residential space and water heating is met via district heating (Hawkes *et al.* 2011). This type of centralized approach to heat supply is an attractive option for a systems-style energy efficient retrofit.

Retrofitting groups of buildings

In addition to the benefits achieved by retrofitting individual buildings, area-wide retrofit schemes can offer further benefits in terms of supply-side technologies, project finances and logistics.

SUPPLY TECHNOLOGIES

Building envelope improvements can be made to a building without regard to location, yet urban areas offer greater cross-optimization opportunities for heating supply in part due to the closely spaced heating demand. There is great potential for reusing waste heat by means of energy cascading using heat exchangers. Kalundborg in Denmark offers one such example based upon industrial ecology (Ehrenfield and Gertler 1997), and Chapter 9 offers a case study of the benefits of distributed heat supply. With respect to power, the voltage may be optimized for large groups of buildings with similar operating voltages. Conservation voltage reduction was first introduced in the late 1970s, and has been discussed since (Von Meier 2006). A famous example took place at Snohomish County's public utility district where line drop compensators in association with capacitors gave definite, although hard to measure, savings (Kennedy and Fletcher 1991). However, optimized heating strategies are much more common large-scale.

Heating multiple buildings by use of either large generation plant or from the waste heat of industry is a significant benefit of considering retrofit at the large-scale. Taking an integrated view of fragmented heat and power supply appears to offer real reductions in the fuel input required to produce the same service.

FINANCE

Financing of systems-style individual retrofits can be expensive and difficult to arrange. In existing commercial buildings, the benefits of retrofit options are usually traded off against their payback period (for a thorough review of retro-commissioning projects in the USA, see LBNL 2009). When residential owners seek to greatly reduce their energy demands, demonstration of the systems approach often costs more than £50,000 per house (e.g. Severn Wye Energy Agency 2011). Even for less ambitious retrofits, financing of individual properties can prove a stumbling block. Aggregating property assets together may allow the appropriate agent to provide the security necessary to assure banks/investors of their expected return. Table 4.5 presents the agents typically involved in the processes of retrofit. The fragmentation of agents along the retrofit supply chain is discussed further in the next section.

Financing is, in part, dependent on the property cycle of the economy. In a depressed portion of the cycle, public agencies could look towards raising bonds, as banks may be looking towards (more secure) public collateral in their investment portfolios. Such a 'retrofit' bond would require an appetite from banks and other financial institutions for perhaps £300 million worth of housing retrofits. In general, debt will be tiered where the cost of finance at a particular stage of a project is taken up by investors with the relevant risk-attitude – mezzanine financiers[5] are willing to take on more risk than a bank for a typical property development. The riskiness of such a bond could be partially offset by public procurement policies, such as minimum company turnover requirements for contractors bidding to take on the work. The point of this discussion is not that large-scale retrofit financing has to be innovative, but rather that the range of options available from the range of agents involved can be innovative where required. Private sector innovation is also a benefit of accessing finance at large scale.

As one of the agents with the most financial leverage, private landlords buying up entire streets of poorly performing building stock allows for serious innovation towards

Table 4.5 Typical agents involved in the process of retrofitting buildings

Agent type	Example
Owner	Occupier
	Public landlord
	Private landlord
Retrofit manager	Contractor
	Engineer
	Estate agent
	House builder
	Owner
	Property company
	Property developer
Finance	Government grant
	Institutional investor
	International grant
	Merchant/clearing bank
	Mezzanine finance
	Municipal bond
	Overseas investor
	Public landlord finance
	Savings/equity
Designer	Architect
	Building services engineer
	Contractor
	Electrical engineer
	Mechanical engineer
	Owner
	Utility
Technology supplier	Builders and trade merchants
	Manufacturers
	Wholesale supplier
Installer	Contractor
	Sub-contractor
	Owner
Other	Commissioning engineer
	Insurer
	ESCO
	Maintenance engineer
	Other consultants
	Planners and regulators
	Sales agent
	Solicitor
	Vested interests

Note: ESCO = energy service company.

the sourcing of their capital. The main benefit to the landlord is that of rental and capital growth in the improved built assets. The business case of refurbishing or regenerating areas of a neighbourhood or city have been known to certain agents for many years (e.g. for institutional investors, see Dixon 2009), particularly those investors who are not afraid of tackling 'hard and dirty' assets as opposed to refurbishing flats in Manhattan or Chelsea! It is this rental aspect of retrofits that is almost always left out of urban energy system modelling. Rent is a prime reason for investors getting seriously involved in the market – see the upwards-only rent review of commercial leases since the 1970s. Mathematical programming may well have a role to play in the capital management of refurbishment/regeneration projects.

LOGISTICS

Economies of scale in retrofittting commercial and/or residential buildings may exist where a contractor and their sub-contractors work on similar jobs in short timescales. This is called 'learning by doing' (for a recent review of cost reductions due to learning, see Weiss *et al.* 2010). If contractors use a systems approach to large-scale refurbishment they can quantify each activity's time and cost as described by the Building Research Establishment's Calibre work group. Retrofit learning rates may be represented by use of a power law $y \propto x^\lambda$, though with caveats (Stumpf and Porter 2012). Quantification and analysis of the learning rates in an optimization framework allows for assessment of alternative plans for manpower allocation, subject to on-site constraints.

An optimization framework applied to the retrofit process can apply similar mathematical methods previously employed in arenas such as chemical engineering and military planning. Indeed the contractors Laing O'Rourke have built a precast concrete factory in Steetley, UK, towards this end. The use of mathematical programming for improved decision-making can increase the throughput, reduce resource inefficiency, and provide details of sensitive parameters on the factory floor. Similarly, for retrofitting of supply-side technologies, much of the contractor's work will be external to the buildings, potentially under the tenure of one owner as in the multi-utility underground pipes or Parisian-style sewer systems.

Summarizing the benefits of retrofitting buildings, well-designed projects can reduce existing fuel and power demands of one-off buildings. Considering retrofits at large scale opens up extra advantages, particularly in terms of shared heat supply, innovative capital management and optimized logistics. However, there are obstacles to achieving effective retrofits at both the individual and the multi-building level. These barriers are discussed in the next section.

4.3.2 Practical obstacles

Having discussed fuel and power demand in buildings, and described the benefits of the multiple technologies available for reducing these demands and supplying them more efficiently, we turn to the question of how these retrofits might be achieved in practice. This section provides an overview of some of the observed obstacles. Literature abounds surrounding the question of why efficient options in building stock are not exercised (e.g Koomey *et al.* 2001, Shove 1998). For ease of navigation, obstacles are divided here into the following three categories: behavioural, technological and managerial.

These categories are arranged in a logical order. First, the barriers as to why a building owner would actively seek to retrofit their building are presented. Knocking these barriers over would then lead to obstacles inherent with the retrofit technologies themselves. Finally, the actual financing, logistics and operation and control of retrofits provide managerial issues.

Behavioural

Why would a building owner seek to retrofit their building, beyond a natural replacement? For an efficient set of technologies with a short payback period, surely the benefits are obvious? Savings can be made on energy bills for a start. Yet, human behaviour is not as simple as that. It is argued here that rational customers do not exist (despite the plethora of energy policies constructed upon the premise that efficient options are chosen on an informed choice of net present value), but rather it is reasonable customers that exist. Reasonable customers make reasonable retrofit investments. Reasonable investment will satisfice (i.e. both satisfy and suffice, Simon 1959) within the context of a given building owner's motivations. Retrofit choices do not need to be optimal for a particular criterion, rather, investments will be made as long as there is nothing wrong with the retrofit technology (Fisk 2005).

The motivations that drive building owners towards retrofits are thought by the author to be most likely a function of the following variables, in decreasing order of importance:

1   The role of the building owner as a social agent in a particular social system. Are they a 'first mover' or do they sit closer to the other end of the adopter curve, i.e. 'laggards'?
2   The computational resources available for assessing retrofit choices. Many cost benefit analyses are computed at the frequency of neural pulses, with spiking frequencies typically around 100 Hz – albeit there are *c.* 10^{14} synaptic connections that potentially could work in parallel leading to 10^{16} potential operations per second (von Neumann 2002), compared with a 2.4 GHz computer processor providing just over 10^9 operations per second
3   The ease of searching for the most satisficing technologies
4   The brand and price of the technology.

For different building owners, the relative importance of each variable entering into a retrofit choice will differ. Nonetheless, while the paradigm of least cost optimization (see Appendix A) may make sense when a government is tendering for a new nuclear plant, it does not usually make sense if modelling a building owner's investment in new technologies. Yet, modelling the motivations and these effects on the sets of technologies available for retrofits is quite an interdisciplinary challenge, as is modelling the future trajectories (or even the historic trajectories) of new technologies in residential and commercial buildings. If the purpose of a model is to determine the best pathways for retrofitting building stock from a *societal* perspective, with the objectives of minimizing fuel and power demand or concomitant greenhouse gas emissions in place, then empirical evidence on the motivations for technologies purchases must be contained in any abstract modelling representation of the retrofit processes.

In general, retrofit technologies will not propagate upwards through social classes. If all the public landlords in a country convened, and decided to install the newest building integrated photovoltaics (BIPV) in their housing stock, it is unlikely that BIPVs would transfer to premium housing and upwards into industrial buildings. This is important, as it limits both the solution space of models (as discussed in section 9.4) and technology options for analysts considering future retrofit pathways. Higher wealth individuals will likely exclude certain technologies because of the lack of esteem attached to certain options. Currently, installing the latest BIPV could be seen as a symbol of status and wealth. Installing cavity wall installation may not have the symbolism. It is not the object here to disparage previous efforts of incentivizing or understanding investments in building innovations, but rather to point out that the social aspect of buildings is often left out of urban energy models.

Technological

Behavioural obstacles notwithstanding, retrofits will continue year-on-year and new technologies will be chosen to replace the incumbents. The second category of practical obstacles is the ability of new technologies to perform as well as, or preferably better than, existing technologies. Renewable technologies in particular are often charged with not being able to provide the same rate of heat supply. Take a simple example. It takes longer to heat water using the sun's radiation than by combusting a very dense compound of carbon, hydrogen, oxygen, nitrogen and sulphur, i.e. coal. A solar thermal collector may be considered less attractive than an electrical immersion heater on account of the rate at which energy is supplied to the water.

As well as issues of available rate of energy supply, technologies often bring with them unwanted by-products. For example an air-source heat pump may create noise levels of 40–50 decibels, need planning permission and may require installation of additional radiators. The by-products of a new technology will often be compared with the incumbent system, which the building occupiers may have grown accustomed to. Poor performance of new technologies is arguably the most detrimental by-product if an effective retrofit is the objective. Technologies form part of a system, and both their stand-alone performance and their role as an element in a heating or power system should be analysed. This is the combined responsibility of the designer and installer.

If the process of retrofitting buildings was broken down into elemental components, and then rebuilt as an integrated system, there would surely be a pinch-point between design and implementation. Whilst the design guides will be based upon heat transfer equations, etc., and installation know-how will be built on tacit experience, there are few codes to ensure technical design and installation produce the intended outcome. It is the interdependent nature of many technologies that proves a stumbling block to optimum performance.

At present, the design, commissioning and maintenance of retrofit components appears unintegrated. It is rare to find the same agents involved in the design, commissioning and maintenance of a retrofit. Energy service companies (ESCOs) are a potential solution, by giving a single entity responsibility for all three stages. Yet, poor systems continue to be designed, built and operated. A recent study of 83 heat pumps in the UK found lower than expected coefficients of performance (CoP) of 2.2–2.4 (versus potential CoPs of 3 and above, Wemhoner and Afjei 2003). The study determined that heat pump performance was sensitive to both installation and commissioning practices (EST 2010). A possible reason for poor performance is that engineers may not get feedback from the contractor on the ground. Full insulation of cavities can be hard to achieve if the insulation is pumped in too quickly, resulting in unfilled cavity pockets. What if a nail goes through a wall's vapour barrier? The problem is not necessarily that simulation models don't consider such worst-case scenarios, but rather that the post-retrofit technologies are rarely analysed to the same level of thoroughness as pre-retrofit. In some cases, the interdependencies of technologies may not be analysed by the designers post-retrofit at all. However, engineers are not the only agents at fault for this lack of recourse, contractors and architects can also be blamed in many instances. The problems of 'silo' thinking may actually lie within the educational syllabus (Banham 1984).

Notwithstanding the cost of production, technological obstacles are largely two-fold: how well the elements work together, and the rates of energy supply available. Assuming the building owner is sufficiently motivated to retrofit a building, and that the technologies available will provide the required service, then the last set of obstacles to effective retrofit is the management of the retrofit process itself.

Managerial

Dependent on the scale and scope of the project, the retrofit process may involve anywhere between one to over 20 sets of agents, as shown in Table 4.5. Considering existing urban areas to comprise energy systems, retrofitting can be described as a process for moving the system from its current state towards a desired future state. The process of retrofit itself can be divided into the following sub-processes:

1 Motivate and/or require
2 Finance
3 Plan and/or consult
4 Design
5 Manage logistics
6 Install/construct and commission
7 Use
8 Maintain
9 Dispose, replace, or upgrade.

Each sub-process has its own agents and set of constraints. The main sub-processes that provide managerial obstacles are finance, logistics and the use phase (i.e. operations, controls and maintenance). These obstacles are inversely mirrored by the potential advantages identified in section 4.3.1. Take finance first: Where do residential building owners get the cash for retrofit? Often the only sources are bank debt or savings. Aggregating private landlords of individual houses may actually increase the risk from a creditor's perspective on account of having to consider multiple credit histories, etc. This is an issue of trust between debtors and creditors. There is also the issue of scale. In Great Britain, finding creditors on a regular basis to retrofit the 17 (out of the total of the 25) million privately-owned residences (DECC 2011b) is a significant obstacle. And this is assuming away the problems of motivation/sociological constraints. For larger projects under one owner however, managing the capital can be easier because of the much larger leverage under a small group of trusted debtors, the retrofit bond being a case in point.

Taking such bonds to be rare, and ignoring equity-led projects, for larger projects the major source of (senior) debt is from banks, which may have different arrangement fees, interest margins, audit fees, non-utilization fees, exit fees, etc. Mezzanine finance can fill in some of the gaps – banks may only lend up to 60 per cent of the gross development value during a depression in the housing cycle. Yet what about alternatives: pension funds, insurance companies or local authority joint ventures? Funding retrofits is difficult if building owners only consider previous lines of credit, but new lines of credit often have confusing fees and may require significant collateral.

Moving from management of capital to management of manpower, other impediments may be set by the rather disaggregate nature of the building industry. Since the modern building professions first materialized in the 1830s, the evolution and expansion of specialized design professions has helped isolate architects (design of buildings) from engineers (design of structures/processes) and from builders (who actually construct buildings) (Gann 2000). The sharing of innovation by builders, in particular, is thought to have been hindered by the fragmentation of the industry. As of 2011, there were over 75,000 construction firms in the UK, of which 64,300 firms have fewer than five employees (ONS 2011). In the USA it is a similar story, with about 10,000 firms in total and fewer than 900 firms having more than 250 employees (Hart 2009). Specialized teams make sense in terms of division of labour, however attempts to re-unify some of the disaggregate firms can be tempered by professional barriers. Both cost and time improvements could potentially be made if these smaller firms were more integrated both within their own profession and across professional barriers.

The final managerial obstacle mentioned here is the management of operations, controls and maintenance. Regular servicing and retro-commissioning are two managerial tasks which residential and commercial buildings undergo on a regular basis. However, in residential buildings, the frequency of maintenance may not be as regular as is optimal. In large commercial buildings there is often a building operator in charge of building automation systems (BAS). In some cases the scheduling, and alarm management, of BAS can be beyond the skills of the building operator. Effective retrofit would require ongoing management of labour training.

The managerial obstacles presented are interdependent to an extent. Without arrangement of sufficient capital, retrofits will continue to be largely one-off affairs without any of the benefits of integrated logistics. Without integrated logistics, the benefits of groups of contractors working together on similar projects will not be gained. And without a systematic analysis of the ensuing management, the results on the ground are likely to be quite different from those modelled.

Taking a systems view of this discussion, the obstacles to effective retrofit mirror the benefits to a large extent. At the individual scale, improvements to the building envelope, appliance demands, control systems and design can be hindered by a lack of understanding of the true aspirations of a building owner, the rate of available heat supply, and the management of the logistics pre- and post-retrofit. At large scale, the problems of financing, managing and supplying fuel and power to efficient buildings are each offset by the greener grass of increased financial leverage, improved organization and planning and efficient use of shared heat supply. With these somewhat symmetrical thoughts in mind, some strategies towards future integrated retrofits are provided.

4.4 Future strategies

The previous sections have outlined the broad principles of energy efficient building retrofits, briefly discussing how fuel and power input to commercial and residential buildings can be reduced while providing the same services. Yet a number of potential obstacles have also been identified. Given the scope of this book, this section focuses on strategies for planning and organizing integrated building retrofits. A particular focus is given to decision-making tools, although strategies for improved government policy and regulations are also offered. The application of the science of decision to retrofits is first discussed.

The process of retrofitting building stock can be thought of as a system of interconnected sub-processes. Each sub-process consists of a number of activities with specific goals, relationships and constraints. A decision-maker managing a sub-process takes decisions in light of single or multi-objectives. Decisions will ultimately consist of 'a statement of the actions to be performed, their timing, and their quantity (called a "program" or "schedule"), which will permit the system to move from a given status toward the defined objective' (Dantzig 1963: 2). When the processes of a system are thought quantifiable they may be represented in the form of mathematical relationships, with symbols representing real-world objects. A model may then be built that reflects chosen aspects of a real-world system under study. Three retrofit sub-processes conducive to representation by mathematical symbols are as follows:

- Finance (capital management)
- Plan and/or consult (capital management)
- Logistics (supply chain and manpower planning).

Mathematical programming applied to the financing, planning and logistics of retrofit may be used to provide the optimal levels of decision variables for the alternative levels available. Mathematical programming is a well-known method and linear programming, a subset of mathematical programming, has been used for decades in many arenas of decision-making. Given a system of linear equality and inequality constraints that represents the relationships and constraints in a retrofit system, and given continuous and integer variables representing retrofit actions to be taken, an objective function may be minimized by the use of linear or mixed integer linear programming (LP and MILP, respectively, see Appendix A for more details on these techniques).

LP and MILP are powerful optimization methods which allow for the solution of problems with large degrees of freedom (i.e. the number of decision variables less the number of binding constraints), much larger than could be computed in an average person's mind or on a handheld calculator. The powerful solution algorithms employed by modern day solvers in parallel with increased memory capacity on relatively inexpensive central processing units has facilitated research interest in the use of optimization methods.

Modelling the process of retrofitting buildings is an exciting field that can build on the previous work of others. Previous work has focused largely on the optimization on the use-phase of retrofitting buildings, for instance by means of minimizing life-cycle costs (Gustafsson 1992) or annualized costs of the technologies involved (Amano *et al.* 2010). A potential strategy then for analysts is directing their research towards mathematical programmes representing the most germane processes (rather than the impacts) of retrofit: arranging the finance, planning retrofit interventions and organizing the logistics. Chapter 9 provides more detail of a MILP formulation and a case study of a planning model. It is expected that in the short-term, mathematical programming will become a more popular decision tool for industry and policy-makers alike.

The fundamental science behind the heat and electrical power demands are well understood and many non-linear relationships can be faithfully solved or reasonably estimated using optimization. However, if such a model is to be used for ascribing policy alternatives, in the mathematical programming field at least, there is a poor representation of the motivation behind retrofit decisions. A number of policy-oriented models minimize the costs to investors in technologies, assuming that real-world behaviour mirrors their models (when it should be the inverse!). This cost-based representation of retrofit decisions may be too narrow a conceptualization of the problem as it ignores the sociological context and satisficing choices that drive retrofitting investments in reality. Whilst solution algorithms, software and techniques have become more sophisticated in recent years, a truly integrated framework for analysing retrofit in buildings has not yet emerged.

So, the initial conclusion of this section is that mathematical programming is an appropriate tool for modelling the financing, planning and logistics of retrofits. The content of integrated optimization models will vary dependent on the objective. Nonetheless, the multidisciplinary nature of programming the best course of action for varying tenures and financing of improved building stock would require an interdisciplinary solution. Depending on the model's purpose, such a model could require input from innovation theory, behavioural science, financial engineering, building services engineering and systems engineering. A fair question is whether or not this is an over-elaborate approach to considering sub-processes in unison?

Models of many of the sub-processes of a retrofit supply chain have been built, and have particular purposes. For example, models focused on the operation do a good job of optimizing operations but have little to say on the effect of financing of refurbishment/regeneration and nor should they. When considering integrated retrofits however, the analyst is forced to be more ambitious, by including what is necessary to represent more than one sub-process of the retrofit. This is not to say conclusively that attempts have not been made in the past but if so, they have not been widely adopted in the literature and there remains great potential for mathematical programming in this area. Debate will likely continue on the optimal content for optimization models of quantifiable processes. Regardless it is clear that optimization models offer a robust framework, that can incorporate physical laws and empirical evidence, towards better understanding and decisions in retrofit projects. Alternative models which are less robust, often neural synapses subject to availability bias, have been the basis of many decisions in retrofit to date.

Similarly, alternative decision-making approaches to the evolution of government policies and regulations of energy demand in the building sector may be explored. With the premise that government policies in this area are more beneficial than detrimental, the author proposes the following strategies going forward. First, innovation theory would suggest the targeting of respected and naturally innovative individuals in the particular system (e.g. those who supply the top-tier of architects, engineers and contractors) for the first adoption of new innovations (Rogers 2003). Taking lessons from the case study of the diffusion of air conditioning units in the USA, incentivizing and mobilizing agents in industrial and premium buildings may well be a stronger policy option than encouraging market pull from the dispersed agents in the general housing sector.

Furthermore, if the exciting technologies of the future, such as remote continuous commissioning, smart appliances and meters, and voltage regulators, are to reach 'tipping points' of penetration in the building sub-sectors, then it surely is better to test new innovations in the sub-sectors who can most afford to and who care the most about their energy bill (Fisk 2003). Focus the policies on the premium and industrial buildings sector, and the expected benefits will accrue to the general housing sector in future years. If policy-makers feel 'something must be done', let them explore methods of speeding up the transfer of technological innovations from one building sub-sector to the next. If regulators are serious about long-term ambitions for reducing fuel and power demand in the building sector, then they should take a patient stance. Regulating the retrofits of industrial and premium buildings more tightly, and regulating towards minimum standards for existing demands in these buildings, should help to push innovations further along the learning curve. Fail early and often, as they would say in Silicon Valley.

The first two strategies going forward are thought to be uncontroversial: the application of mathematical programming as a decision tool for processes of retrofit that are quantifiable; and the application of innovation theory to the intended diffusion of improved technologies for the building sector. These strategies are intended for the analyst and policy-maker, respectively. A brief final offering then for contractors, engineers and architects.

The objective of an integrated approach to retrofitting building stock is more effective understanding, communication and decision-making. Considering the plethora of agents that may be involved in a typical project, there is potential for unnecessary barriers to be formed. Some of these barriers may put in place on purpose: take the example of the restriction of the use of the word 'architect' in the UK (the professional status of architects became formally recognized in the 1830s). The use of such privileges may hinder responsibility and shared learning on retrofit schemes. Architects are often not taught to consider technologies in their design drawings (Banham 1984), almost under the premise that the work that engineers and contractors do is not worth their time. Architects are by no means the sole offender, but the principle remains the same regardless of the profession.

Only considering a certain aspect of the retrofit process entails neglecting the impact of one's decisions on the other aspects. Each profession along the retrofit supply chain has much to learn from one another. Taking an integrated approach allows for contractors, engineers and architect to push towards the same objective in an iterative manner, rather than the mainstream linear approach that neglects feedback. The opening up of professional association's resources to one another would be one tentative step forward in the right direction. Such feedback between the professions is required if well designed, planned and managed retrofits of urban energy systems are to become more commonplace.

4.5 Conclusion

This chapter has attempted to present a thorough introduction to building energy service demands in light of technology retrofits. Technologies, demand profiles, the sociological and historical context, potential benefits, practical obstacles and suggested strategies have been described. The primary intent has been to provide a systems engineering viewpoint of the processes and actions involved in retrofitting buildings.

There are three main conclusions from this chapter. First, the fundamentals of fuel and heat demand in buildings are well enough understood for carefully designed building retrofit to be successful. Such a retrofit design would seek to reduce the thermal transmittance of the building shell and the power demand of non-essential services, and thereafter, to supply heat and power in an efficient manner. Second, the reasons and motivation behind retrofit investments in building stock are poorly understood, using current modelling paradigms as a proxy. This is where integrated urban energy system modelling could play a role. Third, there is great potential for mathematical programming as a decision-making aid to retrofits. Uncovering and quantifying the key variables and relationships would in itself provide insights, and applying a systems framework to the financing, planning and logistics of retrofits could provide robust analyses of the key decision variables and associated obstacles. Retrofitting buildings offers great potential benefits for reducing existing urban energy demands, but to achieve such benefits requires a systematic understanding and analysis of the obstacles present.

Notes

1 It should be noted that fuel and power demand reductions are usually calculated by use of building simulation software. Simulation models are often the best method available for estimating savings post-retrofit, a simple reason being that building owners may not have the inclination (or the discretionary funds) to invest in setting up energy monitors prior to a retrofit taking place. There are various concerns regarding the magnitude of the difference between modelled reductions and actual reductions achieved, but this topic is beyond the scope of this chapter.

2 Passive redesign includes non-mechanical designs that can be implemented in any building, such as optimum use of sunlight by use of thermal mass and overhangs (Randolph and Masters 2008).

3 Demand-side technologies and strategies for changing building occupant behaviour have been omitted from this chapter, as it is strongly believed by the author that it is not the role of an analyst or policy maker to model or regulate how a building occupant should behave.

4 Power is transported from power plants over high voltage transmission lines by means of currents flowing in alternating directions (i.e. alternating current). A sine wave can be used to describe the uniform variation of current and voltage, and the *root mean square* is used to determine the average of the varying quantities. For an induction motor, there is usually a difference in phase angle, θ, (i.e. timing) between the maximum voltage and maximum current. The ratio between the potential (apparent) power and the actual (real) power is referred to as the power factor. The power factor of a large commercial building may cause the bill payer to be billed more by utilities due to the reactive power caused by induction motors. This extra billing charge to the commercial building is because the utility has extra power losses for lower power factors.

5 Mezzanine finance is a form of unsecured debt, typically senior only to common shares. As a result of the higher risk profile of mezzanine debt, creditors require higher rates of return on their investments.

5 Distributed multi-generation and district energy systems

Pierluigi Mancarella

The University of Manchester

5.1 Introduction

The need to fight climate change and meet the challenging environmental targets that have been set by many governments worldwide calls for a rethinking of the traditional ways in which energy systems have been planned, designed and operated. In particular, considering that the general trends in social development all point out a path towards an increasing share of the population that will live in cities worldwide, decarbonization of the energy footprint of urban areas becomes a critical point to address. In this outlook, optimal deployment and integration of locally available multi-energy resources represents a strategic area to enhance the environmental efficiency of the urban fabric. Owing to the recovery of heat that otherwise would be wasted from the thermodynamic cycle, cogeneration (or combined heat and power, CHP) is well known to be able to provide energy savings (Horlock 1997) as well as potential CO_2 emission reductions (Mancarella and Chicco 2008) in given energy contexts, but has been historically limited to industrial or large scale applications. However, growing diffusion of CHP systems has been experienced in the last decade, for smaller scale applications (starting from $1 \, kW_e$[1] in the case of micro-CHP) owing to the technological development of distributed generation (DG) technologies (Borbely and Kreider 2001). However, the interest in DG has been mostly limited to electrical applications and issues (for instance, power system impacts and benefits), and the potential to investigate DG within a comprehensive energy context has often been overlooked. The decarbonization of energy sectors other than electricity, and in particular the heating and cooling sectors (depending on the country), arguably represents an even bigger challenge.

On these premises, recent works have outlined a distributed multi-generation (DMG) framework that can be seen as a conceptual extension of cogeneration and DG at the same time (Mancarella 2006, Mancarella and Chicco 2009a, Chicco and Mancarella 2009a). In this vision, 'multi-generation' refers to the simultaneous production of electricity, hot water, space heating (and in general heat at various temperature levels), cooling power, etc. from one or more fuel sources such as natural gas, biogas, biomasses, etc. DMG technologies may include cogeneration (combined heat and power, CHP) and micro-CHP, reversible electric heat pumps (EHP), various types of chillers for trigeneration (or combined cooling heat and power, CCHP) and electrical and thermal storage. Extensions to entail solar electrical and thermal generation, hydrogen production and storage, etc., can also be considered. From the electrical network perspective, DMG can take place at different decentralization levels (with connection, for instance, at the low voltage or at various medium voltage levels), and can then interact through various types of energy network infrastructure, such as natural gas at the input side or heat networks at the output side. Hence, DMG solutions represent a promising perspective for the development of high efficiency future energy systems, particularly in urban areas, owing to wide availability of suitable multi-energy loads and of electricity and fuel networks. In addition, urban areas usually exhibit high energy density (particularly of thermal loads), so that district networks can be cost-effective (Harvey 2006). At the same time, district energy networks represent an enabling technology for DMG solutions by providing suitable aggregated loads that allow the deployment of larger generation plants that benefit from economies of scale, allowing competition among energy sources and facilitating interaction with centralized energy networks. On the other hand, the complexity of the interactions that take place in such an integrated urban energy system calls for powerful approaches to system modelling and assessment.

The objective of this chapter is therefore to introduce the main technologies and some modelling aspects of DMG solutions and relevant district energy systems, with a focus on electricity, heating and cooling as the most widespread end-use energy vectors in urban areas. In addition, a general analytical framework to assess the environmental performance of DMG options in different energy contexts will be presented and discussed. Emphasis will be set on the role of such systems in evolving scenarios and in potential energy futures.

5.2 Distributed multi-generation: concepts and modelling aspects

5.2.1 *General concepts and technologies*

Distributed multi-generation (DMG) refers to the concept of multi-generation plants interacting among each other through local energy networks (e.g. gas, electricity and heat distribution networks) while also being interconnected with 'centralized' energy transportation networks (see Figure 5.1).

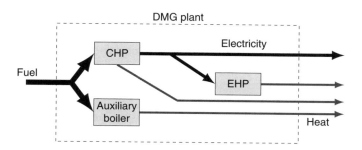

Figure 5.1 A schematic example of a DMG plant for the generation of electricity and heat, with a CHP prime mover, auxiliary boiler and electric heat pump.

Box 5.1 provides a summary of the most common technologies used in DMG systems. For most designs, the key technology will be a cogeneration plant, typically with the combined production of electricity and heat (US EPA 2011, Borbely and Kreider 2001, Mancarella and Chicco 2009a). For small scale applications of about 1 MW$_e$ or below (e.g. small commercial and municipal buildings, sport centres, multi-family and residential buildings, etc.), the most widespread CHP technologies are the internal combustion engine (ICE) and more recently, the micro-turbine. Further solutions, particularly for individual household micro-CHP applications, may adopt Stirling engines. Mid-size CHP applications between 1–20 MW$_e$ include industrial plants, institutional, military and government facilities, large commercial sites and shopping malls and district energy sites. Larger applications up to about 50 MW$_e$ typically refer to industrial sites, college and university campuses, and again district energy systems. All these larger scale applications are typically based on ICEs (up to about 5–10 MW$_e$) or gas turbines (up to about 50 MW$_e$), sometimes in a combined cycle configuration. Applications of various sizes based on fuel cells are also under development and might play an important role in the next decades.

Box 5.1 Distributed multi-generation technologies

A large number of technologies can be used within DMG system designs. This box provides an overview of the most common options, and more details on various DMG technologies can be found in Mancarella and Chicco (2009a).

Internal combustion engines (ICE) that work on the reciprocating principle are a proven technology that can be found in sizes from few kW$_e$ to around 5 MW$_e$ and can typically be fed on natural gas or diesel, as well as biogases. They usually exhibit relatively high electrical efficiencies (more than 45 per cent), good part-load performance, excellent dynamic characteristics, and relatively low costs in the range €500–1000 per kW$_e$ depending on the size (with lower specific costs for larger sizes and for diesel engines compared to gas engines). Local emissions may be a downside for widespread adoption in urban areas.

Microturbines (MT) are available from few tens of kW$_e$ up to 1 MW$_e$ and adopt a variable speed design and usually a recuperator to improve the nominal as well as part-load efficiency (electrical efficiency may reach 35 per cent). A great advantage of microturbines is the relatively low local emissions, while costs are slightly higher than for ICEs, and they can be fed on a number of fuels, although typical applications are based on natural gas. As with ICEs, the technical lifetime is in the order of 50,000 hours.

Stirling engines (SE) are external combustion heat engines operating on a closed thermodynamic cycle. They are cost-effective for small sizes (around €1000 per kW$_e$), are quiet compared with ICEs and MTs, and exhibit excellent dynamic performance. Hence, they are being commercialized particularly for micro-scale cogeneration applications (micro-CHP) at a household level, with capacities from 1 to around 50 kW$_e$. While the electrical efficiency is relatively low (in the order of 10–25 per cent), they are mostly used to replace conventional boilers in domestic applications, and can reach overall efficiencies (electrical and thermal) of up to 95 per cent. Also, as the combustion is external, Stirling engines may be operated using a great variety of fuels, including solar energy and biomasses.

Gas turbines (GT) are the favourite technology above 5 MW$_e$. They are highly reliable with low local emission levels, and can readily produce high-temperature heat for steam-based heat networks. Also due to economies of scale, they typically have longer technical life than ICEs or MTs (even double), and also exhibit lower specific costs (in the order of €500 per kW$_e$). Dynamic characteristics are sometimes worse than for ICEs and MTs, with start-up times in the order of tens of minutes as opposed to tens of seconds.

Fuel cells are reaching maturity for various DG CHP applications from 100 kW$_e$ to a few MW$_e$, although their specific cost is still relatively high (greater than €3000 per kW$_e$). Fuel cells are electrochemical devices that convert the chemical energy contained in the fuel (hydrogen) directly into electricity (at high efficiency, between 40 and 60 per cent) and heat, with no mechanical inter-conversion as for the other technologies. Most fuel cells use natural gas as the source of hydrogen, but the possibility of generating hydrogen from a number of sources, including electrolysis of water using renewable electricity, makes fuel cells promising for a primary role in flexible low carbon multi-energy systems in the future. However, current relatively poor dynamic performance makes them more suitable for constant base-load applications.

Absorption chillers (AC) exploit thermal energy (by burning fuel such as natural gas or supplied in the form or hot water/steam, thus making them suitable for coupling to heat recovered from CHP) to produce cooling (chilled water) in an inverse thermodynamic cycle. In particular, the main difference relative to electric chillers is in the compression stage, whereby the refrigerant is compressed by means of an absorber/generator stage rather than by an electric compressor. A single-effect absorption chiller can be fed by hot water at 80–90°C, and exhibits a heat-to-cooling coefficient of performance (CoP) around 0.6–0.7. While in a single-effect cycle all the condensing heat is rejected to the outside, in a double-effect cycle there are two generators, and the second one is supplied by heat recovered from the condenser of the first cycle. This allows higher performance (CoP around 1.2) but also requires higher firing temperatures (above 100°C) in the form of superheated water or steam. Further cascading is possible, thus leading to triple-effect chillers (with two stages of condenser recovery) that are under commercialization, with CoPs of up to 1.6 and direct fuel firing due to high input temperature requirements. Costs for single-effect absorption chillers are in the order of €200–300 per kW of cooling capacity, while double-effect ones cost 20–40 per cent more. These costs compare to around €150–200 per kW of cooling capacity for electric chillers that are typically operating with electricity-to-cooling CoP between 3 and 5.

Engine-driven chillers are conventional vapour-compression chillers that are driven by a mechanical compressor instead of an electric one, and whose shaft is directly connected to a conventional internal combustion engine. This type of chiller exhibits a fuel-to-cooling CoP around 1.5, and has the advantage that a part of the fuel input can be easily recovered, making it suitable to cogenerate heating and cooling power. Costs are about 50 per cent higher than electric chillers.

Electric heat pumps (EHP) operate along an inverse thermodynamic cycle as for chillers. However, here the purpose is to provide heating rather than cooling, and in order to do so, electrical energy is used to power a compressor to extract heat from a lower temperature source and inject it into a higher temperature sink. Heat pumps are typically categorized according to the temperature source (air, ground, water, geo-thermal, etc.) and the heat delivery means (such as air or water). While performance may be relatively high (typical electricity-to-heat CoP values vary from 2 to 5), it is very sensitive to the temperature lift, so that EHPs may operate poorly under cold winter conditions to provide hot water (CoP below 2) and may need backup heating with a subsequent drop in overall performance. Costs are some 10–20 per cent higher than electric chillers in the same size range, and also depend very much on auxiliary equipment such as the heat extraction system in the case of ground-source heat pumps.

The most used fuel for distributed CHP systems is natural gas, while diesel engines are used as DMG sources, particularly in new applications of traditional on-site backup generators, e.g. at industrial sites, hospitals, supermarkets, etc. Biomass and biofuels also represent new trends of fuels that can be used in CHP systems owing to their limited greenhouse gas impact (Chevalier and Meunier 2005). The cogeneration prime mover is usually backed up by (fuel or electricity-based) auxiliary boilers that are used to top up the heat production at peak times or in the presence of outages and during scheduled maintenance, while electrical backup is usually guaranteed by grid connection. An interesting option to expand the flexibility of a CHP plant is represented by the option of cascading an electric heat pump (Horlock 1997), giving origin to what can be modeled as an *equivalent cogeneration plant* (Mancarella 2009).

A natural extension of cogeneration leads to *trigeneration*, whereby electricity and heat are complemented by the production of another energy vector, which is cooling in most applications, so yielding a so-called CCHP system (Wu and Wang 2006). A trigeneration plant, as well as any generic DMG plant, can be arranged in a number of ways, depending on the equipment that is used along the CHP side. This interest in different schemes and technologies for DMG lies in the flexibility that different

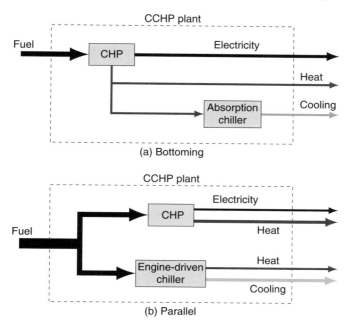

Figure 5.2 Example of *bottoming* and *parallel* generation in a CCHP plant. In the *bottoming* configuration (a), the CHP prime mover is cascaded to an absorption chiller; in the *parallel* configuration (b), the CHP prime mover is separate and parallel to an engine-driven chiller with heat recovery.

multi-generation schemes can provide in pursuit of different objectives, such as profit maximization, network impact minimization, emission reduction. Two basic schemes can be used as a conceptual representation: bottoming generation and parallel generation (Mancarella 2006, Mancarella and Chicco 2009a). These arrangements are described below and shown in Figure 5.2.

Bottoming generation

In bottoming generation, cogenerated heat and/or electricity are used to supply cascaded equipment for further generation. For instance, heat can be used to fire an absorption chiller and produce cooling in the most classical CCHP case. The rationale of this traditional approach is based on the possibility of increasing the utilization of the CHP plant (and thus its economics) throughout the year, by making use of cogenerated heat also in the summertime, when in most applications heat demand is much less than in winter due to lower space heating requirements. On the other hand, high demand of chilled water for space cooling might be required. The 'indirect-fired' absorption chillers used in this trigeneration scheme can be single-effect (usually fired by hot water, and thus suitable also for small-scale applications, coupled to ICEs, MTs, SEs, etc.) or double-effect (usually fired by super-heated water or steam, in which cases larger ICEs or GTs are more suitable). Triple-effect chillers that are recently appearing on the market require higher pressures and temperatures, which could make them more suitable to direct-fired applications (Wu and Wang 2006). Building-integrated micro-scale trigeneration applications can also be realized by coupling micro-CHP systems to emerging heat-fed technologies such as adsorption (Wang and Oliveira 2006, Huangfu *et al.* 2007) and solid or liquid desiccant (Daou *et al.* 2006). A similar form of bottoming generation can be obtained with a reversible EHP, which can be cascaded to the CHP plant to produce cooling as well as further heating from cogenerated electricity. The type of EHP to be used depends on the specific conditions, but most likely water-source and ground-source heat pumps will be used (Harvey 2006). In advanced schemes, EHPs can be coupled to CCHP systems to enhance the temperature level of waste heat recovered from absorption chillers (Havelsky 1999). Mixed schemes with simultaneous presence of more cooling devices are also of particular interest owing to their flexibility (Chicco and Mancarella 2009b).

Parallel generation

In this case, further generation is realized through independent means that are not cascaded to the CHP system. For instance, in the techno-economic assessment of generalized trigeneration systems in Chicco and Mancarella (2006), a direct-fired absorption chiller can be used to generate cooling. Similarly, an engine-driven chiller/heat pump can be used to generate cooling/heating from natural gas (Lazzarin and Noro 2006a), while electricity and heat may still be cogenerated in the CHP side of the multi-generation system.

Another key component of DMG systems is storage. While cascaded or parallel schemes provided *physical* coupling of input/output energy vectors in a DMG system, heat and/or cooling storage represent solutions of increasing importance for *temporal* decoupling of electricity and heat/cooling production from the local needs (Maidment and Prosser 2000, Zhang *et al.* 2007). These solutions allow the creation of an energy buffer to be profitably used for thermal load shifting and control purposes, providing environmental benefits (for instance, less heat may be wasted) as well as economic benefits, particularly by enhancing the potential of the DMG plant to react to real-time prices in a market environment (Houwing *et al.* 2011).

5.2.2 Black-box modelling of DMG systems

When performing energy and environmental assessments of DMG systems, an effective modelling representation of a multi-generation plant can be carried out by considering each component as a *black box*, describing its energy characteristics by means of input–output relations (Mancarella 2006, Mancarella and Chicco 2009a). This approach is convenient in terms of limiting the number of variables to be handled in the analysis and can be scaled-up easily. In particular, the black boxes of various plant components can be aggregated together to build black boxes at a macro-level, for instance, to represent the overall DMG plant. An automatic algorithm to create a matrix representation of the plant suitable for formulating optimization problems is provided in Chicco and Mancarella (2009b). This representation is also consistent with the Energy Hub approach to multi-carrier energy systems (Geidl and Andersson 2007). In fact, DMG systems can also be modelled as interconnected energy hubs. Hence, it becomes straightforward to perform sensitivity and optimization studies while synthetically capturing the main points of the analysis.

Under the black-box approach, energy, environmental and economic indicators can be formulated as functions of some design parameters (such as the first-law efficiencies of the equipment). For instance, let us focus, for the sake of simplicity, on a trigeneration scheme with a CHP plant and a cascaded absorption/adsorption chiller (AC), which will also be exemplified later in numerical terms (as shown in Figure 5.4). The energy performance of CHP prime movers can be synthetically described by means of the electrical efficiency η_W, the thermal efficiency η_Q and the overall efficiency η_y:

$$\eta_W = \frac{W_y}{F_y}, \quad \eta_Q = \frac{Q_y}{F_y}, \quad \eta_y = \eta_W + \eta_Q \tag{5.1}$$

The terms W, Q and F in (5.1), respectively denote electricity, heat and fuel thermal energy, while the subscript y points out cogeneration. Typical figures for various CHP systems for relatively distributed applications are summarized in Table 5.1.

For cooling generation equipment, the energy characteristics of an AC are described by means of the CoP (coefficient of performance), i.e. the ratio of the desired output (cooling energy R, in the form of chilled water for instance at 7°C) to the input (heat Q_R in the form of cogenerated hot water) (Harvey 2006):

$$COP = \frac{R}{Q_R}$$

The above efficiencies depend upon the technology and upon several variables such as the loading level, the outdoor conditions, and so forth. Details on various technologies and DMG applications can be found in Mancarella and Chicco (2009a) and Chicco and Mancarella (2009a).

As shown later in this chapter, it is possible to formulate a methodology for environmental evaluation of DMG systems (Chicco and Mancarella 2008b), which is consistent with the above black-box representation through the concept of an

Table 5.1 Typical CHP performance characteristics

Cogeneration plant	Electrical capacity (MW$_e$)	η_W (%)	η_Q (%)	η_y (%)
Stirling engine	0.001–0.05	8–25	65–80	80–90
Internal combustion engine	0.01–5	25–45	30–45	70–85
Micro-turbine	0.03–1	20–35	40–60	70–85
Gas turbine	5–30	20–40	35–50	70–80
Combined cycle gas turbine	50–100	45–55	5–30	60–80
UK grid electricity (average)		40–45		

Average UK power system efficiency from DECC (2011a).

emission factor. More specifically, the mass m_X of CO_2 emitted to produce the useful energy product X can be defined as $m_X = \mu_X X$, where μ_X is the CO_2 emission factor (specific emissions, in g/kWh) related to the relevant energy product X (e.g. electricity or heat in kWh). As a special case of useful energy product, it is possible to consider the energy F released when burning a fuel and which is used as the input to a thermal plant. In this case, with very good approximation, it is possible to consider the emission factor μ_F related to the fuel thermal energy as a constant depending only on the fuel carbon content and its lower heating value (LHV), that is, on the specific fuel. Since the relations between input and output energy vectors can be modelled through the energy efficiency characteristics, as described earlier, it is straightforward to realize that, once given the fuel type, the CO_2 emission factor μ_X associated to the energy output X can be evaluated as a function of the device efficiency only, as:

$$\mu_X = \frac{\mu_F}{\eta_X}$$

where η_X is the equivalent efficiency to produce the relevant energy output X from the fuel energy input F, as for instance in (5.1) for CHP units. For natural gas, μ_F can be averagely assumed equal to 200 g/kWh$_t$, on an LHV basis. Applications of this concept will be illustrated later. This approach can also be extended to consider other emissions, including local pollutants such as NO_x, CO and so on (Mancarella and Chicco 2009b, 2010).

5.3 District energy systems

5.3.1 *Distributed energy and the benefits of aggregation*

As mentioned above, waste energy recovery (particularly of heat) represents a major opportunity to increase urban energy efficiency. Within the DMG paradigm, heat recovery is naturally performed from distributed CHP systems to supply end-user heat demand (space heating and/or hot water) or for further generation, for instance in trigeneration. However, the necessary condition to set up an efficient DMG system is that there is an adequate energy load. While in the electrical case the power grid ensures the presence of a 'sink' to pool the electricity that might be generated in excess of the local consumption, the same situation does not apply to heat. In this sense, by aggregating several thermal users through heat networks, the so-called district heating (DH) or community heating can represent an enabling technology for the deployment of CHP (Harvey 2006). Indeed, a number of benefits arise from setting up heat networks, among which are the following:

- Residential users tend to exhibit highly variable loads at daily, weekly and seasonal intervals and so the overall operational techno-economic performance of local units such as household EHPs, boilers or micro-CHP plants may be relatively poor due to frequent on–off cycling, part-load operation, low plant capacity utilization, etc.

- The heat demand of various service sectors (such as offices, schools, commercial centres, and so on) typically exhibit complementary thermal load profiles with respect to the residential sector.
- The aggregation of users can therefore provide significant benefits in terms of load diversity, which brings about lower peak demand and better load factors (higher energy infrastructure utilization) than in the case of generation plants installed at individual users.
- Economies of scale in generation generally prompt the adoption of bigger prime movers and larger heat sub-stations.
- Efficiencies and emission profiles of centralized systems are better, owing to economies of scale, steadier operation from network aggregated loads, ease of plant control and economics of local pollution abatement systems.
- Generation systems can thus be generally designed and operated more effectively, with smaller specific plant capacity (per kW of served load) and higher energy infrastructure utilization compared to smaller plant.
- Last but not least, and perhaps mostly important in terms of leaving all future energy options open, DH facilitates the pooling of heat from a number of (low carbon and economical) sources including CHP, industrial waste, refuse incineration, centralized electric heat pumps, centralized solar thermal, and so forth.

From a modelling perspective, and for the purpose of the illustrative analysis discussed here, it is possible to represent district energy systems at first approximation as black boxes characterized by average efficiencies. Focusing on heat networks, the main aspect to model is the heat losses Q_L due to heat transfer between the hot energy carrier (typically water) and the colder external environment (typically the ground). If the DMG plant supplying the DH network is a CHP plant, the heat loss contributions can be expressed with respect to the cogenerated heat as:

$$\varepsilon_Q = Q_L/Q_Y \tag{5.2}$$

Typical values of percentage heat losses ε_Q range between 1 per cent for small networks (few hundreds of meters) and 10–15 per cent for larger DH transportation networks (several tens of kilometers) (IEA 2005).

5.3.2 *The challenges of integrated energy planning*

Based on the above discussions, district energy systems can be naturally set up together with DMG, whereby commercial and sometimes technical arrangements are put in place through a central system (for instance, managed by an energy service company or an energy retailer) to guarantee economic and reliable multi-energy services to a community. In fact, while district energy systems with DMG and local energy networks represent a promising solution to increase environmental performance, the economics of such distributed energy systems need to be thoroughly assessed through cost benefit analysis approaches. For example, the feasibility of heat networks could be hindered by the cost of network infrastructures and energy losses occurring while transporting heat. From this point of view, urban areas lend themselves well to district energy system applications, owing to the high energy density that tends to decrease the cost of energy infrastructure (Persson and Werner 2011).

Economies of scale in both cost and performance of distributed energy generation and distribution add complexity when determining at what level of centralization district energy systems and DH should be optimally deployed. In a DMG framework, in order to pursue economically sustainable solutions and maximize environmental benefits, techno-economic integrated planning of DMG and district energy systems should be carried out. In fact, focusing again on CHP and DH, heat network development could also impact on the cost and performance of the electrical network from the operational point of view (if the electrical network is already developed), as well as from the design point of view (in the case of integrated energy system planning) (Mancarella *et al.* 2011). The shape of power systems could thus be affected by the introduction of district energy systems at different centralization levels of the distribution infrastructure.

For example, in a more *centralized* approach to DH, relatively large DMG systems (for instance, CHP ICEs or GTs from 5 to 50 MW_e) connected to high voltage distribution networks might be used to supply heat through transportation networks to local distribution networks in towns, urban areas and industrial sites (Figure 5.3a). Taking the relevant case of Copenhagen (CTR 2004), whose DH system dates back to the 1970s, the hot water heat transportation network is tens of kilometres long (55 km) and operates at high pressure (in the order of up to 25 bar) and relatively high supply temperature (up to 120°C). Connection to local heat distribution networks at lower temperatures and pressures takes place through heat sub-stations. A number of other examples of real cases CHP-DH energy systems can also be found in Harvey (2006).

On the other hand, in a more *decentralized* approach, DMG plants connected to medium voltage networks and with capacities ranging between a few hundred kW_e and 5 MW_e might be scattered over urban areas and directly supply local heat distribution networks at pressures below 10 bar and a temperature below 100°C (Figure 5.3b). A number of cases of this type are available in the UK (where in contrast centralized schemes are not widespread), including the Southampton scheme (Southampton CC 2010) and the Pimlico and Whitehall scheme in London (LDA 2010). In this more decentralized approach, low temperature systems with plastic pipes directly connected to the users' heating systems might be adopted.

The optimal level of centralization and the network structure therefore depend on a number of factors related to both network side and generation side. Among the others, we can mention electricity and heat density, geographical distribution of main loads and sources, available local generation sources (including waste heat) and fuel networks, users' temperature requirements, projected investment and operational costs.

5.3.3 *Future district energy systems*

In the future, and in some places already today, the integrated energy planning issue will become even more complex due to the possibility of (and need for) various sources of distributed energy generation besides the CHP and boilers already mentioned.

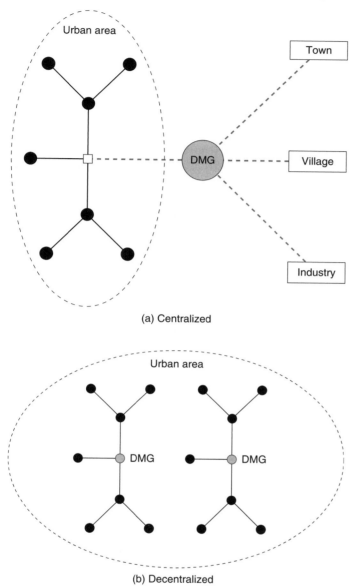

Figure 5.3 Schematics of centralized and decentralized approaches to district heating. The centralized approach (a) features a long transportation network to distribution sub-stations, whereas in the decentralized approach (b) DMG plants are connected directly to distribution networks.

These would include primary inputs from biomass, biofuels and refuse incineration, centralized heat pumps, centralized electrical and thermal solar plants, industrial waste heat, large-scale heat storage, etc. In this sense, the early adoption in some countries, for instance, Denmark and Sweden, of district energy systems with CHP and heat networks, has enabled a number of alternative choices for low carbon heat generation, from industrial waste to urban waste. An interesting case is Stockholm (Friotherm 2005), which has a district heating and cooling system that is also supplied by distributed large-scale heat pumps that utilize sea water as a steady state source of heat throughout the year. This allows excellent performance irrespective of the outdoor conditions.

Following the success of many applications, it is also likely that multi-generation will become more widespread than today, extending classical CHP and DH applications to district multi-energy systems. For instance, district energy systems with heat networks supplied by relatively large CHP or EHP plants might become trigenerative in response to increasing cooling request in various sectors, and particularly for 'anchor loads' such as supermarkets, hospitals, offices, etc. The economics of various schemes should be appraised through systematic approaches that take into account all the uncertainties relevant to multi-generation planning over the whole plant lifetime (Chicco and Mancarella 2006, Carpaneto *et al.* 2011). Optimal management of several DMG resources also needs powerful aggregation concepts that facilitate both technical and commercial viability. The Microgrid (Schwaegerl *et al.* 2011), Virtual Power Plant (Pudjianto *et al.* 2007) and Energy Hub (Geidl and Andersson 2007) concepts are possible approaches that could be adopted to optimally control distributed energy resources in district energy systems while also interacting with the upstream 'central' networks (electricity, first of all, but also natural gas).

5.4 Assessing the environmental performance of district energy systems

5.4.1 *General considerations and comparison with conventional energy systems*

DMG concepts and applications have been introduced in the previous sections, highlighting the potential for setting up more efficient energy systems based on distributed energy. In this sense, it is crucial to have a systematic methodology that is able to quantify the actual emission reductions relative to benchmark technologies and takes into account the interactions with centralized energy generation, particularly of electricity. In fact, in a continuously changing energy framework (particularly with increasing volume of electrical renewable sources), simple and synthetic models are needed to assess in which conditions and to which extent DMG and district energy systems can bring environmental benefits.

A powerful approach developed in recent publications follows the classical studies of Horlock (1997) to assess the energy saving brought by cogeneration plants with respect to separate production (SP) means, namely, electricity produced within the power system and heat produced in boilers. In a similar fashion, DMG has been evaluated in terms of the energy saving that multi-generation can bring compared with benchmark SP means (Chicco and Mancarella 2007). In addition, the assessment methodology has been extended from energy saving to the reduction of global greenhouse gas emissions (Mancarella and Chicco 2008) and local environmental impacts (Mancarella and Chicco 2009a), up to defining a unified approach with structurally similar indicators that can assess the energy and environmental benefits of multi-generation systems compared with conventional (centralized) energy systems (Chicco and Mancarella 2008b, Mancarella and Chicco 2009a). The main objective of these studies was to formulate simple analytical models (based on the black-box approach illustrated above), capable of highlighting the parameters and variables involved in energy and environmental analysis of DMG systems, so as to be able to foresee the potential benefits in different energy contexts. The outcomes from the proposed models and studies that have been carried out highlight the critical role of energy efficiencies and of emission factors (and therefore of primary energy sources) used in both DMG and SP reference systems, as well as of the operational characteristics of the decentralized system. An illustrative application, aimed at analysing the potential CO_2 emission reduction from a DMG district energy system, is shown below.

5.4.2 Emissions assessment of a trigeneration district energy system

In order to compare different energy generation alternatives, it is convenient to establish a reference scenario corresponding to a 'conventional' system and to assess the various alternatives against this reference. For environmental assessment of a district energy system with a CCHP plant, this can be performed through the TCO_2ER (Trigeneration CO_2 Emission Reduction) indicator introduced in Chicco and Mancarella (2008a). The TCO_2ER expresses the relative reduction of the mass of carbon dioxide emitted, due to the use of a fossil fuel-based trigeneration system instead of conventional SP generation means (namely, electricity from power plants, heat from boilers and cooling from electric chillers). Under this assumption, the TCO_2ER indicator is expressed as:

$$TCO_2ER = \frac{m^{SP} - m_F}{m^{SP}}$$

$$= 1 - \frac{\mu_F F_Z}{\mu_F^{SP}\left(W_Z + R_Z/COP^{SP}\right) + \mu_Q^{SP} Q_Z} \tag{5.3}$$

The expression (5.3) applies to a generic trigeneration plant that is supplied by the fuel thermal energy F_z and generates electricity W_z, heat Q_z and cooling energy R_z. The subscript z points out net input–output entries for the overall plant (as compared with the earlier y subscript which applies to a single piece of equipment, see Figure 5.4) and setting $R_z = 0$ in (5.3) leads to cogeneration as a sub-case. As will be shown later, the model can also take into account the presence of distribution networks by black-box modelling of network losses (network losses in SP are intrinsically taken into account by discounting the relevant emission factor). In terms of emissions, m^{SP} is the mass of CO_2 emitted by the production of the same amount of trigenerated energy in SP, which is evaluated through the emission factor μ_W^{SP} and μ_Q^{SP} for separate production of electricity and heat, respectively (emissions from cooling generation are assessed through the reference electricity emissions, and considering an electric chiller with cooling-to-electricity efficiency equal to COP^{SP}). It is also worth highlighting that the heat emission factor is defined generically and thus the model can assess various types of fuels burned in conventional boilers as well as electric systems such as electric boilers or heat pumps. On the other hand, m_F is the CO_2 mass emitted by combustion of the CCHP fuel thermal input F_z with emission factor μ_F.

The model is completely general and allows the user to evaluate the sensitivity of a number of parameters and variables, including the district energy system efficiency and the emission factors of reference generation. This makes it easy to compare a number of possible DMG options in different energy scenarios, in particular with respect to the decarbonization level of the power grid. Positive values of TCO2ER indicate that adopting trigeneration offers environmental benefits relative to conventional energy systems. The maximum positive value of TCO2ER is unity (or 100 per cent), ideally representing a carbon dioxide-free fuel. Negative TCO2ER values (not limited in amplitude) indicate that introducing trigeneration to displace SP is not environmentally beneficial.

5.5 Numerical examples of a trigeneration district energy system

5.5.1 District energy system description

In order to exemplify some of the concepts expressed above, let us consider a district energy system with DMG based on alternative options (including heat networks) to supply electricity, heat and cooling demand. A schematic of the energy system is shown in Figure 5.4, where all the components are represented as efficiency-based black boxes.

In the energy flow model of Figure 5.4, the cogenerated electricity W_y coincides with the overall trigenerated electricity W_z and goes to supply the local user or is injected into a local microgrid or the distribution network (for the sake of simplicity, pumping losses for heat distribution and electricity distribution losses have been neglected). The cogenerated heat Q_y, net of heat distribution losses in DH, splits into two components, namely, Q_z corresponding to the net trigenerated heat output for direct thermal purposes (for instance, domestic hot water generation and space heating), and Q_R to fire absorption chillers that are distributed across the heat networks and are used to generate cooling energy R_z. The relative amount of cogenerated heat that goes to feed the chiller is indicated with the 'splitting variable' α_R. It is easy to check that the TCO2ER indicator (5.3) in this case can be written as:

$$TCO_2ER = 1 - \frac{\mu_F}{\mu_F^{SP}\left(\eta_W + \alpha_R \frac{COP}{COP^{SP}}\left(1 - \varepsilon_Q\right)\eta_Q\right) + \mu_Q^{SP}\eta_Q\left(1 - \alpha_R\right)\left(1 - \varepsilon_Q\right)} \qquad (5.4)$$

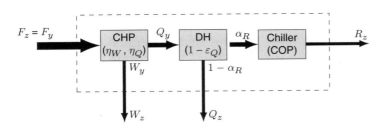

Figure 5.4 Schematic model of a trigeneration system with heat network. See text for explanation of symbols.

The TCO$_2$ER expression in (5.4) yields an analytic formulation of the potential emission reductions in trigeneration as a function of the plant component and network-related efficiencies, the splitting factor (which also corresponds to various relative heating/cooling demand levels), and the emission factors for the input fuel and the SP references. Therefore, it is possible to run various analyses to highlight the role played by the specific entries involved in the study, as shown in the following section. In particular, in the successive analysis, only the parameters of the DMG system and of the reference central production are used, with no specification of the amount of electricity, heat and cooling involved. In fact, the emissions reduction brought by the district energy system depends only on the contribution of DMG in displacing the separate production. That is, since it is in practice impossible to obtain a perfect match between DMG supply and multi-energy demand (Mancarella and Chicco 2009a), the environmental benefits are relevant to the part of the energy demand that is supplied in trigeneration. In this approach, all the trigenerated energy is usefully utilized, and any auxiliary energy that might be needed to balance supply and demand is supplied through the electricity distribution grid or auxiliary boilers (these backup options are not explicitly shown in Figure 5.4), whose characteristics are the same as the ones assumed for the corresponding separate production equivalents. In particular, when $\alpha_R = 0$ there is no cooling demand supplied through the trigeneration chain, meaning that any cooling demand is covered by electric chillers with parameter COP^{SP} supplied by the electrical network with equivalent emission factor μ_W^{SP}. This also corresponds to the case of pure cogeneration with no cooling demand, and with auxiliary boilers that cover the remaining heat demand. Likewise, for $\alpha_R = 1$ there is no trigenerated heat, and the entire heat demand (if any) is conventionally covered by boilers with equivalent emission factor μ_Q^{SP}. The quotas to cover through various pieces of equipment within the DMG system and more generally through various DMG systems depend on the specific control strategies and objective functions that are pursued (Chicco and Mancarella 2009b, Geidl and Andersson 2007).

5.5.2 Numerical case study and results

In the specific case under study here, different district energy systems have been assessed, namely, a micro-turbine and two internal combustion engines of different sizes and efficiency values coupled to single-effect absorption chillers. The average performance characteristics (assumed to be constant and equal to nominal values, for the sake of simplicity) and typical capacities for the equipment analysed are shown in Table 5.2. In addition, average energy penalties due to heat networks of different sizes are considered, with heat losses increasing with the CHP size. All CHP systems are assumed to be fuelled by natural gas.

The TCO$_2$ER indicator is plotted in Figure 5.5 assuming α_R as the independent variable. Two cases are analysed, with particular interest in the central power grid carbon content:

Case 1 The SP emission factor for electricity refers to average UK emissions ($\mu_W^{SP} = 460$ g/kWh$_e$) in 2010, including nuclear and renewables as from DECC (2011a).

Table 5.2 Average capacity and efficiency values for the considered district energy system

	Capacity technology	η_W (kW$_e$)	η_Q	COP	ε_Q
MT	100	0.30	0.55	0.7	0.01
ICE1	1000	0.35	0.5	0.7	0.03
ICE2	5000	0.40	0.45	0.7	0.05

See text for explanation of symbols.

The heat-related emission factor is calculated assuming average boilers with efficiency $\eta_Q^{SP} = 0.8$, fed on natural gas, thus obtaining $\mu_Q^{SP} = \mu_F/\eta_Q^{SP} = 250$ g/kWh$_t$. Finally, for the reference chiller a $COP^{SP} = 3$ has been used.

Case 2   Marginal emissions from the UK power system, estimated around 700 g/kWh$_e$ in Hawkes (2010), are considered, with same conditions for production of heat and cooling as in Case 1.

The results in Figure 5.5 show that for $\alpha_R = 0$, that is, cogeneration mode, the CO_2 emission reduction is the highest relative to the base case references. The environmental benefits also increase with the size of the district energy system, mainly owing to the increasing electrical efficiency in spite of a decreasing thermal efficiency that is also penalized by higher heat losses. With the assumed efficiencies and emission factors, the emission reduction decreases if larger volumes of cooling are trigenerated, but the DMG system is still environmentally effective. A break-even condition is realized for the case with MT and average UK emissions (Case 1) for splitting factors close to electricity-and-cooling cogeneration ($\alpha_R = 1$), while in all other conditions, significant emission reduction can be achieved. On the other hand, in Case 2 the environmental performance is always very good, and emission reductions in the order of 40–45 per cent can be reached for the CHP mode ($\alpha_R = 0$) when DMG is compared with marginal UK emissions.

It is therefore clear that the environmental benefits of the considered district energy system are strongly affected by the selection of the reference scenario. In this respect, apart from the numerical exercise that has been shown here to exemplify the issue, the selection of the correct benchmark (for instance, marginal plants or average mix) for assessment of multi-generation benefits, particularly in the outlook of providing economic incentives to DMG systems, is a critical issue that needs to be solved at the policy level (Chicco and Mancarella 2007, Mancarella and Chicco 2008, 2009a).

5.6 The potential role of district energy systems in a low carbon future

Following up on the above issue of the benchmark selection for DMG assessment, it is natural to enquire about the environmental benefits that thermal DMG plants could bring in a future dominated by low carbon energy sources. From this point of view, it is easy to show, through the models illustrated above, that multi-generation based on natural gas cannot compete with a cleaner centralized energy system, for instance, based on nuclear energy (as in France) or hydro (as in Norway) (Mancarella and

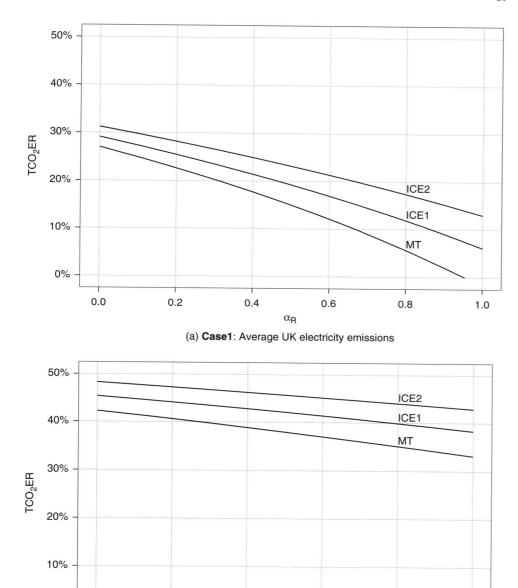

(a) **Case1**: Average UK electricity emissions

(b) **Case2**: Marginal UK electricity emissions

Figure 5.5 CO_2 emission reductions for small-scale trigeneration systems.
$COP^{SP} = 3, \mu_Q^{SP} = 250$ g/kWh$_t$ for both scenarios; $\mu_W^{SP} = 700$ g/kWh$_e$ for the marginal scenario and 460 g/kWh$_e$ for the average scenario.

Chicco 2008). However, if the fuel input to DMG was clean as well, which would be the case of biomasses or biogas for instance, DMG and connected district energy systems might still be a viable option for higher energy efficiency, energy saving and utilization of scarce resources compared with separate production (Mancarella and Chicco 2009a).

Even if the benefits of multi-generation decrease with future cleaner energy systems, there may still be other advantages to the use of distributed energy systems. Among the others, DMG systems, especially if coupled to thermal storage, may represent efficient dispatchable solutions for system balancing and provision of voltage and frequency ancillary services (such as the provision of leading/lagging reactive power and standing reserve). The latter will be particularly critical in a future with less flexible conventional generation (e.g. with an increased role for nuclear plants and carbon capture and storage plants) and more intermittent and unpredictable generation from renewables. The availability of district energy systems connected to relatively centralized DMG plants would facilitate the provision of balancing and ancillary services from distributed energy systems, as it would be easier to control fewer and larger distributed plants rather than more and smaller ones. In addition, there is an intrinsic option value in district energy networks, represented by the possibility of connecting multi-fuel energy sources (including the possibility to switch to those alternatives that become financially and environmentally more viable over different time windows) to local networks and thus also enabling competition at a distributed level (Grohnheit and Mortensen 2003). Hence, district heating that is today connected to efficient CHP plants might later be powered by EHPs when the electrical grid becomes more decarbonized. This energy source switch might not be straightforward or possible at all, without district energy systems.

5.7 Conclusion

In the development of future urban energy systems with enhanced energy efficiency and environmental performance, the exploitation of distributed multi-generation systems based on combined generation of multiple energy vectors, together with district energy systems represent a particularly promising option. This chapter has introduced the general concepts of district energy systems in urban areas, outlining the main technologies and applications on the one hand, and discussing on the other hand, the general issues and models to estimate the potential benefits in different energy contexts. Illustrative numerical applications have shown that the electrical efficiency of the DMG plants, which is also a function of the plant size, plays a key role in the overall assessment.

While significant environmental benefits can be appreciated in various energy contexts based on fossil fuel production such as in the UK, these benefits might decrease with decarbonization of the power grid. However, distributed and district energy systems are nonetheless likely to play a key role in the future owing to the need to increase flexibility in multi-energy supply, which could be facilitated by distributed but relatively centralized systems such as the ones illustrated here. In addition, district energy systems based on DMG will always represent an efficient option to make use of scarce low carbon resources such as biomass, biofuels, and so on, for the production of multi-energy vectors in thermal plants. Overall assessment of future alternatives will have to be based on comprehensive models and analysis that also consider aspects such as electrical network impact, reliability and security of supply, local pollutant emissions (besides global greenhouse gas emissions) and overall system economics. In addition, new urban energy technologies, such as hydrogen generation, storage and transportation and grid-connected electric vehicles, may increase the range of potential distributed options and are likely to be facilitated by relevant aggregation control concepts.

Note

1 Since multi-generation systems have multiple useful products, their rated capacity is commonly given with a subscript to indicate the output of interest. For example, 1 kW_e represents a kilowatt of electricity output and 1 kW_t (or kW_{th}) represents a kilowatt of thermal output.

6 Bioenergy and other renewables in urban energy systems
Potentials, conversion routes and future trends

Antonio M. Pantaleo, Nilay Shah and James Keirstead

University of Bari, Imperial College London

6.1 Introduction

As described in Chapter 2, cities primarily import high-quality raw fuels and export their wastes. This allows them to maintain high degrees of order within constrained spaces, a necessity for urban living. However, that is not to say that cities *only* import energy resources such as natural gas or electricity. Urban renewable energy sources – that is, renewable energy sources that lie predominantly within the urban boundary or the near hinterland – are an option for all cities, particularly those at early stages of development where lower energy demands and a lack of distribution infrastructure may make local resources an attractive option.

This chapter provides an introduction to urban renewable energy resources, with a particular focus on bioenergy. In the first section, a brief overview of different urban renewable energy sources is given to highlight the range of available technologies and their relative performance. However, the bulk of the chapter concentrates on bioenergy as one of the most diverse urban renewable sources and an excellent demonstration of the constraints, benefits and issues that need to be considered when seeking to reduce urban carbon emissions through renewable energy provision. For example, there are several economic, logistic, environmental and technological issues to be addressed when integrating bioenergy into urban energy systems such as transport constraints, storage requirements, energy conversion efficiencies, local air pollution levels and fuel supply costs. Bioenergy resources are also of interest because they provide a unique link with the waste streams of urban areas, offering an opportunity to 'close the loop' of a city's metabolism.

6.2 Urban renewables

For the purpose of this chapter, we can define 'urban renewables' as energy from renewable natural resources within the urban area or its near hinterland. This is a somewhat imprecise definition but it captures both the range of natural resources that are available for energy generation (e.g. sunshine, wind, water, geothermal energy) and the difficulty of drawing a hard urban boundary. For example, many of the resources described here cross the urban boundary (e.g. wind and water) and it would be rather arbitrary to exclude a wind turbine that lies just outside an urban boundary. The key

Table 6.1 Energy densities for various urban renewable energy resources

Technology	Energy density (W/m^2)	Notes/assumptions
Wind energy	2	Average wind speed of 6 m/s (high for urban areas)
Solar photovoltaics	22	20% efficient cells, south-facing roof in UK
Solar thermal	55	50% efficiency
Biomass	0.5	Best performance for Europe, varies depending on crop and other factors
Run-of-river hydropower	0.02	Assuming cities lie mainly in low-land areas
Tidal	3–6	Tidal pool or tidal stream
Geothermal	0.017	Average resource, based on renewable rate of extraction. May be much higher in some locations
Fossil fuel	43	Oil and gas production from Ghawar field, Saudi Arabia
Energy demand	11	London, 2008

Data source: MacKay (2009), GLA (2008) and author estimates.

point is that these are local resources for local use and therefore one potentially useful point of discrimination is whether the technology provides its energy primarily to one location or connects to a national grid or market.

One of the limitations of urban renewables is their energy density. Density – of people, economic activity and energy demands – is a distinguishing feature of a city but urban renewables are largely diffuse, in particular when compared with energy-dense point sources, such as an oil well. This is a major consideration in evaluating the suitability and feasibility of fuelling a city with local renewable energy, and Table 6.1 provides an indicative energy density figure for various technologies. These will vary from city-to-city but should give readers a feel for the relative potential yield of urban renewables.

As noted above, we have focused on biomass, including the waste fraction, since this arises in all urban areas and technologies for energy conversion are relatively straightforward to implement. The other urban renewables discussed here include solar energy, wind energy, hydro power and geothermal energy.

6.2.1 Solar energy

Solar energy, both thermal and photovoltaic, has a considerable role to play in urban areas. An excellent review of the role of these technologies is given by Girardet (2009). He introduces the concept of 'solar cities' as exemplified by cities as diverse as Adelaide and Rizhao City. The latter, in the Shandong Province of China, requires that all new buildings incorporate solar panels. Almost all the households in the central area districts have solar water heaters, and most public lighting is powered by PV (Bai 2007). Girardet makes the point that suburban areas provide plenty of space for solar PV and describes the case of South Australia, where communities bulk-buy 1 kW panels, organize large-scale deployment, and save on both purchase and installation. These should easily provide up to 25 per cent of each household's needs on average. Similarly, Freiburg,

Germany has developed a complete solar value chain including with a zero operational emissions solar panel factory as part of its long-term plans for a solar city.

6.2.2 Hydropower and marine renewables

Although many cities are located on rivers or the coast, there is limited potential for urban hydropower or marine energy. Hydropower relies upon relatively steep topography in order to provide sufficient reservoir volume and head, whereas urban areas are often located in low-lying areas. The value of urban land also means that flooding would incur a significant opportunity cost. Consequently, urban hydropower is limited largely to small opportunistic run-of-river schemes (e.g. on the River Wandle in London).

Marine renewables, including tidal stream, tidal pool and wave energy, are applicable for those cities along the coast. Given the costs of developing in these environments, these nascent technologies are typically constructed at large-scale and connected to the transmission network. They are therefore better thought of as national-scale renewable energy technologies, and not specifically urban.

6.2.3 Wind energy

Wind energy is a technology found both within the immediate urban area and the surrounding periphery. However, the most economic locations are those with steady, strong wind speeds greater than 6 m/s. The built environment interrupts these flows and results in much lower outputs compared with other sites (Bahaj *et al.* 2007). As a result, urban wind is a relatively niche technology, often used in brownfield sites at the urban fringe.

However, large offshore wind farms may be constructed very near to large cities, particularly in shallow coastal areas. An example is the London Array, a 1 GW development in the outer Thames estuary, which should eventually be able to power 750,000 homes. Since many cities are in coastal locations, similar solutions would be expected to emerge over time.

6.2.4 Geothermal

Geothermal energy systems can be divided into two categories depending on the temperature available near the Earth's surface. For example, high-temperature systems, such as those found in Iceland or other seismically active areas, can provide heat at 100°C or more for use in electricity generation or heat provision. Low temperature system, either in the form of open-loop convection-based systems or closed-loop conduction-based systems, offer heat below 100°C. Heat pumps can then be used to extract this heat for use in district energy systems or directly within buildings for heating and cooling provision. A related technology is enhanced geothermal energy systems (EGS), where bedrock is hydraulically fractured to improve heat transfer by convection, which offers a promising option for many cities. However, as the case of

Basel, Switzerland shows, there can be public concerns about induced seismicity events arising from EGS pumping operations within urban areas (Majer *et al.* 2007).

When designing geothermal energy systems, a particular concern is that the rate of extraction should be renewable so that as heat is extracted over the course of the year, the ground temperature shows no net change. For an urban environment, such as a residential neighbourhood, this means that not everyone can use ground-source heat pumps *unless* there is additional recharge of the ground's heat, e.g. via solar thermal systems (MacKay 2009: 152, 302–305). For reviews of related heat pump technology, please see Cantor (2011), Chua *et al.* (2010) or DECC (2012).

6.3 Urban bioenergy

Having provided a brief introduction to urban renewable energy in general, we now turn to bioenergy resources in particular. The discussion is divided into three sections. First, section 6.3.1 presents the potential of biomass resources in urban and peri-urban areas, providing a description of biomass typologies and characteristics, possible end-uses in energy and non-energy sectors, cost ranges, biomass processing and energy conversion technologies for urban energy systems. Section 6.3.2 then reviews the bioenergy conversion technologies suitable to serve the heating, cooling and power demands of urban and peri-urban areas, and the opportunities for integrating these routes with existing energy systems and infrastructures. Some representative case studies of urban biomass supply chains are also reported. The main issues, trade-offs and factors influencing the use of bioenergy in urban areas are addressed in section 6.3.3, and the perspectives of future integration of bioenergy in urban energy systems discussed.

6.3.1 Biomass energy resources and potentials

One of the unique features of biomass energy resources is their diversity. Such resources can come from a range of different feedstocks and offer different energy densities and other characteristics. A broad classification of biomass feedstocks, useful for the purposes of this chapter, is proposed in Table 6.2.

The availability of biomass for energy has been the subject of a great number of studies at global, regional and local scales and these provide a diverse range of resource estimates. At the global level, estimates for the amount of primary energy that might be provided by biomass in 2050 vary from less than 100 EJ/year to over 1100 EJ/year (Hoogwijk 2003, Berndes *et al.* 2003, Yamamoto *et al.* 2000, Smeets *et al.* 2007). The reasons for the large range in estimates include the wide variety of methodologies, datasets and assumptions used to define the bioenergy potential (Slade *et al.* 2010), and differences in key factors such as the availability of land, the yield of biomass and the availability of residues from existing industries.

Methodologically, estimates of biomass potentials can be described as being either resource-focused or demand-driven. The first type of study seeks to compile an inventory of biomass resources based upon assumptions about the availability of supply-side

Table 6.2 Classification of urban biomass feedstocks

Typology	Description
Urban biomass	
Biodegradable municipal wastes (BMW)	Paper/card, food/kitchen waste, textile/wood residues
Organic fraction of municipal solid wastes	As above
Urban wood wastes	Stemwood, chips, branches, foliage from municipal trees and private gardens ('urban green'); demolition wood; industrial wood wastes
Waste vegetable oils	Waste cooking vegetable oils
Sewage sludge	By-product of wastewater treatment processes
Peri-urban biomass	
Energy crops	Short rotation forestry, herbaceous crops, silage crops, oleaginous crops
Forestry residues	Chips from branches, tips and stemwood
Agricultural residues	Straw from sewage crops, pruning residues
Wood residues	Clean and contaminated waste wood
Sawmill co-products	Chips, sawdust and bark
Agro-industrial residues	Wet fermentable wastes, lignocellulosic by-products
Zootechnical residues	Cattle, pigs, poultry manure
Landfill gas	Gas from decomposing biodegradable wastes in landfill sites

Adapted from Hoogwijk (2003), Slade *et al.* (2010), E4Tech (2009).

resources (principally land) and competition between different uses and markets. Demand-driven studies, on the other hand, focus on the competitiveness of bioenergy compared with conventional energy sources or estimate the amount of biomass required to meet specific, exogenously imposed, targets (Berndes *et al.* 2003). A distinction may also be drawn between studies based on their complexity in estimating the future share of cropland, grassland, forests and residue streams available for bioenergy (Smeets *et al.* 2007). Three integrated models have been used to estimate the future potential of bioenergy: the Global Land Use and Energy Model (GLUE) (Yamamoto *et al.* 2000), the Integrated Model to Assess the Global Environment (IMAGE) (Lauer 2009) and the Basic Linked System (BLS) model of the world food system (Fischer and Schrattenholzer 2001). Among others, the Biomass Energy Europe project seeks to harmonize biomass potentials assessment methods and identified 136 studies that have sought to estimate the potential for bioenergy (BEE 2008). Those studies which explicitly identified the resource potentials in urban areas have been included in this review.

Table 6.3 presents the contribution of biomass feedstocks to the primary energy consumption in the UK and Italy. In both cases, biomass represents the largest contribution to renewable energy in input terms. However, for landfill gas, sewage sludge and municipal solid waste (MSW) a substantial proportion of the input energy content is lost in the process of conversion to electricity, so that the same share is not reflected in output ratios.

Most estimates of biomass waste potentials are derived by the economic activity of the main sectors that are responsible for producing these wastes via top-down approaches. These estimates may then be projected into the future, taking into account the influence

Table 6.3 Contribution of biomass to primary energy consumption in the UK and Italy, 2010

	Tonnes of oil equivalent	
Resource	UK	Italy
Landfill gas	1,652	1,215
MSW and other biowastes	200	–*
Sewage gas	303	–*
Animal biomass (meat, bone and farm waste)	530	3,569
Wood and wood wastes	647	–*
Domestic solid biomass	764	–*
Imported solid biomass	729	918
Domestic liquid biofuels for transport	304	521
Imported liquid biofuels for transport	910	608
Total	6,039	6,831
% of total renewable energy supply	81%	53%

Notes:–* = included in preceding figure.
Data sources: DECC (2011a), MSE (2010) and author estimates.

of new legislation or other anticipated changes. This approach is generic to all the biomass potentials studies, although the details of the calculation change for each waste sub-category, and with the inclusion or exclusion of specific waste sub-categories. In EEA (2007), it is assumed that the production of MSW will be driven by GDP growth at national and sectoral level, moderated by the anticipated impact of household waste reduction measures (estimated to be 25 per cent in 2030). Similarly, in BEE (2008) the MSW resource is calculated as a function of the existing resource, moderated by growth rates, recycling rates and availability fractions. However, it now seems that both growing consumption and the trend towards smaller and more households, which have been strong drivers of municipal waste generation in the past, are now decoupling from municipal waste generation. Municipal waste generation per person in the EU-27 stabilized between 1999 and 2007, while consumption expenditure in constant prices increased by 16.3 per cent per person and the number of people per household decreased by 5.6 per cent. However, the total amount of municipal waste generated in the EU-27 over the same period increased slightly to 258 million tonnes mainly as a result of the small growth in population. In 2010, municipal waste generation declined to a level of 252 million tonnes (t) (Eurostat 2012).

Figure 6.1 reports the amounts of municipal solid waste per capita in the European Union for 2010. The average European value is about 0.5 t per capita year, recording a slight decrease since 2000. As regards large urban areas, the range of MSW are of 0.3–0.6 t per capita year (highest values for Rome and London, with 0.66 and 0.56 t per capita year, respectively, and lowest for Stockholm and Tokyo, with 0.3 t per capita year) (Siemens 2009).

MSW consists mainly of household and commercial wastes, which are disposed of by, or on behalf of, a local authority. It is composed mainly of paper/cardboard, plastics,

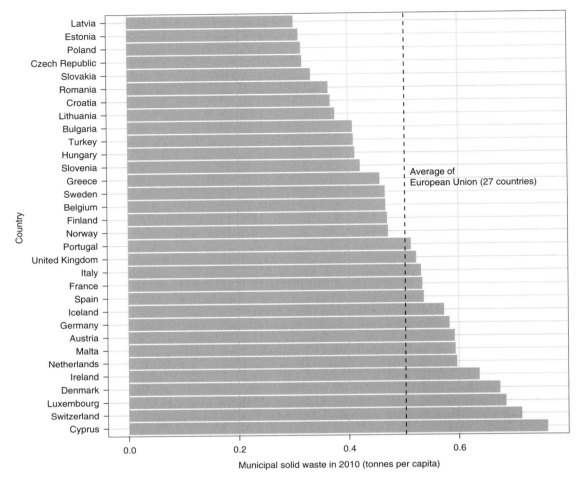

Figure 6.1 Municipal solid waste generated in the European Union (EU-27 plus Norway, Switzerland, Turkey, Croatia and Iceland) (tonnes per capita year, 2010).
Data source: Eurostat (2012).

glass, metals, textiles and food/garden waste. Consequently, the waste contains a high proportion of renewable materials. The good practice hierarchy of MSW management is: prevention or minimization in generation; material recovery; recycling; energy recovery; disposal in controlled landfills (Eighmy and Kosson 1996, Sakai *et al.* 1996). Despite this, landfill discharge remains the prevailing option in many European countries. The Landfill Directive 1999/31/EC (The Council of the European Union 1999) promotes the reduction of wastes that are landfilled and requires that by 16 July 2016, the amount of BMW going to landfill must not exceed 35 per cent of the total amount by weight of the amount disposed in 1995.

Reviews of the typical composition of MSW in various countries are available in the literature (García *et al.* 2005, IEA 2003, Mor *et al.* 2006), and Figure 6.2 shows the typical composition for the UK. These values are strongly influenced by the degree of

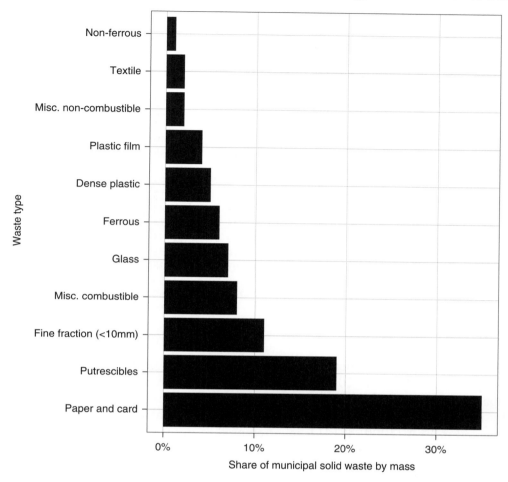

Figure 6.2 Average composition of MSW for the UK (IEA 2003).

separate collection of paper and organic wastes; however, depending on local conditions, eating and drinking habits, climate and the degree of industrialization, between 70 per cent and 80 per cent of MSW consists of BMW (i.e. food and green waste, and paper and cardboard waste).

We now focus on three urban biomass resources: biodegradable municipal wastes, urban wood waste and waste vegetable oils.

Biodegradable municipal wastes

Biomass sources in urban areas are essentially represented by biodegradable municipal wastes (BMW), a fraction of the municipal solid wastes (MSW). The MSW stream consists basically of household wastes and paper and card wastes, that account for 8 and 2 per cent of total wastes at EU level, respectively, according to the EU-27 waste stream outlook. In some cases, wood wastes and animal/vegetal wastes are also included

Table 6.4 Eurostat data for BMW landfilled in 1998

Country	Managed MW (BMW)	Separately collected and recovered BMW (kt per year)	MW incinerated	BMW landfilled	Per capita BMW (annual t per person)
Austria	2,644 (1,745)	791	431	523	0.19
Belgium	5,014 (4,312)	425	1,490	2,397	0.28
Denmark	2,591 (2,560)	641	1,466	453	0.35
Finland	2,100 (1,890)	0	50	1,840	0.33
France	34,700 (27,760)	220	10,352	17,188	0.27
Germany	40,017 (28,700)		8,552	20,148	0.35
Greece	3,000 (2,688)	0	0	2,688	0.25
Ireland	1,550 (1,073)	60	0	1,013	0.27
Italy	24,524 (21,655)		1,572	20,083	0.16
Luxembourg	278 (160)	0	126	34	NA
Netherlands	8,161 (7,280)	2,523	2,192	2,565	0.31
Portugal	3,884 (3,301)		6	3,295	NA
Spain	14,914 (11,633)	2,117	693	8,823	0.31
Sweden	3,200 (2,656)	400	1,300	956	NA
UK	29,000 (21,460)		2,200	19,260	0.31

Data source: EEA (2001).

in the MSW stream. These products account for 3 and 4 per cent of total wastes at EU level, respectively (Eurostat 2012). Total municipal waste and BMW cannot be easily compared between different countries due to differences in the kind of waste collected by different municipalities (EEA 2000). The municipal and household waste survey conducted in EEA (2000) at EU level attempted to improve the comparability of data collected, introducing operational definitions for BMW, the biodegradable fraction of mixed wastes, separately collected mixed wastes, bulky wastes and their biodegradable fraction. The baseline (1998) data for the application of the EU Landfill Directive are reported in Table 6.4. The managed BMW quantity is calculated from managed MW (i.e. collected minus net export) less the non-biodegradable fraction (glass, plastic and metal wastes). The separately collected and recovered BMW is composed of urban green, paper, textile, wood, oil and fat. BMW potentials per capita are also reported, even if the per capita generation of wastes is highly variable on the basis of the economic status of the population (Mor *et al.* 2006).

An overview of strategies and instruments for diverting BMW away from landfill is proposed in EEA (2001), including the production, collection, transport treatment and final disposal phases. Table 6.5 reports the technological options for different typologies of BMW, including energy conversion, while Figure 6.3 describes the pathways for MSW treatment for recovery and recycling. A review of biological treatment processes, and in particular anaerobic digestion (AD) for the treatment of biowastes (both source-separated food wastes and centrally separated organic fraction of municipal solid wastes) is given in Monson *et al.* (2007).

The average biogas yield from AD processes applied to BMW is about 3,200 MJ/t, corresponding to 0.26 MWh per capita year if an average quantity of 0.3 t BMW

Table 6.5 Overview of technologies for the treatment of BMW

Criteria	Anaerobic digestion	Incineration	Pyrolysis	Gasification
Reliable?	Yes	Yes	Partly	Partly
Cost?	Medium–high	Medium–high	Medium–high	High–very high
Processes waste type...?				
Wet household	Yes	Yes	No	No
Dry household	Yes	Yes	Yes	Yes
Urban green	No	Yes	Yes	Yes
Paper and board	No	Yes	Yes	Yes
Food	Yes	Yes	Yes	No
Textile	No	Yes	Yes	Yes
Produces RDF?	No	Yes	Yes	Yes
Energy recovery	3,200 MJ/t	2,700 MJ/t	1,890 MJ/t and energy from char	2,700 MJ/t
End products	Biomethane, heat, power	Heat, power	Heat, power	Heat, power

Note: RDF = refuse-derived fuel.
Data source: EEA (2001).

per capita is assumed. To put it in perspective, these energy potential figures correspond to a range of 4–7 per cent of total energy consumption per capita for heating and hot water in urban areas. The cost of such AD treatments for the Greater London Area would be in the range of 25 €/t (compared with about 79 €/t for mass-burn incineration), according to GLA (2003a). The re-use or disposal of digestate slurry after the AD process is recognized as a key issue for the environmental and economic feasibility of these routes. A great advantage of AD routes is the possibility to upgrade the biogas to biomethane for direct injection into the gas grid (National Grid 2009, IEA 2012, DECC 2009), or for use in the transport system (Monson *et al.* 2007, SBGF *et al.* 2008), in order to increase the overall energy efficiency of the system. Reasonable targets for this route are about 10 per cent of domestic gas customers served by biomethane by 2020 in the UK (IEA 2012). These targets could be achieved including the biomass resources of peri-urban areas, such as manure, brewery, diary and other agro-industrial fermentable wastes, and co-digestion with energy crops.

Other options for energy conversion of BMW, such as gasification and pyrolysis, are described in DECC (2009) and SBGF *et al.* (2008), but are expected to provide some 10–15 per cent lower energy yields than the AD processes. The syngas obtained by biomass gasification processes could also be upgraded through methanation for gas grid injection, even if this route is far less developed than AD biogas.

Urban wood waste

Three major categories of urban wood wastes are available in urban areas:

- Wood wastes disposed of with, or recovered from, the municipal solid waste (MSW) stream ('MSW wood')

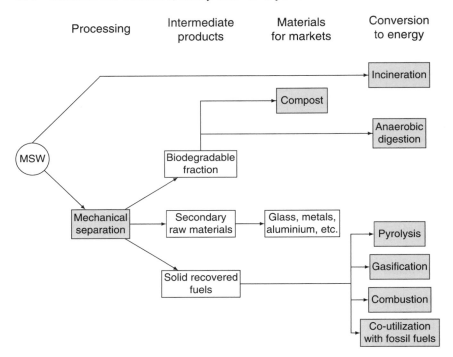

Figure 6.3 Pathways for MSW treatments (IEA 2003).

- Industrial wood wastes such as wood scraps and sawdust from pallet recycling, woodworking shops and lumber yards
- Wood in construction/demolition and land clearing debris.

In Wiltsee (1999) the quantity and typology of urban wood waste in US metropolitan areas is assessed. It is found that significant variations in the methods and costs of urban wood waste disposal and reuse arise, and some of the influencing factors are the levels of landfill tipping fees, access to and regulations concerning rural dumping and burning, public policies that promote waste diversion or recycling, and the proximity of large wood waste users (e.g. power plants, cogeneration plants, pulp and paper mills and medium-density fiberboard plants). The study also proposes predictive tools for the estimation of urban wood waste resources as a function of demographic and economic variables, reporting a range of 0.15–0.8 t wood waste per capita available yearly (average value of 0.3 t wood waste per capita year). This quantity would be sufficient to fire a range of 5–50 MW$_e$ power plant per million people, so producing a range of 0.4–4 per cent of the total electrical energy demand of urban areas (average values of 1–2 per cent).

Slightly lower values are reported for arboricultural residues and wood from civic amenity sites in London (LTOA 1991), with potentials estimated in the range of 150,000 t/year. This wood is dispersed across the city and would be most suited to use in heat producing boilers or relatively small-scale combined heat and power (CHP) schemes. The energy potential is up to 12.5 MW_e and 60 MW_{th} (GLA 2001). Moreover, this urban wood resource can be augmented by means of the integration of woody biomass from peri-urban areas.

A review of solid wood waste resources and potentials in the USA is provided in Falk 1994, with a specific focus on solid wood waste from MSW, demolition wood in constructions, primary timber processing by-products and treated wood resources. The conventional uses of wood wastes and the factors affecting the feasibility of recycling practices are also discussed, and the main drawbacks towards energy recovery result in the dispersion of the resource, the contamination of treated wood and the variability of the resource (with consequent costs of sorting and cleaning), the alternative uses for particleboard industry, and the volatility of market prices for end-products.

In Aiel (2011), the case study of the Province of Padova (North Italy) is explored, and the average wood waste potentials from urban forestry in the whole province (with a total of 0.9 million inhabitants) was found to be about 57 kg per capita year. No correlation was found between wood waste resource and the number of inhabitants of the municipalites included in the study. Moreover, only about 25 per cent of the total wood forestry potentials was considered to be suitable for the production of biofuels (wood chips or pellet), with the other fraction already being used for compost or not suitable for thermochemical conversion processes.

In conclusion, there are several sources of wood wastes in urban areas (urban green, demolition wood, wood processing industry, wood fraction of MSW), and each is highly site-specific and it is not possible to provide reliable ranges of potentials of urban wood wastes in urban areas per capita. Moreover, these resources are usually recovered for other markets, such as composting and particleboard, while in the case of contaminated wood, the energy conversion requires particular regulatory issues. In addition, the dispersion of the resource over the territory and its low energy density (in case of urban green) reduces the profitability of the establishment of a supply chain for energy conversion.

Waste vegetable oils

The waste cooking oils available in urban areas could be used for biodiesel or heat and power production, and there are some interesting case studies in the literature (ISQ 2009, EcoRec 2012). In Innsbruck, Austria, the company Oeli operates a 1 MW_e CHP diesel engine fired by waste cooking oils, with a recovery rate over 2 kg per capita year, while the estimated average quantity of waste vegetable oils is 5 kg per capita year. Moreover, the average potential resource from non-domestic cooking oil consumers (mainly restaurants) is about 240 kg/year (Amsa Gruppo 2009). Other case studies have been explored in Rio de Janeiro (EcoRec 2012), with a potential recovery of Copacabana waste cooking oils in a 450 kW CHP plant or for biodiesel production. The main barrier towards this application is represented by the recovery of the dispersed biomass resource, and for this reason, the use of intermediate collection platforms is one of the best options to achieve high collection rates.

Table 6.6 Classification of biomass resources for urban energy systems

ID	Type	Resulting biofuel/usage
Urban biomass		
1	Urban wood wastes	Chips and pellets
2	Waste vegetable oil	Biodiesel or refined bio-oil
3	Urban wet biowastes	Fermentable residues for AD processes
Peri-urban biomass		
4	Herbaceous residues and energy crops	Chips and pellets
5	Woody residues and energy crops	Chips and pellets
6	Herbaceous energy crops	Anaerobic digestion (AD) processes
7	Lignocellulosic agro-industrial residues	Chips, pellets or direct thermochemical conversion
8	Agro-industrial residues for AD	Fermentable residues for AD processes
9	Oleagineous energy crops	Bio-oil production

6.3.2 *Bioenergy routes for urban and peri-urban areas*

Bioenergy routes are essentially a three-stage process whereby different raw resources (*biomass*) are collected, before being upgraded and processed into *biofuels* and finally used in *energy conversion systems* to provide heating, cooling or power. In Table 6.6, a classification of biomass resources is proposed, including those produced in urban environments and in the neighbouring areas (Jablonski *et al.* 2008, Wit and Faaij 2010).

Because of the high energy intensity of urban areas in comparison to their biomass potentials, it is estimated that 5–10 per cent of the total heat and power energy demand of urban areas could be reasonably served by urban biomass, while in the case of integration with peri-urban biomass and rural communities, this percentage could increase up to a range of 30–50 per cent. These percentage ranges are also highly influenced by the energy demand typology of urban areas and their size, as discussed in the following sections. Urban areas located near the premises of local biomass potentials (i.e. forestry or wood processing residues) can take advantage of this opportunities to develop biomass-fired district heating (DH) or combined heat and power (CHP) plants (Freppaz *et al.* 2004). In the case of biomass imports over long distances, the sustainability assessment of the whole bioenergy conversion chains is a crucial issue and is a specific eligibility requirement for subsidies in several countries (e.g. EU 2009).

A set of biomass treatment and upgrading processes are required to obtain high energy density biofuels, which can be easily transported, stored, and that are suitable for high efficiency energy conversion processes, possibly at the premises of the energy demand. In Table 6.7, the commercially available and most promising biomass treatment processes are described to produce solid, liquid and gaseous biofuels. In most cases, these processes are implemented near to the biomass production sites, in order to minimize transport costs, facilitate trade on the market and storage issues. However, when integrating bioenergy routes into UES, the specific logistics, economic and environmental constraints of urban areas imply locating these processing facilities in industrial areas outside the town, eventually decoupling them from the final energy conversion of biofuels near

Table 6.7 Biofuel typologies and treatment processes for urban energy systems

ID	Biofuel	Treatment	Input biomass IDs
Solid biofuel			
1	Pellet	Chipping-drying-pelletization	1, 4, 5, 7
2	Torrefied pellet	Torrefaction-pelletization	1, 4, 5, 7
3	Chip	Chipping-drying	1, 4, 5, 7
4	Torrefied pellet	Hydrotreatment-drying/dewatering (wet lignocell biomass)	3–5, 7, 8
Liquid biofuel			
5	Bio-oil	Mechanical or chemical refining	2, 9
6	Pyrolysis oil (BTL)	Pyrolysis and thermochemical processes on lignocell biomass	1, 4, 5, 7
7	Biodiesel	Esterification of fatty acid methyl esters	2, 9
8	Biodiesel-FT	Gasification coupled to FT biodiesel process	1, 4, 5, 7
9	Bioethanol	2nd generation process from lignocellulosic biomass	1, 4, 5, 7
Gas biofuel			
10	Syngas	Gasification of lignocellulosic biomass	1, 4, 5, 7
11	Biogas	Anaerobic digestion	3, 6
12	Biomethane-AD	AD and biogas upgrading	3, 6
13	Biomethane-FT	Gasification+syngas upgrading	1, 4, 5, 7
14	Biohydrogen	Dark fermentation–AD processes	3, 6
15	Biohydrogen-FT	Catalytic synthesis from FT processes	1, 4, 5, 7

Notes: AD = anaerobic digestion, FT = Fischer–Tropsch process, BTL = biomass-to-liquids.

to the loads. Moreover, locating these processes in industrial areas could facilitate the implementation of bio-refineries approaches and the integration of multiple processes (Fatih Demirbas 2009).

The most promising biofuels for UES are pellets and in particular torrefied pellets with higher low heating value (Van Der Stelt *et al.* 2011, Hoekman *et al.* 2011), bio-oils (both from fatty acid methyl esters and second generation thermochemical processes on lignocellulosic biomass) (Tyson *et al.* 2004, Bridgwater 2011) and biomethane (from anaerobic digestion biogas upgrading or second generation Fischer–Tropsch processes on lignocellulosic biomass) (Ryckebosch *et al.* 2011, Der Meijden *et al.* 2010).

The biofuels can be converted into energy by means of several technologies, as shown in Table 6.8. Heat production tends to be the cheapest and most profitable conversion system for solid biomass and in the absence of specific incentives for bioelectricity. The district heating option is interesting in the case of high heat density (Zinko *et al.* 2008, IEA 2008b), new buildings or refurbishment of existing ones, and possibility to increase network load factors by district cooling with adsorption chillers (Rentizelas *et al.* 2008). The CHP option with solid biomass can be attractive in the case of high electricity costs, incentives for biomass electricity, favourable rules for on-site generation and net metering, the presence of suitable heat/electricity demand, and possibilities to manage the logistic constraints of the biomass transports and storage (Siemens 2009). Specific technological options are ORC (organic Rankine cycle) plants of up to 1–2 MW_e (Chinese *et al.* 2004, Rentizelas *et al.* 2009), and steam turbines, possibly in co-firing,

Table 6.8 Biofuel energy conversion technologies in urban areas and size ranges

ID	Technology	Size range	Input biofuel IDs
Heat technologies			
1	Multi-biomass boiler	0.02–20 MW	1–6, 10–11
2	Pellet stove boiler	0.02–20 MW	1, 2, 4
3	Biogas boiler	0.1–20 MW	10–13
4	Bioliquids boiler	0.5–20 MW	5–9
CHP technologies			
5	OCR	0.5–2 MW$_e$ / 2–10 MW$_{th}$	1–8, 10–13
6	ST	3–40 MW$_e$ / 9–120 MW$_{th}$	1–6
7	ICE diesel	0.1–15 MW$_e$ / 0.15–15 MW$_{th}$	5–8
8	ICE spark	0.1–15 MW$_e$ / 0.15–15 MW$_{th}$	9–13
9	MT	30–500 kW$_e$ / 50–500 kW$_{th}$	6, 9–13
10	GT	0.5–30 MW$_e$ / 0.5–25 MW$_{th}$	5–13
11	IGCC	15–40 MW$_e$ / 15–30 MW$_{th}$	1–6, 11
12	SOFC	10–100 kW$_e$ / 10–100 kW$_{th}$	10, 12–15
13	SE	100–500 kW$_e$ / 300–1000 kW$_{th}$	1–8, 10–13
Co-firing/dual-fuelling technologies			
14	EF-CCGT	15–800 MW$_e$	1–6
15	EF-GT	1–30 MW$_e$	1–6
16	Doubly fed GT	5–15 MW$_e$	5–13
17	ST co-firing	50–500 MW$_e$	1–6
18	Doubly fed ICE	5–15 MW$_e$	5–13

Notes: ORC = organic rankine cycle, ST = steam turbine, GT = gas turbine, CCGT = combined cycle gas turbine, SOFC = solid oxid fuel cells, ICE = internal combustion engine, MT = gas microturbine, SE = stirling engine, IGCC = integrated gasifier combined cycle, EF = externally fired.

for larger sizes (Berndes *et al.* 2010). In the case of liquid and gaseous biofuels, the options of internal combustion engines (ICE) and gas turbines (GT) (Lazzarin and Noro 2006b), also in co-firing mode with natural gas, are available and minimize the biomass transport, storage and air emission constraints which are typical of large solid biomass boilers and make their diffusion difficult in urban areas. The use of small-scale ICEs, microturbines (MT) (Kaikko and Backman 2007) and fuel cells (SOFC) (Lin *et al.* 2007), fired by high quality biofuels (bioethanol, biomethane, biohydrogen) for CHP (Saxena *et al.* 2009, Hamelinck and Faaij 2002) could be a very promising option for UES, in particular if connected to a centralized biofuel distribution network, and integrated with the gas network.

6.3.3 Bioenergy in urban energy systems and bioenergy

The use of bioenergy in urban energy systems can be affected by many different factors, which may be broadly grouped as pertaining to the urban area type (climate, energy demands, building fabric, etc.) or energy framework (existing infrastructure, financial incentives, environmental regulations, etc.). These are now briefly discussed

Table 6.9 Specific factors of biomass integration into urban energy systems

Factor	Description
Bioenergy-specific factors	
Environmental/planning constraints	Air quality issues, noise levels, landscape constraints, space availability for storage, biomass/biofuel transport congestion
Biomass resources availability	Urban and peri-urban biomass availability
Biomass actual use	No use, old low efficiency boilers, new boilers
Infrastructures	
Biomass transport	Presence of ports, rail networks, roads for long distance biomass supply; traffic congestion for biomass transport in urban areas to energy loads
Energy/biofuel transport	Presence of gas/district heating networks for integration of bioenergy
Storage infrastructures	Presence of storage infrastructures for imported biomass and seasonal biomass
Upgrading/processing infrastructures	Availability of biomass treatment and upgrading platforms or bio-refineries outside the urban areas to produce high quality biofuels for high efficiency conversion near to urban energy loads

and Table 6.9 presents some specific factors to consider for biomass integration into urban energy systems.

Urban area factors

The first category contains factors relating to the urban area and associated energy demands. The local climate is one such driver, as the load factor for residential–tertiary heating can vary between 800 and 2,200 hours per year according to climatic zone (Persson and Werner 2011). In cold weather areas, high load factors can therefore be very favourable to district heating (DH) applications. In hot climate areas, the presence of cooling demands (mainly air conditioning for residential–tertiary sector) is a key factor for the profitability of biomass trigeneration systems using adsorption chillers. The load factor of DH networks can be increased by combining heating and cooling distribution systems.

The energy efficiency of buildings also has a significant influence on the heat demand and profitability of DH schemes. In particular, differences arise in:

- new buildings or refurbishment of existing ones
- the type of buildings (single family dwellings or multi-dwelling blocks)
- the energy efficiency of buildings (specific heat load varies in the range 0.03–1.2 kW per m^2 of heated building)
- the specific residential building space (commonly variable in the range 30–60 m^2 per capita).

(Persson and Werner 2011)

The presence of anchor loads – i.e. large concentrated heat/cooling loads such as hospitals, sport and leisure centres or large supermarkets – is important to facilitate the location of CHP plants (Entec 2010). This is because thermal storage in DH–CHP systems is still an expensive option (IEA 2008a, Tveit *et al.* 2009) and therefore, thermal energy must be consumed when produced, while excess electricity produced by CHP plants can be fed into the grid. Similarly, the heat:electricity ratio in energy demand and the relative penetration of demand-side management techniques to shift loads and shave peaks can influence the choice of CHP system (IEA 2008b, Tveit *et al.* 2009).

The feasibility of building and operating biomass DH networks will depend on load density, which can be expressed as network length per dwelling, which varies in the range of 4–20 m per dwelling for high/low heat density areas. The corresponding linear heat density is calculated on the basis of climatic area and building energy efficiency, as well as the factors listed below:

Population density Typical values are below 1,000 inhabitants per km^2 for rural areas, in the range of 1,000–5,000 inhabitants per km^2 for average urban areas, of 10,000–80,000 inhabitants per km^2 for city districts.

Urban area The size of urban area and number of inhabitants also plays a relevant role in the logistics of biomass supply and may range from a few km^2 for small borough, to 100–200 km^2 for average urban areas, and up to 2,000–5,000 km^2 for urban regions.

Energy framework

The baseline energy framework also plays an important role in the choice of urban bioenergy system. In particular, the cost of fossil energy and tax levels, the baseline energy efficiency, the subsidies available (i.e. tax exemption levels, direct subsidies to biomass plants, rules for on-site generation, net metering options), and the possibility to integrate biofuels into existing plants (both DH and CHP), are some of the most important factors to investigate. Moreover, in some cases, biomass fuels are already used in urban areas, but with very low efficiency levels and poor logistic networks, so improvements can be introduced.

Other factors, such as biomass/biofuel transport congestion, space availability for storage, landscape constraints, noise and air emission levels (Jonsson and Hillring 2006) can be major bottlenecks, in particular when dealing with low energy density fuels and poor conversion processes. The potential to integrate biofuel transport systems into existing urban infrastructures is also relevant, as shown in some ongoing and existing projects (e.g. integration of bio-methane from urban biowastes into gas network (National Grid 2009); large DH networks with biomass co-firing (Copenhagen Energy Summit 2009); integrated network distribution of bio-oils (Kumar *et al.* 2004)).

A summary of the main trade-offs in the integration of bioenergy routes into urban energy systems is provided in Box 6.1. Several of these research problems

Box 6.1 Trade-offs and optimization problems of bioenergy routes for urban energy systems

Optimization modelling can be helpful for many different aspects of bioenergy systems design, particularly within an urban context. Here are some example applications from the literature:

Small distributed versus large centralized heating plants Small distributed heating plants present higher biomass transport costs, air emission levels and local storage issues, but lower heat distribution losses and the optimization is based on the energy demand heat density, transport and storage constraints of the specific urban area, energy losses of heat distribution and relative costs and conversion efficiencies of small/large-scale biomass plants (Kaikko and Backman 2007, FIPER 2011).

Decoupling of processing and biomass conversion systems The decoupling of processing and energy conversion steps offers the possibility to produce energy near to the loads and minimize the biofuel upgrading costs with large centralized facilities (eventually integrated into biorefineries) that serve distributed gensets. The biofuel transport and storage costs can also be minimized when high quality and high energy density biofuels are produced; however, energy conversion near to the biomass processing could make available heat and power for the conversion process itself, so reducing the biofuel production cost (Bridgwater *et al.* 2002, Uslu *et al.* 2008, Wright *et al.* 2008).

Transport modes: pipeline, network or road transport? Various typologies of biomass and biofuels transport modes can be optimized on the basis of fixed and variable transport costs, distances, local constraints and energy balances of the different options (Kumar *et al.* 2006, Pootakham and Kumar 2010).

Interconnected DH network with distributed generators versus decentralized DH systems Interconnected DH systems with distributed gensets are typical of existing DH networks with increased loads connection rates. In view of potential expansions of the DH network, and on the basis of the heat demand density, heat distribution losses and investment costs, it is possible to select the pipeline diameter, network pressure and temperature levels that minimize energy costs. In general, smaller pipeline diameters imply lower investment and energy losses, but with limits for further network expansions (Overgaard *et al.* 2005).

CHP versus stand-alone heating/cooling The selection of CHP configurations is strongly based on the value of electricity produced (subsidies and/or cost of electricity) and presence of heat/electricity demand in case of on-site power generation for autoconsumption. Again, these trade-offs can be handled in an optimization model.

Sizing of CHP and back-up boilers The optimal size of CHP plants is based on factors such as heat load patterns, levelized electricity generation costs and costs of heat generation from back-up boilers. The thermal power of CHP plants is commonly fixed in the range of 50–75 per cent of peak load, in order to ensure at least 5,000 operating h/year; the back-up boilers produce about 20–25 per cent of the annual energy delivered to the load (Overgaard *et al.* 2005, Benonysson *et al.* 1995, Vallios *et al.* 2009).

Optimal operation of CHP–DH plants CHP plants can operate baseload, peak shaving or on–off cycles, on the basis of technical plants characteristics and operational flexibility, levelized costs of electricity and hourly selling prices (Biezma and Cristobal 2006).

Dedicated biofuel plants versus dual fuel/co-firing options The selection of optimal configurations is based on the quantities of biomass available, increased efficiencies of co-firing, economies of scale of larger sizes achievable with mix of various fuels and the possibility to repower existing plants.

Definition of biomass supply strategies On the basis of relative seasonality of biomass availability and heat loads, and storage costs, it is possible to optimize multi-biomass supply strategies (Verbruggen 1982, van Dyken *et al.* 2010).

Location of processing and energy conversion plants The optimal location of storage, processing and energy conversion plants is a trade-off between transport costs and constraints, available infrastructures, scale economies and relative efficiencies of coupled processes (i.e. pelletizing integrated into CHP plant to use excess heat for biomass drying) (Ebadian *et al.* 2011, Reche-López *et al.* 2009, Dunnett *et al.* 2008, Keirstead *et al.* 2012b).

are modelled and solved by means of mathematical programming techniques and MILP tools (Weber and Shah 2011, Keirstead *et al.* 2012c), and can be classified into strategic, planning or operational problems, according to the temporal horizon and the specific objective function (i.e. technology selection, location, coupling versus decoupling biomass upgrading and energy conversion, strategies for biomass purchase costs optimization, operational mode of CHP plants).

Promising bioenergy routes for urban energy systems

Table 6.10 summarizes the most promising bioenergy routes for urban energy systems. For each option, the size range, conversion efficiency and biofuel supply options are reported. Moreover, the possible configuration of the logistics of biomass upgrading and energy conversion, the main constraints when integrating into UES, the technology

Table 6.10 Most promising bioenergy routes for stationary applications in urban areas

Route	Size	Efficiency (%)	Biofuel	Logistics	Constraints T	S	E	Reliability	UES type/Key factors
Pellet stoves/small boilers	20–100 kW$_{th}$	75–85$_{th}$	Pellets, TOP, chips	Centralized pellet/TOP plant + road distribution to small plants	+++	+++	+++	+++	Peri-urban rural/low heat density
Boilers + DH	0.1–5 MW$_{th}$	80–90$_{th}$	Solid biomass	Centralized pellet/TOP or biomass storage + road distribution to plants + heat distribution to loads	++	++	++	+++	High heat density/no gas available/ existing DH networks
ORC-CHP + DH	0.25–1 MW$_e$	15–25$_e$	Solid biomass	Centralized storage/upgrading + road biofuel transport to plant + heat distribution to loads	++	++	++	+++	
ICE-CHP + DH	0.1–15 MW$_e$	30–45$_e$	Bio-liquids; bio-gas; gas co-firing	Centralized upgrading + gas/ bio-liquids networks to plants + distribution to loads	+	+	+	+++	High heat density/biofuel transport infrastruct. / bioelectricity incentives

(Continued)

Table 6.10 Cont'd

Route	Size	Efficiency (%)	Biofuel	Logistics	Constraints T	S	E	Reliability	UES type/Key factors
GT-CHP + DH	0.5–15 MW$_e$	30–38$_e$	Biomethane/biogas/gas co-firing	Centralized upgrading + gas/biogas network to distributed GT	+	+	+	+++	Large urban areas/existing DH network/co-firing in existing plants
ST-CHP + large DH	5–100 MW$_e$	28–38$_e$	Solid biomass/gas co-firing including CCGT cycles	Centralized energy conversion + DH to loads	++	++	++	+++	
MT-CHP	30–500 kW$_e$	30–35$_e$	Bioethanol/biomethane gas co-firing/bio-hydrogen	Centralized upgrading + gas/bioethanol/hydrogen network to distributed MT/SOFC + on-site heat and power	–	–	–	+	High energy density and cost/bioelectricity incentives/on-site generation schemes/biofuel infrastructure
SOFC-CHP	10–100 kW$_e$	35–45$_e$	Biohydrogen/biomethane		–	–	–	–	

Notes: TOP = Torrefied pellet, $_{tb}$ = thermal, $_e$ = electrical, (+) high level, (–) low level, T: transport, S: storage, E: environmental constraints (air emission, noise).

maturity level and the typical urban areas suitable for each route, are also described. We can therefore draw the following conclusions:

- High quality biofuels should be used in urban areas to minimize transport, storage and environmental issues.
- Decoupling of biomass upgrading and biofuel energy conversion near to the loads is required in most cases.
- Small boilers are suitable for rural areas and low heat density zones.
- DH is feasible with high energy density loads or when cooling distribution can be introduced to increase the network load factor.
- Integration with existing infrastructures is a key factor (i.e. possibility to use existing gas networks for biomethane).
- CHP in urban areas is more promising with high quality fuels such as liquid or gaseous biofuels, potentially integrated with natural gas.
- Solid biomass CHP implies large storage, transport and air emission issues, integration with DH schemes and localizaton in peri-urban areas.
- Large CHP plants should be located where possible on brownfield sites and use co-firing options to maximize energy conversion efficiencies while limiting the amounts of biomass required.
- The most reliable technological option currently available for small scale biomass CHP in urban and peri-urban areas are ORC plants fed by solid biofuels and ICEs fed by liquid or gaseous biofuels.
- Promising technologies for small-scale on-site biofuel CHP are microturbines and fuel cells.
- The economic competitiveness of bioenergy routes in CHP schemes is strongly influenced by the subsidies available for bioelectricity, while biomass heating and cooling can be, to some extent, competitive with fossil fuels, even without incentives.

6.4 Conclusion

Although cities get most of their energy resources from outside the urban boundary, urban renewable sources are often an important element of plans to reduce greenhouse gas emissions from urban energy use. This chapter has provided an overview of the main renewable energy resources available within urban areas and highlighted in particular, the low energy density of such resources. While certain forms of solar energy and fossil fuels could theoretically provide sufficient energy on a per metre-squared basis, in practice most cities do not wish to produce large quantities of fossil fuels within their local environment and the cost of installing extensive rooftop solar energy systems is unfavourable compared with the use of imported fuels.

However, certain renewable energy resources can be attractive, in particular bioenergy resources. Bioenergy resources are widely available in urban areas and can be collected from urban waste streams which otherwise would incur disposal costs. Such resources are highly diverse and can be used in a range of applications including the provision of heat, power and transport. The choice of any particular bioenergy route will therefore depend on the specific technologies available, financial conditions, local environmental impacts and many other factors.

7 Urban transport technologies

Aruna Sivakumar, Salvador Acha and James Keirstead

Imperial College London

Urban areas offer significant opportunities for innovations in efficient transport technologies and services (Banister 2008). Emerging trends include the alternative fuel vehicles market (e.g. electric vehicles, hybrids, biofuels), co-operative vehicle infrastructure systems (e.g. automatic crash avoidance, fleet logistics management), real-time information and intelligent transport systems, and smart mobility services such as car and bike-sharing. This chapter will present an overview of the major trends and technologies that have the potential to create a significant impact on the urban transport system. In this chapter, we focus exclusively on technologies and services that are directly related to transport; there are a number of land use policies and developments influencing the design of the urban area that also have a significant impact on the transport system.

7.1 Introduction

Urban transport systems have seen a wide range of technologies and services evolve over the years, from the automobile (based on the internal combustion engine) and the steam-powered railway locomotive to the futuristic Segway personal transporter[1] and high-speed maglev.[2] While the innovations of the nineteenth and early twentieth centuries were driven primarily by the need to meet the mobility demands of people and freight on a large scale, in the last 40–50 years, transport technologies and policies have been driven increasingly by the need to provide efficient, clean, safe and sustainable mobility. Accordingly, there has been a flurry of policy directives and environmental legislation in recent years establishing aggressive personal and industrial carbon targets nitrogen oxides (NO_x) and particulate matter emission targets and noise abatement targets, and so on (Banister *et al.* 2000, Santos *et al.* 2010, Rodrigue *et al.* 2009). For instance, the UK has adopted a path towards an 80 per cent reduction of CO_2 emissions by 2050. These policy directives in conjunction with high petroleum prices, increasing congestion and health concerns, form the main drivers behind the innovations in transport technology and services in urban areas.

Sustainable transport policy comprises a wide range of solutions. Figure 7.1 presents a sample of current practice in demand management policies, which range from charging and taxation to information and communication technologies (ICT) substitutions, such as e-commerce and telecommuting, to smart mobility measures encouraging lifestyle

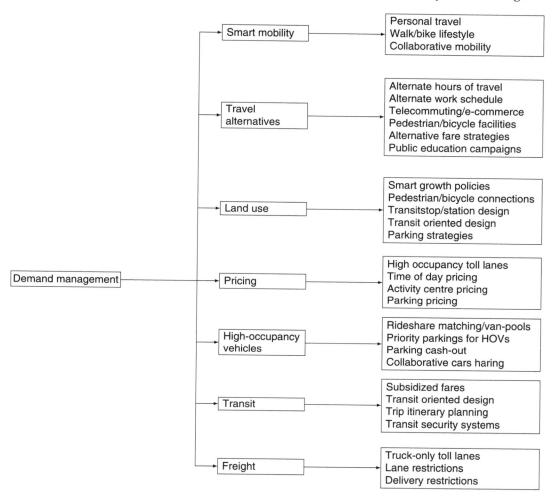

Figure 7.1 Demand management technologies and services for vehicle congestion. Adapted from US FHWA (2005).

changes such as increased walking or biking. Cairns *et al.* (2008) summarize the impacts of smart measures from seven different studies, and these estimates suggest that soft policies may reduce traffic levels by between 4 and 26 per cent.

While these demand management and smart mobility measures are expected to play a big role in the future of transport systems, real innovations in transport technologies and services will clearly be required in order to meet aggressive policy targets. Examples of technological innovations include the electrification of mobility (electric and hybrid vehicles and public transport modes) and development of alternative, cleaner fuels such as biofuels. The role of electric mobility in cleaning up the environment depends largely on the electricity generation process, which is still heavily based on coal and natural gas worldwide, but electric vehicles (EVs) are expected to play a major role in the future.

Efficient deployment of electric mobility relies on commercial fleets, car-sharing services and innovative service models aimed at maximizing the strengths of electric vehicles, while minimizing the weaknesses of range limitations and recharge time.

Looking ahead, we are poised on the brink of a future of sustainable and efficient personal transport technologies (such as Segway), and lifestyle changes based on walking and cycling-friendly neighbourhoods, which also rely on ICT and service innovations that enable easy access to shared bikes and provide real-time information on air quality and congestion. Further, looking beyond passenger transport, there have also been significant advances in freight transport technology, e.g. vehicle-to-vehicle communications that increase the efficiency of freight logistics.

In this chapter, we discuss a variety of these emerging transport technologies and services that influence urban passenger and freight mobility. Section 7.2 focuses on vehicle and fuel technologies; section 7.3 presents intelligent transport systems and the role of transport telematics; while section 7.4 deals with emerging smart mobility policies and services. Note that this chapter is not meant to be a comprehensive review of emerging trends, rather we have picked a few key technologies and services that are expected to play a significant role in shaping the future of urban mobility. Section 7.5 concludes with a discussion of the implications for urban energy systems of the up-and-coming innovations in transport technologies. We also present a case study which analyses the impact of future vehicle fleets on urban energy systems in the UK.

7.2 Vehicle and fuel technology

Governments worldwide rely on low emissions vehicles (based on hybrid technology and lean-burn engines) and alternative fuels (such as compressed natural gas, liquid petroleum gas, biodiesel, electricity and hydrogen) to meet the aggressive climate change policy targets (Hickman and Banister 2007). Vehicle and fuel technology is believed to offer a carbon reduction potential of 11–27 MtC, compared with the carbon reduction potential of, e.g. pricing regimes, which is expected to be around 1.1–2.3 MtC. A report by Element Energy on the future of electric vehicles in the UK and Ireland presents the emissions trajectories in Figure 7.2 under various scenarios of EV uptake. The analysis assumes a relatively low EV uptake of 0.5 per cent of the car stock in 2020 and 5 per cent in 2030 under the business-as-usual (BAU) scenario; a higher level of EV uptake in the Extended scenario of 5.5 per cent of the car stock by 2020 and 18 per cent by 2030; and a Stretch scenario with 13 per cent of the car fleet being plug-in vehicles by 2020 and 75 per cent of the car fleet being plug-in by 2030.

Studies of the transport sector in future urban energy systems require consideration of the different alternatives available to mitigate carbon emissions either be it via using various fuel sources, adopting new power train architectures or by relying ever more on fuelling vehicle power through the electrical grid. This section has the goal to review the state of the art in such areas in order to provide the reader with a grasp of basic concepts that are worth considering regarding the decarbonization of transport systems.

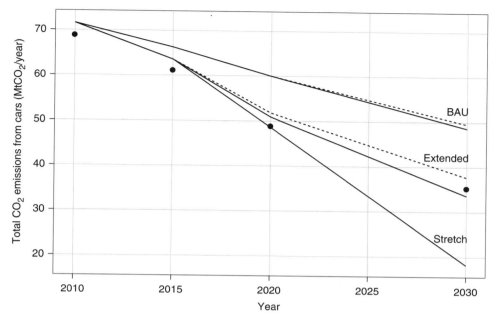

Figure 7.2 Projected emissions from UK cars to 2030. Dashed lines represent scenarios without electric vehicles. Scenarios: BAU = business-as-usual, Extended = medium EV uptake in-line with the UK Committee on Climate Change recommendations, Stretch = upper bound on EV uptake. Solid dots represent emission milestones derived from the UK Committee on Climate Change recommendations.
Data source: Element Energy (2010).

7.2.1 Fuels

Even though today, petrol prices to fuel conventional vehicles are on the rise and environmental legislation promotes vehicle fuel efficiency, technology displacement will only occur gradually. Rapid changes will only become the norm as alternatives become more attractive to deploy than conventional fuels.

Available fuels will be of different kinds as the twenty-first century progresses, however they can be categorized into two broad groups: conventional fuels and alternative fuels (FURORE 2003). Conventional fuels consist of petrol and diesel resources that have powered vehicles for most of the past century and they are not expected to change much; however, most likely, regulations will specify that they reduce their sulphur content. Alternative fuels ranging from natural gas, biogas (i.e. derived via anaerobic digestion), biofuels (i.e. ethanol or vegetable oil based), electricity from the grid and hydrogen are the options currently being considered to displace petrol demand. Major issues pending to make these options feasible solutions deal mainly with production costs and the development of large-scale facilities to produce, store and distribute these fuels; hydrogen is the most technically constrained of the fuel options. Meanwhile, the use of electricity as the prime mover of vehicles can only be achieved when batteries are cost-effective in the market, otherwise the technology will only be used in niche applications and markets, e.g. where there are strong environmental standards as in California, USA. Technological breakthroughs are indeed needed in alternative fuel development in order to achieve economies of scale that no longer leave these options marginalized. Until this occurs, low carbon fuels will be trapped in specific applications and isolated, thus only serving niche markets and surviving via governmental support. Nevertheless, there are always exceptions and, in most parts of the world where growth of biofuels has taken place (e.g. sugarcane-based ethanol in Brazil), it has been mainly due to ambitious government policies that are keen on both minimizing energy security concerns and revitalizing rural economies (IEA 2011b).

The International Energy Agency has identified a series of actions that should be undertaken by governments wishing to meet greenhouse gas emission reduction targets by deploying alternative fuels. These include (IEA 2011c):

- Creating a stable and long-term policy framework that increases investor confidence that aims at producing biofuels in a sustainable manner
- Short-term funding (e.g. 10 years) to promising biofuel technologies so that commercial competitiveness can be achieved
- Introducing requirements that reduce land use changes and hence promote sustainability
- Linking financial support for biofuels to its sustainable performance in reducing life-cycle greenhouse gas emissions
- Incentivizing companies and households to use residues and wastes as biofuel feedstocks
- Researching land availability and feedstocks with the objective of identifying the most promising locations for future biofuel production
- Reducing trade barriers that diminish the practice of sustainable biomass and biofuel trade
- Supporting international collaboration on knowledge creation and technology transfer, thus promoting sustainable biofuel production globally.

7.2.2 Power train

Another substantial stream of research that can help lower carbon emissions and improve fuel consumption efficiency of light-duty vehicles is by developing more robust and efficient power train technologies. At present, much attention is given to plug-in hybrid and electric vehicles, definitely promising technologies by their own right. However, reviews indicate that internal combustion engine (ICE) technology – both spark ignition and diesel engines – will still account for a substantial amount of vehicle usage by 2050 (IEA 2011d). Each propulsion system has its own set of challenges going forward that need to be addressed. For ICE units, the main concern will be reducing NO_x levels and becoming compatible with natural gas fuels and hydrogen. EVs will instead require batteries which are less heavy, more energy dense, with a longer life expectancy and at a lower cost (e.g. be it either lithium-ion, fuel cell based, etc.) (FURORE 2003).

Similar to the debate of using high or low carbon intensity fuels, the evolution of vehicle propulsion systems will be constantly evolving in order to meet policy and

Box 7.1   Will consumers buy electric vehicles? The role of stated preference modelling

Technological innovations can only stimulate the market and offer alternatives to the consumer; the ultimate level of penetration of new technologies such as electric vehicles depends on the dynamics of the consumer market. There have been several studies in the last 30–40 years that have attempted to understand and predict this market and the resulting demand for EVs. Most of these use a technique known as stated preference modelling (Train 1980, Beggs *et al.* 1981, Calfee 1985, Brownstone *et al.* 1996, Dagsvik *et al.* 2002, Caulfield *et al.* 2010). Stated preference models (also known as stated choice experiments or multi-attribute contingency models) are commonly used to estimate the demand for new products and services, and are designed to present survey respondents with a variety of alternatives based on various combinations of attribute levels (e.g. EVs of varying configurations in terms of vehicle category, electric range, recharging time, capital cost, etc.) much like the alternatives presented in automobile comparison sites. The choices made by the respondents in these hypothetical experiments can then be statistically analysed to extract the implicit willingness to pay for increased range or lower recharge time, and further to predict the demand for EVs based on future scenarios of EV configurations.

While many of the early studies highlight range anxiety issues among consumers (i.e. the fear that a vehicle has insufficient range to reach its destination) and a high sensitivity to EV costs, more recent studies show a reduction in range anxiety and a higher willingness to pay. Some of these changes are attributable to current economic conditions and increased fuel prices, but studies show that there is a more fundamental reason for this change in behaviour. Essentially 'multi-attribute choice models applied to the early phase of a product's life cycle are particularly vulnerable to ill-structured perceptions of the new choice object and its proper place in the larger set of products with which it competes' (Wilton and Pessemier 1981: 163), and as consumers become more familiar with a product they are able to process the choice experiments more realistically. A number of recent EV adoption studies therefore attempt to quantify the impact of real experience in using an EV on the stated demand by undertaking before and after surveys (Sivakumar *et al.* 2012).

environmental regulations, while customer preferences can also play a significant role in influencing the marketplace (see Box 7.1). For example, in the short-term it is difficult to massively deploy plug-in hybrid electric vehicles (PHEVs) because they are much more expensive than conventional vehicles, while customers also need to reduce their 'range anxiety' and charging infrastructure is scarce in densely populated areas. Nevertheless, cities such as Paris have been the first to develop innovative EV-sharing schemes that have the intention to create a virtuous cycle that can begin addressing some of the limitations delaying PHEV penetration (Niches 2010). Thus, progress is being made and the IEA suggests that, by the year 2050, sales for 100 million of these vehicles is achievable and car manufacturers such as Nissan, Tesla, Chevrolet, BMW and others are launching models for residential and commercial customers (IEA 2010b).

The introduction of PHEVs, which obtain their fuel from the grid by charging a battery, suggest that the electrification of the transport sector is imminent and that PHEVs provide a good opportunity to reduce CO_2 gases from transport activities. However, this assumption can be deceiving. This is because the emissions that might be saved from reducing the consumption of petrol could be off-set by the additional CO_2 generated by the power sector in providing for the load the vehicles represent. Therefore, PHEVs can only become a viable effective carbon mitigating option if the electricity they use to charge their batteries is generated through low carbon technologies (RAE 2010). In addition to this environmental issue, these unique vehicles also bring techno-economic challenges for utilities, since its intrinsic mobile characteristic means loads can appear and disappear in different parts of the electricity network. Addressing this issue will require thorough research into driving patterns and vehicle use, thus providing utilities with the capacity to predict when and where these vehicles may require energy for their travelling needs and whether they are capable of supplying this new stochastic demand whenever required. Electric vehicles will have great load flexibility for two key reasons. First, they are idle 95 per cent of their lifetime making it easy for them to charge either at home, at work or at parking facilities (Brooks 2002). Second, most marketable batteries exceed the 40 miles per day average urban travel observed in surveys, implying that the time of day in which they charge can be varied with minimal disruption to drivers (Slater and Dolman 2009). Thus, if set up correctly, the above conditions allow PHEVs to adopt flexible tariff schemes letting them charge when electricity is more accessible and cheaper. Consequently, as renewable energy sources become prominent (e.g. with increased wind and solar power penetration) and intelligent communication infrastructure more abundant (e.g. smart grid implementation), these mobile loads should seek to take advantage by charging whenever electricity is at its lowest cost and the generation fuel mix of the region or area is less carbon intensive (Acha *et al.* 2011a).

7.2.3 *The role of the electricity grid in transport systems*

Since no electric car is carbon free while the electricity used to charge its battery is generated in power plants that produce CO_2 emissions, metrics are fundamental to assess the trade-offs this new technology brings. Careful studies are needed to analyse and compare the efficiency of different vehicle models on a so-called well-to-wheel (W2W) basis. W2W analysis is popular within the literature and follows the energy content of the fuel from its original source up to its point of consumption, quantifying the distance that a car can travel per unit of energy used (measured in km/kWh). For example, a Tesla Roadster if charged with electricity from coal has a slightly better energy efficiency than a Toyota Prius charged on petrol. It is also possible to compute the W2W emissions of automobile technologies. In this manner, the environmental impact of replacing petrol with diverse power generation can be estimated and be compared on a like-for-like basis; this is done by dividing the carbon content of the fuel used (measured in $kgCO_2/kWh$) by its W2W energy efficiency. Once again, if a Tesla Roadster using coal power is compared with a Toyota Prius using petrol, the Roadster would emit almost three times more carbon per kilometre, thus confirming that the environmental impact of electric vehicles must take into account the source of the electricity. Therefore, it would be ideal for these new types of automobiles to fill up their batteries when the carbon emissions from power generation are at its lowest (Eberhard and Tarpenning 2006).

It is for these reasons that the power system sector needs to comprehensively understand how PHEVs can impact low voltage networks and re-think some of the assumptions that have become common in the literature, since the infrastructure will most likely become proactive in order to adapt to ever-changing conditions and hence its studies and forecasts should accommodate these features (Acha *et al.* 2010). The ability to determine ideal charging times of PHEVs in local electricity distribution networks, considering temporal and spatial issues, is paramount in developing an effective and robust smart-grid in urban areas. Hence, much work in academia and industry is to be done in merging transport-related modelling concepts from EV travel with traditional optimal power flow issues in order to identify ideal EV charging strategies that enhance power network performance (Acha *et al.* 2011b).

Box 7.2 demonstrates these transitions in fuels, power trains and the electricity grid with a case study of future vehicle fleets in London.

Box 7.2 Case study: Analysis of the impacts of future vehicle fleets on UES

This chapter presents a wide range of emerging, new and future technologies and services that could have a significant impact on urban transport systems. The natural next question then is: what are the likely impacts of these technologies and services? In this case study, we present an analysis of one such trend – changing vehicle fleet characteristics – using an agent and activity-based demand modelling tool developed at Imperial College London, illustrating one of the many methods that could be used for such an analysis.

Aim   To estimate the energy consumption and greenhouse gas (GHG) emissions from passenger transportation within London to 2050, based on assumptions of future vehicle and transportation fleet characteristics.

Methodology   An agent-based microsimulation model of urban activities and a synthetic population for London in 2050 (see Chapter 11 for more details of the modelling tool) was used to build a bottom-up estimate of the activities undertaken and therefore the implications for energy consumption and GHG emissions.

The composition of the vehicle fleet in 2050 was assumed to be similar to existing vehicles, except for their energy consumption and emissions standards, which are projected into the future based on expert opinion. Three climate policy scenarios were used: a business-as-usual (BAU) scenario, as well as 450 ppm and 550 ppm CO_2 stabilization scenarios. A set of 37 'vehicle types' or modes were identified, including non-motorized modes such as cycling and walking. For simplicity, only one vehicle type was used for the bus, medium- and heavy-duty vehicle categories, and less common car and light commercial vehicle (LCV) technologies were also removed, leaving a final set of 22 modes. Average vehicle occupancy rates were assumed as follows (obtained from Transport for London where possible): car = 1.36, LCV = 1.00, bus = 41.30, taxi = 2.00, other (including medium- and heavy-duty vehicles) = 1.00.

The detailed specifications of the final vehicle types considered are presented in Table 7.1. In addition to the headline statistics presented here, we also had details on the electrical energy consumption of each vehicle type and a breakdown of the well-to-tank (WTT), tank-to-wheel (TTW) and tailpipe emissions. Further assumptions were made regarding the forecasted carbon intensity of the electricity grid from 2010 through to 2050, based on data from the Committee for Climate Change (CCC 2008), as presented in Figure 7.3.

The data in Table 7.1 was used to determine the energy consumption and emissions profiles of each vehicle technology from 2008 to 2050. For the car and light commercial vehicle categories, the data describe the estimated fleet mixes through to 2050 and therefore the only emissions improvements shown in our analysis come from the reduced carbon intensity of grid electricity. However, for the other vehicle categories, some assumptions were made regarding likely performance improvements over time. For medium- and heavy-duty vehicles, a 10 per cent improvement in fuel efficiency and carbon performance is assumed by 2050. For motorcycles, it is assumed that efficiency can be improved by 30 per cent (though no electrification has been assumed here). For rail and tube, a 10 per cent improvement in performance was assumed by 2050. Buses and taxis are assumed to improve in-line with the transitions observed within their vehicle categories (i.e. the current diesel taxi transitions to an electric vehicle).

Analysis and results   The demand modelling tool was first implemented to predict the activities undertaken by the agents in the synthetic population (designed to be representative of the real population) in great detail over time and space, including predictions of the mode used for travel. The predicted demand for each mode is then combined with the data outlined above to estimate the energy consumption and GHG emissions. Figures 7.4 and 7.5 present the results.

The advantage of the approach shown here is that it allows transport activity to be disaggregated to the individual and household level, providing much greater detail on the activities which ultimately create transport demand.

Several assumptions and simplifications were made during the course of this analysis that could be revisited or improved.

1   A simple model of total population growth was used to estimate mobility patterns through to 2050, and this could be enhanced by bootstrapping our original sample to ensure that future populations include a representative mix of individuals and hence trip activity. However, substantial mode shifts might not be picked up by this modification, as the underlying choice model is still based on 2001 data. The Mayor's draft transport strategy has a more detailed breakdown of hypothesized mode shifts in future, which forecasts a significant shift to public transport by 2031.

2   The present work does not include a vehicle ownership model. A significant research investment would be required to incorporate this additional detail, as the likely adoption patterns for new vehicle technologies are unknown and potential consumers would have to be surveyed using stated preference questionnaires or similar techniques (see Box 7.1).

3   The model treats urban freight and commercial transportation simplistically. In other words, a detailed model of individual consignments, vehicle choices and routes would be expected to give improved results.

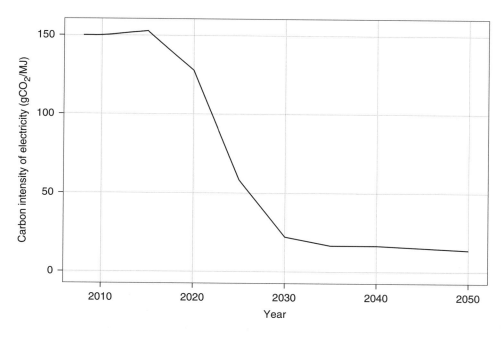

Figure 7.3 Forecasted carbon intensity of UK grid electricity.
Data source: CCC (2008).

Table 7.1 Key parameters of vehicle types for base year 2008

Technology	Fuel type	Energy consumption (MJ/pass-km)	Tailpipe emissions (gCO$_2$/pass-km)	Description
Car	Petrol	1.67	123.4	Baseline NA PFI no VVT
	As above	1.31	97.1	Downsize NA mild hybrid
	As above	1.00	74.0	PFI VVT series-parallel full hybrid
	As above	0.70	45.9	PFI VVT series-parallel PHEV 40 miles
	Diesel	1.5	108.26	Baseline Euro 4
	As above	1.15	84.3	Euro 6 DS mild hybrid
	As above	0.87	64.0	Future DS parallel hybrid + energy recovery full hybrid
	Electric	0.52	0.0	Electric vehicle 100 mile range
Light commercial	Petrol	2.95	176.3	Baseline NA PFI no VVT
	Diesel	3.01	220.8	Baseline Euro 4
	As above	2.41	176.8	Euro 6 DS mild hybrid
	As above	1.93	141.2	Future DS parallel hybrid + energy recovery full hybrid
	Electric	1.06	0.0	Electric vehicle 200 mile range
Medium duty	Diesel	6.63	486.4	Baseline Euro 4
Heavy duty	Diesel	12.37	907.0	Baseline Euro 4
Bus	Diesel	0.31	23.0	Baseline Euro 4
Rail	–	0.58	0.0	Estimated from TFL data
Tube	–	0.69	0.0	From TFL
Taxi	Diesel	1.00	73.6	Assume same as Baseline Euro 4 diesel car
Motorcycle	–	1.56	104.4	4-cylinder engine
Bicycle	–	0.00	0.0	
Walk	–	0.00	0.0	

Notes: NA = naturally aspirated, PFI = port fuel injection, VVT = variable valve timing, DS = downsized, PHEV = plug-in hybrid vehicle.
Data source: Chester and Horvath (2009) and author estimates.

7.3 Intelligent transport systems

Intelligent transport systems (ITS) are a broad range of diverse technologies applied to transportation to make systems safer, more efficient, more reliable and more environmentally friendly, without necessarily having to physically alter existing infrastructure. The term 'transport telematics' is also often used to refer to the integration of ICTs into transport systems. ITS make use of a variety of advanced technologies, including computers, communications, sensors, collision warning systems and vehicle-sensing technologies in order to secure, manage and control transportation systems (Goldman and Gorham 2006). ITS technologies can be broadly divided into intelligent infrastructure and intelligent vehicles. Each of these is discussed further in the following sections. Moreover, ITS plays a big role not just in passenger transport networks but also in freight and urban logistics. We also discuss the role of ICT specifically in enabling new transport products and services.

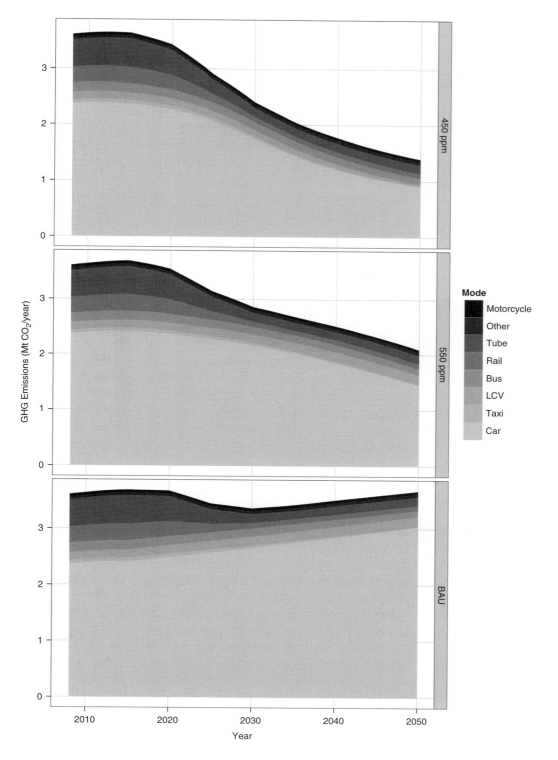

Figure 7.4 GHG emissions from London passenger transport to 2050 by transport mode. LCV = light commercial vehicle. Panels show three climate policy scenarios: business-as-usual (BAU), 450 ppm stabilization and 550 ppm stabilization.

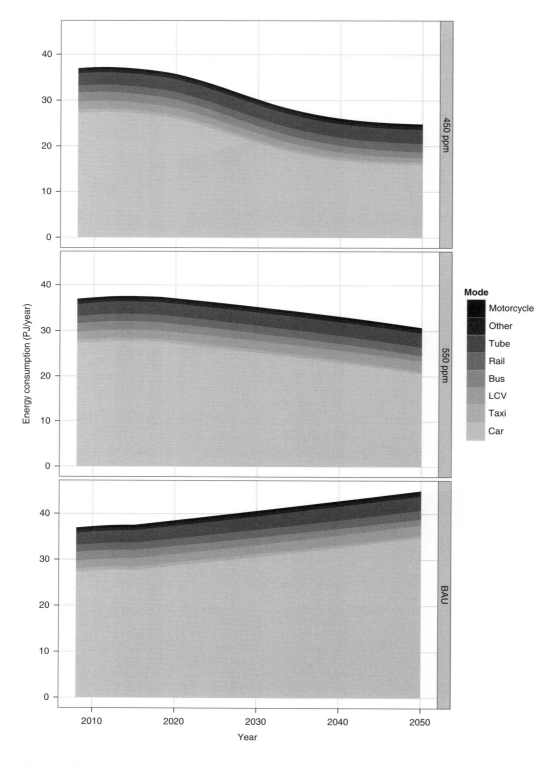

Figure 7.5 Energy consumption of London passenger transport to 2050 by transport mode.
LCV = light commercial vehicle. Panels show three climate policy scenarios: business-as-usual
(BAU), 450 ppm stabilization and 550 ppm stabilization.

7.3.1 Intelligent infrastructure

Intelligent transport infrastructures are wired to transmit real-time information (e.g. variable message signs on roads and public transport networks), enforce traffic rules and charges more efficiently (e.g. automatic number plate recognition, speed cameras, monitor applications such as security CCTV systems), improve safety and enable intelligent traffic regulation and efficient signal control systems, including bus priority networks. More advanced ITS applications integrate live data and feedback from a variety of sources, for instance, in advanced route and parking guidance systems, weather information and bridge de-icing systems (Skinner *et al.* 2003). Moreover a range of short-term predictive modelling techniques have been and are being developed to enable the advanced modelling underlying these ITS applications, drawing on historical baseline data where available (Vlahogianni *et al.* 2004).

In a future of high electric mobility, intelligent parking systems and advance parking reservations can contribute to the development of business models that will enhance parking management for the operators while offering drivers the opportunity to eliminate their uncertainty of finding a parking place with an available charging post and assisting them with valuable guidance and pricing information. Online reservation systems and call-ahead technologies have the potential to improve the customer's experience and satisfaction (WSA 2008). Intelligent parking systems can potentially also interact with in-vehicle navigation systems to provide information about capacity and availability and automatically accept or reject parking requests.

Another less dramatic but very effective technological improvement in transport networks is the introduction of integrated public transport fares that work seamlessly across buses, light rails/trams, commuter rails and underground systems within an urban area. This has been made possible by the modernization of fare payment technologies and the use of smart chips. Such fare integration technologies cut barriers to transit access, encourage participation in monthly pass programmes and potentially serve as new revenue sources for transit agencies (WSA 2008). There are several such systems worldwide, for instance the Oyster card in London, UK, the SmartRider in Perth, Australia and the EZ-Link card in Singapore.

7.3.2 Intelligent vehicles

Advanced sensor technologies and global positioning systems (GPS) have played a big role in the development of advanced in-vehicle technologies such as smart navigation, collision alert and parking assist systems, and smart active suspension systems that are environmentally efficient and respond to the occupancy and load in the car. As with intelligent infrastructures, intelligent vehicle technologies also rely on ICT advances such as cloud technology and the invisible and all-pervasive Internet and are poised to exploit these resources much further.

In-vehicle technologies enable efficient implementation of charging schemes such as tolls or congestion charges through the use of electronic toll collection tags, such as the E-ZPass in the USA, the Via Verde in Portugal and the AutoPass in Norway. At the other extreme are newly emerging driverless vehicles or autonomous cars, such as the

Figure 7.6 Maglev-based public transit system with personal pods.
© skyTran LLC (http://www.skytran.us). Reprinted with permission.

Google car, which combines information gathered from Google's Street View maps with artificial intelligence software and inputs from video cameras and sensors mounted on the car. Such an innovation could have far-reaching impacts on the efficient utilization of road networks and open the doors to a competitive market in 'personalized' public transit offered by systems such as skyTran^{TM} – a revolutionary personal rapid transit system that travels at speeds of up to 150 miles per hour and operates using maglev (see Figure 7.6).

7.3.3 *Urban logistics*

ICT, cloud computing technology and related innovations have already played a significant role in improving urban logistics by introducing major efficiency gains and new service models. For instance, ICT is the key element in container management systems at ports, keeping track of the numbers of empty and full containers in the warehouses, and matching them in real-time against the containers carried into and out of the port by container ships to determine the most efficient sequence of loading, unloading and distributional activities. Cloud technologies also play a key role in shipper–carrier interactions, relaying shipment status information, vehicle tracking and fleet management. These innovations in urban logistics have been facilitated by a wide range of optimization tools for routes, network and assets. These mobile asset management technologies improve asset tracking and inventory control, reduce loss or damage of goods in transit, reduce labour cost and human error, and generally streamline business processes, thus leading to overall gains in efficiency.

The rapid growth in urban freight traffic as a result of e-commerce has in turn, facilitated the rapid growth of the small package delivery business. Increasing numbers of trucks on the road are below capacity or running empty, because of rapidly changing logistic and supply-chain pressures, such as just-in-time deliveries, as well as asymmetrical patterns of trade. This is inevitably where future innovation in technologies and services for urban logistics lies through the development of cooperative distribution systems and the use of clean vehicle, connected fleets. For example, Nuremberg and Freiburg in Germany and Bristol in the UK have encouraged (and in some cases subsidized) voluntary, private cooperative systems among shippers or major retailers to coordinate deliveries. A number of Dutch cities have set up licensing schemes whereby participating firms receive certain privileges (e.g. longer delivery hours) in exchange for performance commitments (e.g. using only electric vehicles, and exceeding minimum loading standards). Freiburg's programme reduced truck journeys by 33 per cent and truck operating times by 48 per cent (Goldman and Gorham 2006).

A model sustainable city in Stockholm (the Hammarby Sjöstad development) attempts to take city logistics a step further. As in other programmes, deliveries to the community will use electric vehicles coordinated through a privately operated logistics centre. Furthermore, this centre will operate a web-based service featuring 15 local businesses and 300 local farmers, from which it will deliver everything from food to dry cleaning services. Suppliers or shops pay the centre by the shipment; participating households pay a monthly subscription fee. Another emerging trend in urban logistics and commercial fleets is cooperative vehicle networks,[3] touted as a future killer application for ad-hoc networking, adding extra value to the car industry and network operator services (Goldman and Gorham 2006).

7.4 Smart mobility initiatives

Even as markets innovate and governments encourage new technologies and services, it is acknowledged that a change in attitudes and mobility lifestyles will be needed in order to meet the various climate change and carbon reduction targets, and to ensure a sustainable transport system for the future. This has been the motivation for a number of smart mobility initiatives, many of which are feasible only because of the advances in ICT.

On the one hand, there have been several creative new technologies and business models to provide competitive alternatives to the private automobile. These new mobility strategies, such as car and bike-sharing systems, enable the most efficient use of mobility assets and are facilitated hugely by ICT networks designed for customers to search for an idle vehicle and make quick bookings from anywhere, enabling keyless access through the use of smart cards, allowing operators to track member usage and bill them on a pay-as-you-drive basis, and geo-fencing the vehicles to monitor and regulate usage (e.g. Australia's Charter Drive). Car sharing, and more generally the cooperative mobility business, is a fertile area for innovation. For instance, they present a good context for the deployment of electric vehicle fleets.

On the other hand, ICTs play a big role in making real-time information easily accessible to travellers, including driving times between specific locations, live bus or metro arrivals, real-time congestion levels and availability of parking at specific lots. Innovative services are being built around all this data, so that it is easily accessible on the internet and on mobile devices. There are even mobile applications that can pinpoint the locations of taxicabs on the road networks, and mobile applications to chart walking routes with the least congestion or the lowest pollutant levels.

Transit agencies are increasingly rebranding themselves as integrated mobility service providers, working with customers to help them plan their travel needs, using a variety of journey planning tools to design door-to-door travel plans that are efficient and sustainable. The RATP public transport operator in Paris offers such a seamless mobility service, integrating traditional transit offerings with bike rental, paratransit, car-sharing and car-pooling services. In its stations, it provides amenities to minimize the inconvenience of waiting and will soon provide mobile telephone and Internet service throughout its network.

A more subtle but nevertheless significant trend is the substitution of virtual activities for travel, such as teleworking, e-shopping, online banking and social networking. While some of these may substitute travel, the substitution patterns are generally far more complex. For instance, an online banking episode may clearly substitute a trip to the bank. But an online shopping episode might merely serve as a means of comparison shopping, followed by a trip to the store for the actual purchase; social networking activity online may serve as a scheduling tool for a meeting later in the evening at the theatre; time freed up due to teleworking may be spent in additional leisure travel and so on. While this trend of online activities is not directly related to the transport system, it is important to be aware of the potential knock-on effects. Similarly, the role of mobile phones and smart phones in real-time scheduling, through the use of maps, text messages and location-based applications, has very far reaching implications for how the transport system will be used. Location-based applications, in particular, have the ability to produce completely unplanned, spur-of-the-moment activity driven by offers and discounts or online reviews.

7.5 Implications for urban energy systems

This chapter describes a wide range of emerging and future technologies, grouped into vehicle and fuel technologies, Intelligent transport systems and smart mobility initiatives that could potentially have direct and indirect impacts on the urban transport system. We have also presented a case study demonstrating an analysis of future vehicle fleets and their impacts on resource demands and the environment. While we have attempted to cover a wide range of topics, it must be noted that this chapter is by no means comprehensive in its discussion of emerging and future transport technologies or services. Rather, it is meant to stimulate thought and to encourage the analysis of potential future scenarios as a means of understanding and predicting the future of urban energy systems.

The primary objectives of most new urban transport technologies and services are to increase efficiency, safety, convenience and, increasingly, sustainability. Telematics technologies have enabled shared mobility systems (such as car clubs and bike shares), which maximize the efficient use of resources. Camera and sensor-based speed and law enforcement services are far more efficient and less expensive than traditional approaches that require a large commitment of police and judicial resources and often exacerbate traffic conditions on busy streets.

A widespread network of driverless networked electric vehicles that can move people and/or goods safely from A to B without intervention from the user would dramatically change the way transport systems work. This would enable the traveller to use their travel time more productively while being conveyed to their chosen destination in privacy and comfort. Implemented correctly, it could bring a step-change in safety and open up the market for personalized public transit systems as a seriously attractive alternative to private cars. Infrastructure maintenance in such a system could be implemented far more easily as the vehicles can be automatically directed away from roadwork sites. In addition to driverless pods, Singh (2009) outlines futuristic possibilities for urban public transport including zeppelins, backpack helicopters, slidewalks and electric bicycles.

On the other hand, there is also danger of inefficiency until suitable services and innovations are put into place. For instance, while an increase in home shopping has the potential to reduce passenger transport trips, if it is not organized efficiently, it could increase the amount of kilometres that freight vans and lorries make to deliver to peoples' homes. Moreover, the true socio-economic impacts of new transport technologies and services are very difficult to assess.

According to the rhythmic historical model developed by Ausubel *et al.* (1998), a new fast transport mode should enter about the year 2000. Their analysis suggests that, in the past 200 years, the transport system has embraced a new means of transport roughly every 50 years (i.e. barges, trains, autos, planes), and the time is ripe for a new mode of transport. All bets seem to be on magnetic levitation systems that can operate at speeds of up to 600 km per hour. However, only one high-speed maglev system is currently in operation – the 30 km Shanghai maglev train. As Ausubel *et al.* claim, given the current pace of innovation, it is very likely that the next revolutionary change will in fact be a package of revolutionary changes.

Notes

1 The Segway personal transporter, a two-wheeled self-balancing battery-powered electric vehicle produced by Segway Inc., is marketed as a zero emissions mobility solution: http://www.segway.com/
2 Maglev or magnetic levitation technology is based on the principles of electromagnetic propulsion and is considered a promising means of designing very high speed public transport modes (e.g. trains, personal transport pods, etc.).
3 See, for instance, the CVIS project: http://www.cvisproject.org/download/Events/validation_ws/CVIS_UK\%20TS_Validation-workshop_20100521.pdf.

Part III
Analysing urban energy systems

8   Modelling urban energy systems

Nilay Shah

Imperial College London

8.1 Introduction

The importance of computer modelling in urban energy systems can be seen as two-fold. First, it allows one to analyse and understand the current state of these systems. But also, and perhaps more importantly with regards to sustainability, it allows one to 'predict, prescribe and invent' the urban systems of the future (Batty 1976). Clearly however, 'exact' modelling of urban systems, on the scale of building-to-building or finer detail, is not feasible, as an enormous amount of information would be required to construct a complete urban model. It would also likely not be desirable, as it would limit the applicability of the derived model (e.g. to a specific city), with little scope for generalization. Therefore any urban energy systems model developed will be an idealization or simplification of reality; the extent depends on model complexity and the approaches used. Models can range from the highly aggregated to disaggregated, partial to general, static to dynamic, and so on, depending on their intended use.

In this section, we describe the ideal components of a toolkit for urban energy system modelling and in section 8.2, we will review some of the important work in the literature to date. Section 8.3 describes our 'SynCity' integrated framework.

8.1.1 Approaches to modelling urban energy systems

Before getting into the detail of specific modelling approaches, it is worth pausing to consider the general process of model development. Rosen (1991) offers one popular explanation, which draws a distinction between the natural system being modelled and the formal system we use to represent it (e.g. a set of mathematical equations or computer software). Each of these systems are self-consistent and the modeller's task is to first encode the internal logic of the natural system into the formal system (i.e. model building) and then decode the model results to assess their meaning in reality (i.e. model interpretation).

A similar generic approach to model development was outlined by Batty (1976) in Figure 8.1. Here the first, and perhaps most challenging, stage is model theory development, as it must be decided precisely what has to be modelled and how. Data collection, calibration and testing would serve to 'fine-tune' the model (assuming that

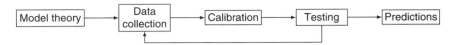

Figure 8.1 A generic approach to model development.
Source: Batty (1976)

the underlying theory is sound), and model predictions would be the final desired result of the process.

It should be noted that there are few current studies which utilize a general approach to urban energy systems modelling; however, several commonalities can be identified in the model theories. Below we describe the main elements of urban energy systems modelling.

Energy demand modelling

Energy usage patterns and the potential for optimization vary depending on urban sectors (or land-use types). Thus, these are typically disaggregated: at the highest level into domestic, commercial, industrial and transport sectors, with a number of further subdivisions possible within each sector to indicate specific activities and user types.

The method commonly used to model energy demand is to obtain energy consumption data, and use regression to devise correlations. However, this approach is contingent on a sufficient body of available data. A good example is seen in Brownsword's (2005) work on modelling urban energy supply and demand. An alternative approach is suggested by Richter and Hamacher (2003), where the demand for energy services is used to infer useful energy demand, which when combined with information about technologies in use (i.e. efficiencies), would provide final energy demand. This is perhaps more useful as the energy service demand can be calculated from information regarding building/consumer types and distributions, economic activity, etc., leading to a more theoretically based and robust model.

Energy supply modelling

The energy supply side also presents opportunities for urban energy system optimization. The Brookhaven Laboratory's Reference Energy System (RES) methodology by Beller (1976) provides a starting point: comprehensive approaches track energy from primary energy extraction through conversion and distribution to end-users. Provisions are made for the introduction of alternatives at each of these stages; Bose and Anandalingam (1996), for example, allow for increases in technology efficiency and changes in consumption patterns, such as shifts in vehicle type preferences. The downside of RES-based approaches (including descendents such as the popular MARKAL energy model) is their aggregate nature. In urban settings, there is a need for geographical (i.e. spatial) disaggregation to effect true optimization because many interactions take place at the local rather than aggregate level.

Environment modelling

The environmental impact of activities is a main consideration of almost any definition of sustainability (Egger 2006, Giddings *et al.* 2002); this is reflected by its inclusion in optimization models. In many studies considered, greenhouse gas emissions are usually used to quantify this impact. There is a need for other measures of impact (e.g. local air quality via photochemical ozone creation potential).

Modelling approaches used include those which produce aggregated outputs which calculate, for example, the tonnage of pollutant generated based on energy consumption, and more complex dispersion models which also consider the spatial distribution. The first is possibly more suited for optimization purposes, while for visualization and detailed impact assessment, the dispersion models are ideal.

Optimization

Energy demand, energy supply and environmental impact form the basic considerations in optimization calculations, along with financial cost. Supply and demand optimization does ultimately relate to cost and environmental considerations: e.g. a calculation to represent the introduction of a CHP system to replace a traditional generation system would introduce a capital cost, a probable decrease in operating costs and an overall decrease in emissions. Other options commonly considered for supply and demand optimization include introduction of energy-efficiency measures, renewable energy, new generation, conversion and distribution technologies, and so on.

Multi-objective optimization methods are typically used to identify ideal energy systems. Linear programming is used to constrain one or more variables while minimizing another. For example, one may constrain greenhouse gas emissions (in order to satisfy legislation) and minimize the total costs. This introduces 'hard' constraints and a single objective function; a different approach is goal programming, which can consider a range of objectives and trade-offs between them.

8.1.2 Modelling dynamic evolution of urban areas

A limitation of many urban energy systems models is that they tend to describe and generate optimal solutions for a system at a particular cross-section in time, i.e. they are steady-state models. The continuous evolution of urban systems with time and the resulting impacts on energy systems is not considered. The usability of this approach is therefore limited within the broader requirements of urban energy systems modelling tools. In other words, the goal should be to develop strategic solutions which are time-phased and applicable over an extended period of time, as opposed to some end state, which would possibly only be optimal within a constrained time period.

These issues have also been a concern in the urban transport planning literature where system dynamics are inherently part of the exercise. This type of urban modelling is a well-established field with a large body of research available, beginning with the seminal works of Alonso (1964), Lowry (1964), Batty (1971, 1976) and others, and continually being developed to present day.

Barredo *et al.* (2003) point out that the current theoretical approach to urban systems regards them as complex systems, displaying characteristics such as emergence, self-similarity, self-organization and other non-linear dynamics. Tobler's (1970: 236) assertion that 'everything is related to everything else, but near things are more related than distant things' emphasizes spatial relations between urban features (neighbourhood effects). These behaviours can be mathematically formulated as systems of differential equations.

It is therefore unsurprising to see that the current state-of-the-art approaches utilize variations of cellular automata (CA) to model urban land use and evolution: Barredo *et al.* (2003: 150), based on Stephen Wolfram's (1984) work, describe CA as 'spatial idealisations of partially differential equations' – they are capable of replicating the spatial relations and complex behaviours/dynamics which define urban systems. In simple terms, CA can be described as a lattice of cells; each cell can take on any number of different states; the state of each cell at a time step $t + 1$ is determined by the state of its neighbourhood cells at time t (as well as its own state), according to a set of transfer rules (Torrens 2000). In an urban context, cells represent an urban area, states represent land-use types (perhaps specific building types, depending on how extensive the set of states are), and the time steps in conjunction with transfer rules simulate dynamic evolution of land use.

Care is needed however in defining available cell states, transition rules and initial cell states to ensure that realistic simulations are produced (e.g. obeying certain constraints on aggregate behaviour), as opposed to fanciful 'game playing' scenarios. The literature suggests that the introduction of appropriate constraints can be used to guide simulations. These constraints may, for example, represent physical and policy-based land-use restrictions, the suitability of land for particular types of use, or the costs associated with developing new land or changing from one land use to another (Barredo *et al.* 2003, Li and Yeh 2000). Geographic information systems (GIS) have been used in an attempt to tackle these issues of model calibration by a number of researchers (Clarke and Gaydos 1998, Batty *et al.* 1999, Li and Yeh 2000), coupling CA models with GIS datasets to generate transition rules and constraints. GIS also provides the opportunity to generate highly detailed model visualizations.

A clear weakness of cellular automata is their immobile nature: while infrastructure, buildings, etc. need not be mobile and thus can be modelled with CA, transport systems, which form an important part of an urban (energy) system, may not be treated satisfactorily using this type of model. Detailed investigation of transport systems has not been undertaken using CA; however, the potential for using agents (which can be mobile by definition) for transport modelling has been investigated, again by Batty (2005).

8.1.3 Integrated land-use transport models

The preceding review highlighted the two broad areas of urban energy modelling and dynamic urban modelling, and discussed their importance in the context of developing comprehensive, realistic models of urban energy systems. It is clear that none of the approaches on their own would lead to a satisfactory result: the urban energy models considered do not take into account the importance of urban dynamics; the dynamic urban models do not specifically deal with energy. A more satisfactory approach would incorporate major elements of these areas to create a dynamic urban energy system model and this is the focus of much current research in land-use transport modelling (for a more detailed review, see Chapter 11).

Land-use transport (LUT) models, from the early static models to the current state-of-the-art micro-simulation models, provide a means of estimating travel demand in response to changes in land-use. In other words, LUT models are predictive models that provide a description of the state of the transport system under different policy scenarios. The accurate assessment of policy impacts on the transport system, and the key to modelling energy consumption by the transport and indeed other sectors, therefore requires the application of a behaviourally realistic land-use transport model.

The earliest land-use transport models were essentially static models, driven typically by gravity formulations or input–output formulations (see e.g. Lowry 1964). The static models, by their very nature, cannot realistically capture urban spatial processes and their effects on the transport system. These early models are therefore not very responsive to policy analyses. Since then, however, the development of improved modelling methodologies such as entropy-based interaction, random utility theory, bifurcation theory and non-linear optimization, together with significant computational advances has paved the way to several dynamic land-use transport model systems. These model systems can be broadly classified, based on the unit of analysis and the operational theory, into general spatial equilibrium models (e.g. MEPLAN, see Hunt and Simmonds 1993) and agent-based micro-simulation models (e.g. UrbanSim, see Waddell 2002).

The main approaches to land-use transport modelling can therefore be as follows:

Static models Basic static models are typically entropy-based models and are linked directly to a four-stage transport model (i.e. trip generation, trip distribution, mode choice and route choice). By their very nature, all static models treat land-use and transport systems as being exogenous to each other.

Spatial equilibrium models are typically spatially aggregate models with fully integrated land-use and transport elements. The interactions between these elements are determined by input–output analysis or discrete choice models, and these interactions are used to derive the demand for transport. General spatial equilibrium model systems are, therefore, based on random utility theory and theories of competitive markets.

Agent-based micro-simulation models (see Chapter 11) are activity-based models with the individual (one person, household, firm or any other agent in the urban system) as the unit of analysis. Hence, these models are intuitive in their formulation and capture the interactions between land-use and transport systems to the greatest extent possible.

In understanding energy consumption *vis-á-vis* transport, it is important to take into account not only the short-term decisions such as destination and mode choice, but also the medium and long-term decisions made by individuals with respect to residential and workplace location, vehicle ownership, labour force participation, and so on.

Therefore, the land-use transport model must take into account all these elements and the interactions between them.

8.2 Meta-analysis review

The discussion above reflects the breadth of modelling potentially relevant to the analysis of urban energy systems. Of course, within each tradition there are a variety of specific modelling tools and it can be difficult to get an overview of the relative popularity of each approach. Keirstead *et al.* (2012a) have therefore undertaken a comprehensive review of the literature related to urban energy systems and used this to develop a categorization of the field and to identify areas of strength, gaps and opportunities. This section provides a summary of their key findings.

The review paper began with the theoretical definition of an urban energy system model presented in Chapter 2 and then searched the literature for research matching relevant key words. It identified 219 papers and each was scored based on the spatial and temporal scale of the analysis, the modelling techniques used, and the extent to which the supply and demand sides of the energy system were considered. A cluster analysis was then performed which identified five key areas of practice: technology design, building design, urban climate, system design and policy assessment. It was additionally noted that, although not represented in the initial review, land-use and transportation modelling should be added as an important and emerging part of urban energy systems modelling. Table 8.1 summarizes the attributes of these six modelling approaches.

This section is particularly concerned with the 'system design' cluster. A review of the work in that field is given below.

8.2.1 Review of urban energy 'system design' literature

These studies are characterized primarily by their use of optimization techniques (72 per cent of the 39 studies in the cluster). The typical problem definition in these

Table 8.1 Medoid characteristics of main paper types described by the review. The 'transportation' models were not part of the main clustering analysis, but their typical features are listed here for comparison

Category	Spatial	Temporal	Method	Supply	Demand	n
Technology design	Technology	Monthly	Simulation	Endogenous	None	33
Building design	Building	Annual	Simulation	None	Endogenous	56
Urban climate	Sub 1 km	Hourly	Simulation	None	Endogenous (indirect)	36
System design	District	Static	Optimization	Endogenous	Exogenous	39
Policy assessment	City	Static	Empirical	Exogenous	Exogenous	55
Transportation	District*	Dynamic	Econometric	Endogenous	Endogenous	–

*The spatial unit of analysis for transportation models is typically the transportation analysis zone (approximately 3,000–5,000 people) but can be as detailed as the precise latitude and longitude coordinates of a single point.

studies is: given an exogenously-specified pattern of energy service demands, determine the combinations of capital equipment and operating patterns to meet some economic or other objective subject to supply and other constraints (e.g. minimize cost of the system subject to meeting heat and power demands within a carbon budget).

Given that most urban energy systems in place today will still be in place (at least in some form) in the near to long-term, an important subset of this category are models concerned with designing the retrofit of urban energy systems. Retrofit is defined here as 'a planned action intended to improve upon existing energy infrastructure with the provision of appropriate technology and methods'.

Manfren *et al.* (2011) have previously reviewed studies of this kind, focusing particularly on distributed generation techniques. Indeed most of the studies identified by the review do consider combined heat and power and district energy. Early work in the field was undertaken by Gustafsson *et al.* (1987) who were the first identified to use optimization techniques to assess the impact of district heating rates on the life-cycle cost (LCC) of retrofit strategies for multi-family buildings. An LCC typically consists of conflated capital, installation and operational costs.

System design studies typically consider either the entire city or a specific sub-district, although similar techniques can be applied to buildings (Lozano *et al.* 2009). District scale models have been the prominent choice of spatial optimization since at least the year 2000 (e.g. Bojic *et al.* 2000), while others have focused on building stock models (e.g. Zavadskas *et al.* 2004). It is reflective of the recent advances in both programming software and computational hardware that some researchers have increasingly pushed from district scales towards urban scale models.

The temporal resolution of these models has also changed. Within Keirstead *et al.*'s sample, 51 per cent of the models performed static analyses, although multi-period assessments at hourly, monthly, annual and decadal resolutions were also seen. For example, Rolfsman (2004) describes a model for minimizing the total cost of a municipal energy system, by either investing in new plant or retrofitting buildings (or both), wherein he divides a year into periods of 3 hours duration. Overall, a larger trend can be seen where static programming models in the early 1990s are increasingly superseded by temporally disaggregate models in the late 1990s. This can be seen in the comparison between two static models, as applied to a district heating network in Italy (Adamo *et al.* 1997) and to a district supply system in Germany (Bruckner *et al.* 1997), and to the annually optimized design for the addition of solar heating systems to housing stock in Germany (Lindenberger *et al.* 2000).

8.2.2 Land-use transport models for urban energy systems

As explained above, land-use transport (LUT) models are critical to integrate within the discipline of urban energy systems modelling. From the urban energy modelling perspective, these tools offer a behaviourally-realistic means of simulating consumption activities, and therefore resource demands. LUT model systems are typically a suite of interconnected descriptive and normative models that can jointly predict urban processes and activities. The embedded models are usually either micro-econometric and based on random utility maximization principles (e.g. CEMDAP, Bhat *et al.* 2004), or based on decision heuristics (e.g. ALBATROSS, Arentze and Timmermans 2000). Some models combine heuristic rules with econometric models (e.g. TASHA, Roorda *et al.* 2008).

The state-of-the-art LUT models, regardless of the underlying model types, are implemented as agent-based micro-simulation systems with the activities of all the agents in the study area being simulated. These models are also highly disaggregate with respect to time and space, with some of the models operating on a continuous (second-by-second) time scale with parcel-level spatial detail. The flip side of such descriptively rich models is the quantity of data and computational time (effort) required to validate them for operational modelling, with some models taking up to 36 hours to simulate a single day's travel. On the other hand, once operational, such models can be excellent test beds for a variety of policy scenarios, engineering and technological solutions. As integrated models, they effectively capture both direct and indirect effects of the scenarios of interest.

A key feature of the state-of-the-art LUT models is the underlying models of individual and group behaviour. These models, unlike typical engineering models, which are employed in the supply components of UES model systems, acknowledge the stochasticity of human behaviour and the intrinsic heterogeneity in this behaviour, which results in both the same individual, and observationally-identical individuals, making very different choices. The earliest models of agent-level behaviour focused principally on predicting the choice of specific facets of individual trips (such as mode or route location) and tended to be deterministic in nature (typically assuming that behaviour was driven solely by considerations of cost or travel time minimization). From the 1970s onwards, these approaches were gradually replaced by models which, at the conceptual level, consider travel decisions explicitly as part of the broader context of an individual's programme of activity participation and at a methodological level, treat decision-making as a stochastic (rather than deterministic) process.

The current state-of-the-art is represented by techniques based on the random utility formalism, which can accommodate a wide variety of decision-making contexts including both individual and group decisions, decisions regarding both discrete and continuous outcomes, static and dynamic decisions, decisions with single or multiple expressed outcomes, decisions made under uncertainty and those influenced by qualitative as well as quantitative factors. These methods can also be used as a means of integrating data both from real market outcomes ('revealed preference data') and data from hypothetical market studies ('stated preference data').

8.2.3 *Challenges in UES modelling*

Keirstead *et al.* (2012a) concluded from their survey that current modelling practices only partially address the key features of an urban energy system. The deficiencies include:

- Studies often focus on specific aspects of energy use, with only the system design models looking at the full set of 'combined processes' within an energy system, although often excluding the transport components of a city.
- Studies often rely upon exogenous input data, for example, user-supplied electricity demands in the case of many system design models. Only a small number of

integrated assessment models for policy considered both supply and demand endogenously.

Based on the review, a number of challenges were identified as described below.

Complexity of the modelling domain

It was observed that a range of spatial and temporal scales are used but, particularly in older studies, the resolution and fidelity of the models can be limited by data availability and computational performance. For example, 44 per cent of the 219 studies analysed district or coarser spatial scales and 58 per cent dealt with annual or greater temporal resolutions. Optimization techniques may attempt to incorporate some of these complexities with a technique like multi-objective optimization (Alarcon-Rodriguez *et al.* 2010) or sensitivity analysis (Sundberg and Karlsson 2000). Yet all models tread a fine line between tractibility and model performance. For example, modern models can take between several days to several hours and shorter, dependent on the method, scope and computing power at play. Related to this, over-simplifying or over-detailing a model may obstruct the true value of the output from the model's internal relationships.

Data availability and quality

Many of the papers use modelling techniques that require large amounts of data such as econometric models of transportation or consumer behaviour, detailed GIS and 3D mapping of cities, and hourly profiles of energy consumption demands. Acquiring good quality data for urban energy modelling faces at least one intrinsic challenge, namely that cities are open systems and defining the boundaries of the urban energy system can be difficult. The review studies primarily used data gathered from administrative or district boundaries and only a few included life-cycle assessments that extend beyond the city boundaries. This creates a challenge for the modeller, who must assess the quality of the available data source and determine whether or not it is sufficient to answer the research question at hand. In developing countries in particular, acquiring any urban energy statistics can be difficult although there were some exceptions within the dataset including the energy consumption of bakeries in Nigeria (Ekechukwu *et al.* 2011), gasoline consumption in Mexico (García *et al.* 2010), and industrial combustion in Brazil (Lucon and Santos 2005).

There are some technique-specific data availability issues as well. For example, LUT models are essentially descriptive models that need data in order to be calibrated prior to implementation. As the models become more complex – in terms of integrating land use, transport and the environment, operating at the level of the decision-making agent (e.g. individuals, households, businesses), assessing detailed spatial and temporal scales – the corresponding data needs to grow very quickly. However, this need not be a stumbling block in the development of such models as researchers have explored and exploited mathematical approaches of pooling data from different sources.

The second major issue is data uncertainty. Observed data may be uncertain due to measurement error, the need to use proxy data sources or calculation adjustments (e.g. for downscaling national-level data to a city scale on a per-capita basis). Ideally, such parameters would be expressed along with their uncertainty but the majority of papers surveyed did not explicitly describe methods of dealing with the uncertainty of parameters. Deterministic optimization models in particular are guilty of this, whereas stochastic algorithms such as genetic algorithms do provide a probabilistic solution set. Other relevant optimization techniques include two-stage stochastic programming, parametric programming, fuzzy programming, chance constraint programming, robust optimization techniques and conditional value-at-risk. More generally there is a need to describe both the uncertainty of collected input data and the way in which this feeds into the modelling methodology.

Lack of model integration

The reasons for this are not covered by the scope of this chapter, but it is clear that the urban energy systems modelling community develops models for distinct purposes and unique audiences, and so less effort appears to have been focused on creating larger integrated modelling systems. There appears to have been very few attempts to integrate models that span across multiple sectors or disciplines. Depending on the goal of the analysis, this may not be an issue. However, if urban energy systems modellers are to tackle the complexity of their domain, then model integration is a sensible strategy. Integration raises both practical and theoretical questions about how models might fit together. As a simple example of the potential interactions, consider the problem of the urban heat island. In these cases, a building model may be able to determine indoor climate given orientation, solar exposure, and so on. A technology model might then determine the best air conditioning system for the building, but the rejected heat from this air conditioner would of course be rejected into the environment thus changing the building's thermal performance (particularly if groups of buildings are located in close proximity). A more significant example where model integration seems to be missing is between urban land-use and transportation models and energy system models.

Policy relevance

Integrated model systems are arguably necessary to address the fourth challenge, policy relevance. Specific models may be able to answer narrowly-defined policy questions (e.g. building standards) but urban energy systems can be shaped by both direct and indirect effects of policy interventions, such as land-use planning restrictions on density which affect both building energy consumption and transportation options. For decision-makers looking to improve the performance of the overall urban energy system, many of the models identified were applied to a relatively narrow set of policy problems – evaluating transport pricing policies, estimating the impacts of new low-energy technologies, examining the environmental impacts of demographic evolution, evaluating the social equity of a specific policy, and so on. This is limiting, as the narrow perspective often fails to account for indirect effects of policies on the urban system. It also fails to account for the conflicting effects of different policies. As urban systems face tighter integration through developments such as electric mobility, it becomes increasingly important to account for the combined effects of policies (e.g. transport and energy policies in the case of electric mobility). The challenge for urban energy system modellers, then, is to create tools that explicitly capture some of the linkages between energy systems and other aspects of urban policy. At the very least, modellers should be aware of these connections so that modelling results can be presented in a policy-relevant fashion with caveats acknowledged.

8.3 The SynCity modelling framework

Much progress has clearly been made in the field of urban energy systems modelling, but it remains to be seen whether the multiple challenges identified above can be overcome within a single modelling framework. We have attempted this by creating a model platform known as SynCity (short for 'Synthetic City'). The main objective of the toolkit is to capture both the demand and supply aspects of urban energy systems. We believe that the toolkit's capability to look at the overall energy system of a city in this bottom-up, integrated way is unique. It contrasts with other modelling approaches which tend to look at broader geographies (country, regional) and consider supply and demand independently and largely from a top-down perspective.

The SynCity platform consists of a Java code library, a database of energy technologies and fuels, and four connected submodels. Together these components enable users to explore the role of energy in urban areas at key design and operational stages, namely: the master planning of a new development or redevelopment (the land-use or layout model); the daily activities of individuals within a city (the agent-activity model); the design of energy supply strategies (the resource flow model); and the detailed operation of energy service networks (the service networks model). Each component can be run separately or as a single integrated analysis and the toolkit has been applied to case studies in the UK, China and elsewhere. Figure 8.2 provides an overview schematic of the framework and the respective submodels are discussed in detail below.

8.3.1 *Layout model*

This is an optimization-based approach to organizing the city layout. The model is provided with basic information about the city, its residents, boundary conditions, desired activities and so on, and uses a combinatorial optimization technique to develop alternative city layouts, where the degrees of freedom are the allocation of different land uses to zones, given the need to allocate sufficient land to each activity. The expected passenger transport flows through the layout are also calculated, from which transportation energy is estimated. Different performance measures may be used for the optimization including total energy consumption or cost. The key variables include the allocation of a land use to each cell and for the residential cells, the number of residents expected and type of housing used.

The model is described in greater detail in Keirstead and Shah (2011). However, in Figure 8.3, we illustrate a simple case study of a proposed eco-town in the UK. The developers had prepared an already highly-efficient land-use plan, making particular use of efficient building types to deliver approximately a 70 per cent reduction in

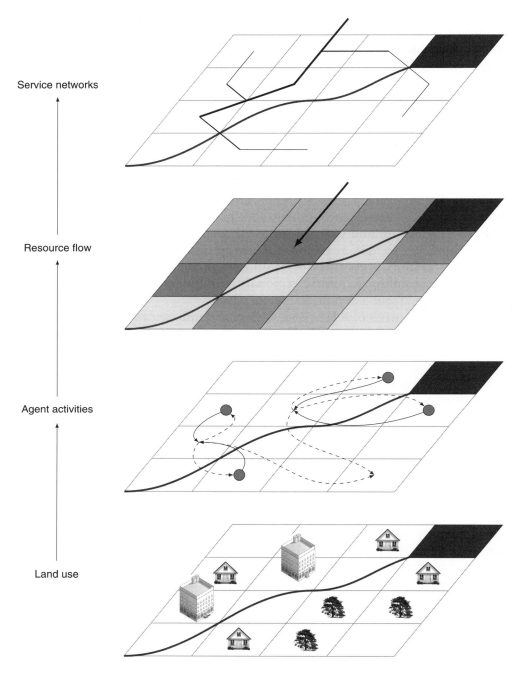

Service networks

Resource flow

Agent activities

Land use

Figure 8.2 The SynCity hierarchical modelling framework. Briefly, the land-use layer determines a low energy spatial plan for the city, the agent activity layer simulates the behaviour of citizens and firms and hence their energy requirements, the resource flow layer calculates an optimal energy strategy for a given pattern of demands, and the service networks layer analyses the detailed operation of urban energy supply networks. See acknowledgments for clip-art sources.

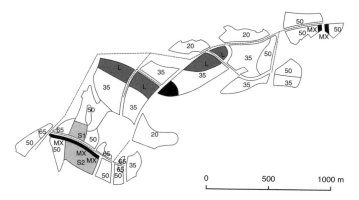

(a) Original land-use plan, 24 GJ/capita

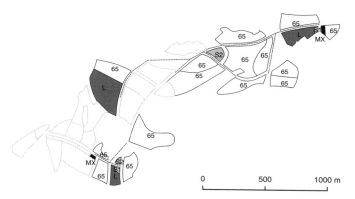

(b) Minimum energy layouts with no constraints, 11 GJ/capita

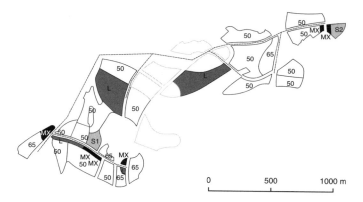

(c) Minimum energy layouts with constraints, 17 GJ/capita

Figure 8.3 Alternative optimized layouts for a UK eco-town. White cells with black borders are labelled with the housing density in dwellings per hectare. Activity cells are shaded as follows: light grey, schools (S1 and S2); medium grey, open space/leisure (L); black, mixed use (MX). Road networks are shown with dashed line (omitting connections to individual cells for clarity). For reference, London's energy consumption is 79 GJ/capita.
Source: Keirstead and Shah (2011).

energy consumption compared to current practice in London. This was then compared to two cases. First, the model was run with minimal constraints in order to identify a low-energy benchmark layout; this layout represents the minimum possible energy consumption that still satisfies basic constraints on adequate housing and activity provision. The second scenario then adds constraints to reflect the realistic pressures facing developers: e.g. not using exclusively high-density housing or providing sufficient greenspace.

The layout model therefore enables analysts to quickly develop alternative low energy designs and, perhaps more importantly, to identify a minimum energy benchmark layout against which the ambition of other plans can be judged. It should be noted that this is a highly stylized model, intended for developing ideas at early stages of the design process. In practice the connections between urban form and energy consumption are much more complex (Echenique *et al.* 2012).

8.3.2 *Agent activities model (AMMUA)*

This model takes the city layout and establishes where the agents are at any given time, what activities they are involved in, and their transportation mode choices when they move between locations. This information can then be used to infer time-dependent resource demands associated with the built environment (which can vary depending on building designs and standards) and the transport system.

We call this tool the agent-based micro-simulation model of urban activities (AMMUA) and it forms the disaggregate demand modelling component of SynCity. It incorporates important features of agent-based micro-simulation modelling and state-of-the-art activity-based transport demand models. AMMUA therefore operates on individual agents (such as households, citizens, businesses) with heterogeneous properties. The AMMUA model system is comprised of a number of sub-models such as activity-generation models, scheduling models to determine spatial and temporal elements of demand, and an urban goods and services models. Together, this framework enables the detailed assessment of the direct and indirect effects of technological change and policy interventions. More details on AMMUA can be found in Chapter 11.

8.3.3 *Resource flow model (TURN)*

Once the demands for energy services are known, either from AMMUA simulations or observed data, the next step is to design an energy system that supplies these demands efficiently. Here, we build upon the 'resource technology network' concept from process systems engineering in order to determine how to supply the time- and space-dependent resource demands in an optimized fashion. We call the model TURN (Technology and Urban Resource Networks) and its formulation and a case study are described in detail in Chapter 9.

The TURN model uses mixed integer programming in order to find an optimal solution for all resource demands simultaneously. It can therefore automatically identify opportunities for integration across energy networks, a feature which is particularly helpful for analyses of urban bioenergy systems which may have multiple energy conversion pathways (Keirstead *et al.* 2012b). The model's decision variables include:

- The choice of resources to import into the city
- The choice of resource conversion, transportation and storage technologies to use in the city
- The design of any networks for resource flow through the city
- The destination of wastes (including waste heat).

The objective function can be defined in a flexible fashion to incorporate capital cost, life-cycle cost, fossil energy consumption, GHG emissions, criteria pollutant emissions, and so on.

A number of case studies have been used to demonstrate the speed, robustness and flexibility of the TURN model. For example, the UK eco-town described above was analysed to examine the potential for converting waste into energy, rather than landfilling (Kostantinidis *et al.* 2010). Another study considerd a large eco-city in China (Lingang, near Shanghai) and the potential use of various heat pump and trigeneration technologies to provide electricity, heating and cooling. Energy systems were designed for minimum cost, minimum primary energy consumption and minimum CO_2 emissions (Liang *et al.* 2012). A final example is the use of the TURN model to generate marginal abatement cost (MAC) curves. Traditional expert-derived MAC curves treat each intervention (e.g. wall insulation, combined heat and power, etc.) as independent, assuming that their respective benefits are additive (Kesicki 2010). However, model-based approaches, for example, using the TURN framework, can capture the interactions between technologies (Philippen 2011).

8.3.4 Service networks model

The TURN model generates a series of resource networks that indicate the flow of resources between interconversion technologies and final demand points. It sets the target conditions for the actual service networks that must meet a variety of technical performance measures. The service networks model is therefore concerned with the design of robust urban networks that embrace heterogeneity of generation and conversion and which incorporate the state-of-the-art in the particular network type (power, gas, heat, etc.). Given that we anticipate greater integration in the future, the interdependence between networks must be quantified and operational feasibility guaranteed.

Research has therefore focused on assessing the detailed energy flows that occur in urban energy systems through key infrastructures such as electrical and natural gas networks. After establishing the principal guidelines of the time-coordinated optimal power flow (TCOPF) programme – a modelling framework that considers the interactions and interdependency between conventional networks and embedded distributed energy resources (DERs) – work has focused on assessing the storage features of key embedded technologies. As a consequence, combined heat and power (CHP) units with thermal storage and plug-in hybrid electric vehicles (PHEVs) with battery storage have

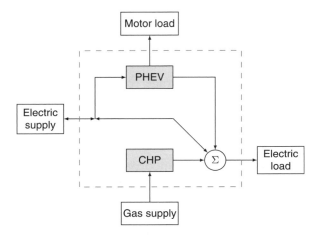

Figure 8.4 Schematic showing the role of plug-in hybrid electric vehicles (PHEVs) and combined heat and power (CHP) in linking both electricity and gas networks. Demand for electric power within the home can be satisfied by combining electricity flows from the grid, CHP and PHEV units.

been included in the model. The connections between these DERs and the underlying energy networks is shown in Figure 8.4.

In these settings, DERs such as CHP and PHEVs provide a great deal of flexibility in power provision. The TCOPF tool can therefore function as a global coordinator to simultaneously evaluate the performance of natural gas and electrical networks, dispatching DERs according to stakeholder requirements. The TCOPF programme effectively takes the role of an intermediary, similar to an ancillary service provider, which can send operating signals based on grid conditions and the status of the connected embedded technologies. Furthermore, the modelling framework includes the feature of vehicles giving power back to the grid (V2G) if required (Acha 2010).

Modelling these innovative interactions between DERs and energy infrastructures leads to questions of optimal system operation, such as 'How will DERs be used under different operating strategies?' The TCOPF programme was therefore tested using the following strategies:

Plug and forget refers to a status-quo scenario in which the network operators concentrate only on supervising the technical conditions of its assets. PHEVs are only plugged in for charging when at home.

Fuel cost dispatches generation units such that the total fuel cost for consumers is minimized while satisfying operational feasibility constraints.

Energy loss focuses on reducing the power losses incurred in the networks by dispatching the embedded technologies whenever it is necessary, while again meeting all operational requirements.

Energy cost uses the day-ahead natural gas and electricity spot market prices to reduce total energy costs incurred in the system while meeting all the demanded technical needs.

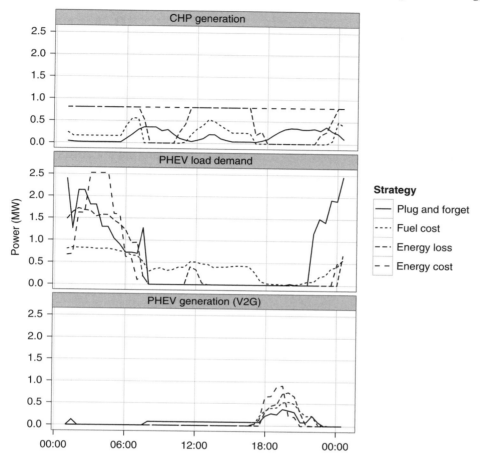

Figure 8.5 Load profiles for CHP and PHEVs under different TCOPF operating strategies. See text for description of strategies.
Data source: Acha (2010).

Figure 8.5 illustrates these different scenarios, showing when CHP units are generating electricity and when plug-in hybrid vehicles are both drawing power from, and returning power to, the grid. Note that in most scenarios PHEVs delay their charging until the early morning hours when other demands on the grid are lowest; similarly PHEVs operate in a vehicle-to-grid mode, providing power back to the grid, at times of peak demand (early evening).

Using Repast Simphony,[1] an agent-based model representing the behaviour of owners of plug-in electric vehicles in a city has also been developed, which can be connected with a power flow optimization model of the electricity distribution grid. The link between these two models showcases how different infrastructures in a city are related (in this case, road and electricity networks) and allows for spatial analysis of vehicle charging and the effects on the distribution grid (Acha *et al.* 2011b).

8.4 Conclusion

This chapter has provided an introduction to the computer modelling techniques that can be used to analyse urban energy systems. It was shown that a wide range of methods must be used to fully capture the complexity of citizen behaviour, technologies, infrastructures and their interactions. In particular, the role of optimization in system design and agent-based micro-simulation for understanding behaviour were highlighted. A review of current literature showed that there have been only limited attempts to connect these different modelling paradigms and so the SynCity modelling toolkit was introduced as a way of exploring these linkages and hopefully providing a more complete assessment of potential improvements in urban energy performance.

In the next four chapters, we will describe some of the models discussed here in greater detail and present sample application case studies. Uncertainty and sensitivity analysis techniques are also presented as a vital part of any modeller's toolkit.

Note

1 An open-source toolkit for agent-based simulation, available at http://repast.sourceforge.net/

9 Optimization and systems integration

Nouri Samsatli and Mark G. Jennings

Imperial College London

9.1 Introduction

Optimization models of urban energy systems (UES) are argued to provide the advantages that mathematical modelling offers in any field (Williams 1999). As with any model, the internal structure and relationships (not the data) are what determine its usefulness to a decision-maker. The construction of a UES model may uncover relationships not initially known or well understood. Once built, a model allows for quantification and discussion of plans and strategies not initially conceived. Also an optimized mathematical model of UES permits experimentation which may be too costly or risky for investment in practical terms. These advantages are thought to be embedded in major UES optimization models.

However, there may be skeptics who point out that models cannot quantify many real decisions in UES. Yet even qualitative decisions will likely be made on the basis of internal reasoning that includes an implicit trade-off analysis. There is thought to be value in bringing decisions, such as the planning of retrofit or distributed generation investments, out into the open by representing them as optimization variables. Should the decision-maker then elect not to implement a decision determined by the optimization, at least the disadvantages of this choice will be known. The strength of UES optimization models therefore lies in their impartial optimization, and they offer an invaluable tool for any urban decision-maker or researcher.

Details of the theory behind mathematical programming, optimization and the science of decision-making are beyond the scope of this chapter, and have been extensively covered by others (e.g. Taha 1982, Dantzig 1963). Optimization is defended as being a stronger method than others, for instance, due to its capacity for considering alternative pathways over time. The output from optimization models is a programme for the best decisions towards achieving a stated objective, i.e. prescriptive output. Simulation models on the other hand provide descriptive output by imitating or mimicing the real world so that its actual behaviour may be analysed.

In general, researchers appear to favour simulation models for technological and building design, and construct econometric models for transportation models (see Chapter 8). Optimization is the method most applied to UES modelling, one could argue, on the basis of its powerful description of systems. The recent increase in computing power and more efficient solution algorithms have helped popularize the use of optimization models as planning tools for energy systems, particularly on a bottom-up basis (for a recent review, see Hiremath *et al.* 2007). A brief overview of their structure is given now.

Optimization models all involve the minimization or maximization of an objective function, the value of which represents some quantity of interest (such as total cost or pollutant emissions). There has been a great deal of interest in what constitutes an 'optimal' objective function. The debate of single objective versus multi-objective functions is particularly interesting (see White (1990) for bibliographical survey of multi-objective methods in earlier years, and Ehrgott (2005) for a description of techniques). One way to solve multi-objective optimizations is to define the objective function to be a weighted sum of objectives and to perform a sequence of single-objective optimizations with different weights to produce a pareto-optimal set (Diwekar 2008). The next element of importance in an optimization model is that of the modelled relationships: how one represents the complexity of the UES through suitably defined variables and constraints.

Most formulations of UES optimization models can be classified as linear programming (LP), mixed integer linear programming (MILP) and mixed integer non-linear programming (MINLP) problems; Appendix A provides a more detailed discussion of optimization models and their solution procedures. Alternative formulation and solution strategies are provided by meta-heuristic methods such as those of evolutionary algorithms (of which genetic algorithms (GA) are a subset). The main advantage of meta-heuristics is that, based upon previous experience of the problem, they will likely provide a reasonable solution to a problem while their main disadvantage is that they cannot ensure that an optimum solution will be found. Table 9.1 gives an outline of the typical characteristics of optimization models.

9.1.1 Optimization modelling in urban energy systems

The use of optimization techniques for UES modelling has grown in recent years, as recent reviews detail (Keirstead *et al.* 2012a,c). UES modelling examples of LP, MILP, MINLP and GA are now briefly described.

Table 9.1 Characteristics of optimization models and energy system applications. See text for relevant citations and abbreviations

Category	Benefits	Drawbacks	Energy system examples
LP	Global optimum	Linearizes all relations	Deeco
	Shadow pricing	No discrete variables	Modest
MILP	Global optimum if all nodes searched	Longer solution time	EnerGIS
	Integer variables	Combinatorial explosion	Markal
MINLP	Local optimum	Difficult to solve	Frombo *et al.* (2009)
	More realistic functions	Global optimality not guaranteed	
GA	Reasonable solutions	Many more function evaluations required	Ooka and Komamura (2009)
	May avoid inferior local optima	Optimality not guaranteed	

Deeco (dynamic energy, emissions and cost optimization) is an LP framework for determining the relative utilization of the (pre-specified) available energy technologies at each time interval and has been applied to a UES in Germany (Bruckner *et al.* 1997, 2003). It is based on a previous modelling framework called NEMESS (Groscurth *et al.* 1995). Deeco was formulated with cost-effective system integration synergy in mind. The model considers, by default, 1 year using intervals of 1 hour, achieved by optimizing the supply of energy for each interval independently of the others.

Another generic LP tool for optimizing energy supply systems is Modest (Henning 1997, Henning *et al.* 2006). It can be used from the local level to the national level, at a number of sectors at various temporal resolutions. For example, it has been applied to a municipal utility in Sweden, which uses a number of fuels and cogeneration plants, and the model includes diurnal and monthly changes to demand. However, the model lacks any endogenous spatial disaggregation, and also linearizes investment costs.

When designing a UES, a number of discrete decisions must be made, such as the existence of a particular facility in the design or the existence of a gas pipeline connecting two different locations. Problems such as these are often modelled using discrete variables and therefore LP may not be appropriate for a number of investment decisions. Integer variables, in combination with continuous variables, are thus of use when considering both new UES networks and retrofit of existing ones. EnerGIS (Girardin *et al.* 2010) uses such an MILP formulation in combination with a GIS (geographical information system) to design district heat and cooling networks in Geneva. Annual demands for space heating, hot water, cooling and electricity are provided to the model from GIS data, and heating/cooling loads are then calculated for various building types based on outdoor temperatures. A number of supply-side technologies are then selected to minimize investment cost.

Investment costs are often included in the objective function, as in the model of Frombo *et al.* (2009). Here, an MINLP optimizes the investment in biomass conversion processes (pyrolisis, gasification and combustion), biomass extraction and transportation costs less the power sales. The model determines what type and size of energy production plant should be chosen, and where the biomass is obtained. The non-linearity is brought in by the product of a binary variable (whether or not a technology exists in a location) and a continuous variable (the annual biomass quantity harvested for a technology). The model is intended for long-term planning, hence only annual energy demands are considered. Other drawbacks include: only considering thermal demand, with excess heat production being converted to electricity and sold to the grid, and drawbacks typically associated with MINLP problems (e.g. the feasible region may be non-convex, such that a local optimum may not be the global optimum).

The NLP sub-problems of MINLP are usually computationally expensive to solve, such that alternative solution strategies may be sought. Ooka and Komamura (2009) developed an MINLP model of a hospital with a floor area of 40,000 m^2 using demands at 1-hour intervals for a typical day. The MINLP formulation allowed them to retain the non-linear relationship between efficiency and load factor, which must otherwise be approximated when developing an MILP model. Rather than using a gradient-based solution procedure, they used a genetic algorithm. This is an approach, used by a number of researchers in many different areas, that aims to avoid being trapped in a local optimum by employing stochastics as part of the solution procedure. Yet, as mentioned above, meta-heuristics cannot be sure that their solutions are globally optimal. Furthermore, their model only considered a single day, and they note that 1 or 2 months would be required to optimize their formulation for an entire year.

It can be seen from above that computational limits, in part due to the combinatorial complexity of UES models and in part due to the formulation/solution strategies, are as much of a concern as the correct abstraction of the real-world energy systems within cities. There are also weaknesses in modelling of investment decisions, particularly with regard to building systems where non-economic factors may be significant and there are gaps in knowledge with regard to the true limits of optimization techniques as UES decision tools. Yet, there has been great progress made in optimization modelling in recent years, and models continue to become more representative of the UES networks they seek to reflect.

9.2 An MILP model for UES design

A novel MILP formulation has been developed in order to be easily extensible and applicable over different spatial and temporal scales.

There are three main dimensions that need to be considered when developing a UES model: spatial, temporal and the description of the technologies that interconvert energy resources.

The spatial dimension is addressed by dividing the region of interest into a number of zones, which may be any shape and size. Each zone may then have time-dependent demands for various resources, plants for converting resources, storage facilities and connections with other zones by which resources may be transported. In addition, certain zones may be allowed to import and export resources from outside the city. The spatial resolution will have a direct effect on the tractability of the model because of combinatoric explosion in the number of decision variables needed to describe flows between zones.

Similarly, the temporal resolution needs careful consideration. Most UES models discretize the time domain, resulting in a finite number of time intervals that are almost always of the same duration. Since the number of intervals also limits the tractability of a model, it is usual to consider either detailed operational aspects, on an hourly basis, or investment decisions on a yearly or decadal time scale *but not both simultaneously*, since to do so would result in a very large number of time intervals. A more efficient approach is required to consider various time scales simultaneously. The model presented in this chapter uses an efficient hierarchical non-uniform time discretization that enables a number of time scales to be considered, simultaneously or separately, within the same model.

Finally, a flexible framework is required in order to describe processes that convert primary energy resources into forms that can satisfy the energy demands of the urban area. This model employs the *State-Task Network* (STN) representation first developed by Kondili *et al.* (1993) and later extended by Pantelides (1994). The STN was developed to represent recipes for batch chemical processes as part of an optimal batch-plant scheduling formulation. The states represent distinct material states and the tasks

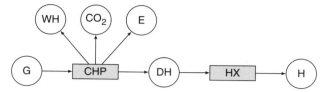

Figure 9.1 An example technology and urban resource network (TURN). The resources (circles) are: natural gas (G), waste heat (WH), carbon dioxide (CO_2), electricity (E), district heat (DH) and heating (H). The technologies (rectangles) are: combined heat and power (CHP) and heat exchanger (HX).

represent chemical processes that convert a set on input states to a set of output states; these are connected to form a directed graph known as the STN. In this model, the states are used to represent resources (e.g. heat, electricity, water and also wastes such as CO_2) and the tasks are used to represent the technologies that convert one set of resources to another (e.g. power stations, PV cells, wind turbines). In this way, the model is completely flexible and can therefore be used for a wide variety of existing and future technologies. To avoid confusion with Pantelides's Resource Technology Network (RTN) formulation (Pantelides 1994), networks for UES are called Technology and Urban Resource Networks (TURN). A simple example of a TURN is shown in Figure 9.1, where natural gas is converted in a CHP plant to electricity, district heat, some waste heat and CO_2. Demands for heating can then be met by connecting to the district-heat network using heat exchangers.

The TURN is not only useful for representing resource-conversion technologies but also for transport and storage of resources. Transport technologies have two sets of input and output resources: one for the source zone and one for the destination zone, and it 'converts' resources in the source zone to resources in the destination zone. Storage technologies are represented using three tasks: one to store a resource, one to maintain it in storage and one to retrieve it for consumption. Thus, it is easy to apply different efficiencies and resource requirements for these three phases of resource storage.

The details of the mathematical formulation are presented next. For readers who are interested in the applications of this modelling framework, see section 9.3 below.

9.2.1 Hierarchical non-uniform time discretization

The efficiency in the time discretization comes from decomposing the time domain into a number of different sub-domains of varying granularity: hourly intervals, h, daily intervals, d, seasonal intervals, s, and yearly intervals, y. The hourly intervals may have different durations, Δh_h (such that $\sum_h \Delta h_h = 24$), so that daily demand profiles can be represented more efficiently: more short-duration intervals during peak time, to capture the dynamics, and fewer longer-duration intervals during off-peak times where the demand is relatively static.

Further efficiency is gained by only using as many day intervals as there are days with *distinct* demand profiles (e.g. weekday and weekend). To model the demand for a whole week, each day type is allowed to repeat a number of times. For example, one could model the demands for the whole week as a sequence of repeated profiles: five repetitions of a weekday profile, followed by two weekend profiles. The number of *sequential* occurrences of a day type in a week is specified using the parameter R_d, subject to $\sum_d R_d = 7$.

Each year in the time horizon comprises $\bar{\bar{s}}$ seasons[1] of W_s identical weeks each, such that $\sum_s W_s = 52$.

Finally, the yearly interval is used to represent the longest time frame, over which investment decisions might be made. Each interval y has a duration of Y_y identical years.

The above structure provides a rich variety of temporal resolutions. One can focus on the details of a single day by selecting only 1 day type, one season and 1-year interval and using a very high resolution for the hourly intervals, e.g. 15-minute intervals. Instead, one might choose only to consider long-term investments and define only 1 hourly-interval of 24 hours' duration, 1 day type and one season (meaning the demands would be the annual average) and then use a large number of yearly intervals. An example of a more balanced approach would be to look at the next two decades using 6 hourly intervals per day, two day types (weekday and weekend), four seasons (of 3 weeks each) and four yearly intervals (of 5 years each). This would result in a total of $6 \times 2 \times 4 \times 4 = 192$ intervals, which is far fewer than the $24 \times 365 \times 20 = 175{,}200$ intervals required to model the whole time horizon at a 1-hour resolution.

9.2.2 Indices

A number of indices are used to denote the dependence of variables and parameters on position, time and types of technology. The main indices used in the model are given in Table 9.2; others will be introduced as required.

Table 9.2 Main indices used in the TURN model

Index	Description
h	Hourly interval
d	Day type
s	Season
y	Yearly interval
i	Zone
r	Resource
p	Resource conversion technology
m	Mode of operation for resource conversion technology
t	Resource transport technology
ς	Resource transport infrastructure
σ	Resource storage technology
μ	Key performance indicator (KPI)

9.2.3 Resource balance

The central constraint in the model is the resource balance. It states that for each resource r in each zone i at each time interval $hdsy$, the net production rate P_{rihdsy}, the net inflow from all other cells Q_{rihdsy}, the rate of import from outside the city I_{rihdsy}, and the rate of consumption from storage S_{rihdsy} must be equal to the demand D_{rihdsy}, plus the rate of export E_{rihdsy}. This is written mathematically as:

$$P_{rihdsy} + Q_{rihdsy} + I_{rihdsy} + S_{rihdsy} = D_{rihdsy} + E_{rihdsy}. \tag{9.1}$$

D_{rihdsy} is the demand for resource r in zone i during hourly interval h of day type d in season s of yearly interval y. Demands must always be met and can be positive (e.g. for heat, cooling and power) or negative to represent wastes being generated in the urban area that must be treated or disposed of in some way (e.g. municipal solid waste).

The demands (if positive) can be met by any combination of net local production of the resource ($P_{rihdsy} > 0$), net inflow from other cells (i.e. the resource is produced in another zone and transported here so that $Q_{rihdsy} > 0$), import and consumption of stored resource ($S_{rihdsy} > 0$). Any surplus resource can be exported, if allowed.

The net production of resource in each zone, P_{rihdsy}, is due to the activity of all resource-conversion technologies. Some resources will overall be consumed by these activities, hence P_{rihdsy} will be negative; others be produced overall, resulting in $P_{rihdsy} > 0$. Q_{rihdsy} and S_{rihdsy} can similarly be positive or negative. The constraints defining these variables will be discussed in detail in subsequent sections.

Note that the domain of equation (9.1) should be $\forall\, r, i, h, d, s, y$ but for simplicity, all indices that are to take all possible values will not be shown in the domain specification.

9.2.4 Imports and exports

Each zone may import or export any resource at any time. The rates or import and export of resource r at time $hdsy$ in zone i are represented by the *positive* variables I_{rihdsy} and E_{rihdsy}, respectively.

Clearly, these variables will be subject to a number of constraints, such as maximum rates, limiting the locations where imports and exports can take place and so on. Since these constraints are all simple to implement, in the interest of space they are not shown here.

9.2.5 Resource conversion technologies

The net rate of generation of resource r in zone i at time $hdsy$ is P_{rihdsy}. This variable accounts for the activity of all technologies operating in the zone that inter-convert resources (within the same zone). It is positive if the result of all conversion technologies is a net increase in the resource; it is negative if there is a net consumption.

Resource-conversion technologies are characterized by their rate of operation and a set of coefficients that define how much of each resource is generated per unit rate of operation. The *total* rate of operation of all technologies of type p in zone i at time $hdsy$

is the *positive* variable $\mathscr{P}_{pihdsy}$. The parameters α_{rp} define how much of each resource r is generated per unit rate of operation of technology p; they are negative for resources that are consumed. The total rate of generation of resource r can now be defined by:

$$P_{rihdsy} = \sum_{p} \alpha_{rp} \mathscr{P}_{pihdsy}.$$

Since $\mathscr{P}_{pihdsy}$ is the total rate of operation of all technologies of type p, it will be limited by the maximum rate of a single technology, $\mathscr{P}_{phdsy}^{\max}$, and the number of technologies present in zone i during yearly interval y, N_{piy}. The former are, of course, parameters and the latter are *integer* variables and are degrees of freedom in the model. The total rate of operation is therefore bounded by:

$$\mathscr{P}_{pihdsy} \le N_{piy} \mathscr{P}_{phdsy}^{\max}. \tag{9.2}$$

By convention, α_{rp} is set to one for the main resource produced by the technology; then $\mathscr{P}_{phdsy}^{\max}$ is the 'nameplate' capacity of the technology.

In the most general case, the maximum rate of a technology is a function of time. This allows one to specify periods when certain technologies may not be used and to model renewable technologies. For example, the maximum output from a single wind turbine will depend on the strength of the wind, which typically varies with time of day and season. If these are known, then the maximum power generation for each type of wind turbine can be pre-computed and used to set the values of $\mathscr{P}_{phdsy}^{\max}$.

The number of technologies present in any zone during any yearly interval, y, is subject to change due to investments and retirements. Therefore there needs to be a technology balance to track the number of technologies present over time. It is assumed that investing more frequently than on a yearly basis is not necessary. The technology balance is:

$$N_{piy} = N_{piy-1} + N_{piy}^{+} - N_{piy}^{-} \qquad \forall \, y > 1,$$

where N_{piy}^{+} is the number of technologies of type p built in zone i in yearly interval y and N_{piy}^{-} is the number of technologies retired.

9.2.6 Transport technologies

As has been mentioned before, transport of resources is also modelled using tasks. The tasks behave very similarly to the conversion technologies but, in this case, transport tasks have *two* sets of conversion factors: one for the source zone and one for the destination. These are denoted $\beta_{rt}^{\mathrm{src}}$ and $\beta_{rt}^{\mathrm{dst}}$ for the source and destination zone, respectively. Setting $\beta_{rt}^{\mathrm{src}} < 0$ and $\beta_{rt}^{\mathrm{dst}} > 0$ causes the resource to be consumed in the source cell and generated in the destination cell, effectively transporting it. Losses can be modelled by setting $|\beta_{rt}^{\mathrm{src}}| > |\beta_{rt}^{\mathrm{dst}}|$. Resources other than the transported one may also have non-zero values for β_{rt}^{src} and β_{rt}^{dst}, which allows one to model resource requirements and emissions, for example, transporting biomass by road requires some fuel and results in emissions. These interactions often depend on the distance between the two zones ($l_{ii'}$), and so there is a second set of conversion factors that describe how much resource is required/emitted in each zone *per unit distance*: these are denoted γ_{rt}^{src} and γ_{rt}^{dst}.

The rate of operation of a transport technology between zones i and i' is defined as $\mathcal{Q}_{tii'hdsy}$, a *positive* variable. The net rate of increase of resource r in zone i due to the activity of all transport tasks is given by:

$$Q_{rihdsy} = \sum_{ti'} \left(\beta_{rt}^{\text{src}} + \gamma_{rt}^{\text{src}} l_{ii'}\right) \mathcal{Q}_{tii'hdsy} + \sum_{ti'} \left(\beta_{rt}^{\text{dst}} + \gamma_{rt}^{\text{dst}} l_{i'i}\right) \mathcal{Q}_{ti'ihdsy}.$$

The rate of operation of any transport technology is limited by its maximum rate and also the existence of a transport infrastructure between the two zones of interest. In general, infrastructures can support more than one transport technology, so a constraint is also required to ensure that the maximum capacity of the infrastructure is not exceeded.

The utilization of a transport infrastructure ς from zone i to i' will be equal to the sum of the rates of all transport technologies on that infrastructure:

$$\mathscr{I}_{\varsigma ii'hdsy} = \sum_{t} U_{\varsigma t} \mathcal{Q}_{tii'hdsy} \leq \mathscr{I}_{\varsigma}^{\text{max}},$$

where $\mathscr{I}_{\varsigma ii'hdsy}$ is the utilization of intrastructure ς from zone i to i' at time $hdsy$, $\mathscr{I}_{\varsigma}^{\text{max}}$ is the maximum capacity of intrastructure ς and $U_{\varsigma t}$ is a parameter specifying the utilization of intrastructure ς per unit rate of transport technology t.

Finally, the existence of an infrastructure ς between zones i and i' is represented by the *binary* variable $X_{\varsigma ii'}^{\mathscr{I}}$, which is equal to one for a connection. The rate of each transport technology is therefore also constrained by:

$$\mathcal{Q}_{tii'hdsy} \leq \sum_{\varsigma | U_{\varsigma t} \neq 0} X_{\varsigma ii'}^{\mathscr{I}} \mathcal{Q}_{t}^{\text{max}},$$

where the summation selects the infrastructure, ς, that supports technology t by using the requirement that each transport technology must only be associated with one infrastructure, so that $U_{\varsigma t}$ must have exactly one non-zero value for each t.

The $X_{\varsigma ii'}^{\mathscr{I}}$ variables are *directional* in that a connection from zone i to zone i' does not imply a connection in the reverse direction. A number of simple constraints can be added to the model to account for connections that are bidirectional by default, optionally bidirectional or unidirectional.

Finally, if connections are permitted between all zones, the number of $X_{\varsigma ii'}^{\mathscr{I}}$ variables will be proportional to the square of the number of zones. This combinatorial explosion can be relieved by defining a set of allowed connections, Γ, and removing all binary variables associated with disallowed connections from the model. This can be achieved by using the following constraint:

$$X^{\mathscr{I}}_{\varsigma ii'} = 0 \qquad \forall\,(i,i') \notin \Gamma,$$

which will cause these variables to be removed at the pre-solve stage of the solution algorithm.

9.2.7 *Storage technologies*

Most MILP models have a fairly simplistic approach to resource storage. This model uses the STN approach to model storage using tasks. As can be seen in Figure 9.2, tasks are used to model the process of placing a resource into storage, holding it there, and retrieving it. This allows one to define different efficiencies for each phase, e.g. there may be no loss of energy when adding or removing heat from storage, but energy will be lost over time while the heat is held in storage. The use of tasks also allows one to

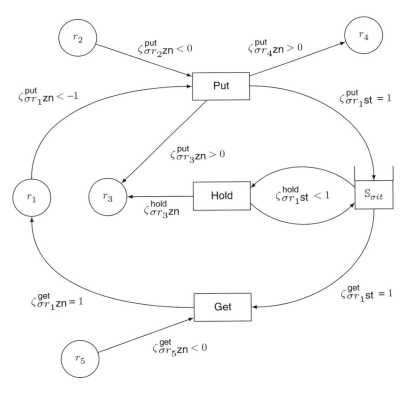

Figure 9.2 An example set of storage tasks to store resource r_1. The 'put' task transfers r_1 from the zone to the store, requiring some r_2 and producing some wastes r_3 and r_4 (e.g. waste heat and CO_2). The 'hold' task maintains r_3 in storage, but at less than 100 per cent efficiency, the losses being converted to r_3. Finally, the 'get' task retrieves r_1 from storage and delivers it to the zone, requiring some r_5.

define resource requirements for each phase, e.g. the use of electricity to power pumps when moving to and from storage, keeping LNG liquefied, and so on.

The three tasks defined for each storage technology, σ, are called 'put', 'hold' and 'get'. Stored resources are considered distinct from resources in the zone (i.e. resources that are immediately available for use by technologies or to satisfy demand): the actions of the 'put' and 'get' tasks are, respectively, to place excess resource production into storage and make stored resource available for immediate consumption; the 'hold' task keeps a given resource in storage, with a possible drain on resources in the zone. The rate of operation of the 'put' task is represented by the variable $\mathscr{S}^{\mathrm{put}}h$, its effect on the resources in the zone is given by the user-defined parameter $\zeta_{\sigma r zn}^{\mathrm{put}}$, the effect on the stored resource is specified through $\zeta_{\sigma r st}^{\mathrm{put}}$ (which must be set to 1 *for the stored resource only, all others being zero*) and the maximum rate of the task is given by the parameter $\mathscr{S}_{\sigma}^{\mathrm{put\,max}}$. With similar variables and parameters defined for the other tasks, a wide variety of situations can be modelled, some examples of which are shown in Figure 9.2 and discussed below.

The operation of all three types of storage task has an effect on the resource balance through the variables S_{rihdsy}, which are the net rates of increase of resource r in the zone due to storage activities, and also on the amount of resource in storage: $\mathbb{S}_{\sigma ihdsy}$. Their effect on S_{rihdsy} is given by:

$$S_{rihdsy} = \sum_{\sigma} \zeta_{\sigma r zn}^{\mathrm{put}} \mathscr{S}_{\sigma ihdsy}^{\mathrm{put}} + \sum_{\sigma} \zeta_{\sigma r zn}^{\mathrm{hold}} \mathscr{S}_{\sigma ihdsy}^{\mathrm{hold}} + \sum_{\sigma} \zeta_{\sigma r zn}^{\mathrm{get}} \mathscr{S}_{\sigma ihdsy}^{\mathrm{get}}. \tag{9.3}$$

The three terms represent the effect of all 'put', 'hold' and 'get' tasks on resource r in the zone (hence the subscript 'zn' in the $\zeta_{\sigma r}$ parameters). The rate of the 'put' task, $\mathscr{S}_{\sigma ihdsy}^{\mathrm{put}}$, multiplied by the conversion factor, $\zeta_{\sigma r zn}^{\mathrm{put}}$, gives the rate of addition of resource r to the zone when placing into storage the resource that technology σ stores. For the stored resource, $\zeta_{\sigma r zn}^{\mathrm{put}} \leq -1$ because it is being 'removed' from the zone in order to be placed into storage; the equivalent conversion factor for the store ($\zeta_{\sigma r st}^{\mathrm{put}}$) should be set to 1. Values of $\zeta_{\sigma r zn}^{\mathrm{put}} < -1$ represent losses, e.g. heat dissipation when charging a battery could be modelled by setting $\zeta_{\mathrm{batt\,ele\,st}}^{\mathrm{put}} = 1$, $\zeta_{\mathrm{batt\,ele\,zn}}^{\mathrm{put}} = -1.05$ and $\zeta_{\mathrm{batt\,wh\,zn}}^{\mathrm{put}} = 0.05$.

The 'get' tasks are almost identical to the 'put' tasks but now $\zeta_{\sigma r zn}^{\mathrm{get}} = 1$ for the stored resource and $\zeta_{\sigma r st}^{\mathrm{get}} \leq -1$ *for the stored resource only, all others being zero.*

For the 'hold' tasks, the efficiency of the storage device is specified by setting $\zeta_{\sigma r st}^{\mathrm{hold}} \leq 1$ for the resource being stored, with all others being zero. $\zeta_{\sigma r zn}^{\mathrm{hold}}$ represent losses, emissions and the resource requirements to maintain the stored resource in storage. For example, a resource requirement might be electricity to maintain the stored resource in a refrigerated state and the emissions would be waste heat from the refrigeration cycle. Another example would be the storage of hot water: if the efficiency is less than 1, then some of the stored heat is lost and becomes waste heat.

Equation (9.3) uses the $\zeta_{\sigma r zn}^{\star}$ parameters to determine the net rate of increase of each resource in the zone due to storage activities. $\zeta_{\sigma r st}^{\star}$ play an equivalent role in defining how the inventory levels, $\mathbb{S}_{\sigma ihdsy}$, of the storage devices, σ, change. This is given by:

$$\frac{\mathbb{S}_{\sigma ihdsy}}{\Delta h_h} = \sum_r \zeta_{\sigma rst}^{put} \mathscr{S}_{\sigma ihdsy}^{put} + \sum_r \zeta_{\sigma rst}^{hold} \mathscr{S}_{\sigma ihdsy}^{hold} + \sum_r \zeta_{\sigma rst}^{get} \mathscr{S}_{\sigma ihdsy}^{get}; \tag{9.4}$$

$$\mathscr{S}_{\sigma ihdsy}^{hold} = \frac{\mathbb{S}_{\sigma ih-1dsy}}{\Delta h_h} \qquad \forall\, h > 1; \tag{9.5}$$

$$\mathscr{S}_{\sigma i1dsy}^{hold} = \frac{\mathbb{S}_{\sigma idsy}^0}{\Delta h_1}. \tag{9.6}$$

These equations determine how the inventory level changes on an hourly basis *for each day type in each season in each yearly interval*. (Note that there is only one non-zero term in each sum over r: that belonging to the resource being stored.) By substituting equation (9.5) into equation (9.4), with the hold efficiencies set to 1, it can be seen that these equations are a finite-difference approximation to a simple energy balance:

$$\frac{d\mathbb{S}}{dt} = \eta_{in}(\text{Flow in}) - \frac{1}{\eta_{out}}(\text{Flow out}).$$

Equation (9.6) represents the initial condition, $\mathbb{S}_{\sigma idsy}^0$, for each day type d, season s and yearly interval y. Although there are $\bar{\bar{d}} \times \bar{\bar{s}} \times \bar{\bar{y}}$ initial conditions, only the initial condition of the first day type in the first season of the first yearly interval, $\mathbb{S}_{\sigma i111}^0$, need be specified because all of the others are coupled.

To determine the initial condition for the next day type, it is necessary to calculate the change in inventory over each day:

$$\delta_{\sigma idsy} = \mathbb{S}_{\sigma i\bar{\bar{h}}dsy} - \mathbb{S}_{\sigma idsy}^0.$$

These can then be used to calculate the change in inventory *over 1 week* for each s and y by multiplying the change in 1 day by the number of occurrences in the week and summing over all day types:

$$\delta_{\sigma isy} = \sum_d R_d \delta_{\sigma idsy}.$$

Finally, the change in inventory over a year is just the sum of the changes in each season, which is the change over a week multiplied by the number of weeks in each season:

$$\delta_{\sigma iy} = \sum_s W_s \delta_{\sigma isy}.$$

These variables are then used to couple the inventory profiles for each day type by defining all of their initial conditions. This is done by the following constraints.

$$\mathbb{S}^{0act}_{\sigma idsy} = \mathbb{S}^{0act}_{\sigma id-1sy} + R_{d-1}\delta_{\sigma id-1sy} \qquad \forall\, d > 1. \tag{9.7}$$

$$\mathbb{S}^{0act}_{\sigma i1sy} = \mathbb{S}^{0act}_{\sigma i1s-1y} + W_{s-1}\delta_{\sigma is-1y} \qquad \forall\, s > 1. \tag{9.8}$$

$$\mathbb{S}^{0act}_{\sigma i11y} = \mathbb{S}^{0act}_{\sigma i11y-1} + Y_{y-1}\delta_{\sigma iy-1} \qquad \forall\, y > 1. \tag{9.9}$$

$$\mathbb{S}^{0}_{\sigma idsy} = \mathbb{S}^{0act}_{\sigma idsy} + \frac{R_d - 1}{2}\delta_{\sigma idsy} + \frac{W_s - 1}{2}\delta_{\sigma isy} + \frac{Y_y - 1}{2}\delta_{\sigma iy}. \tag{9.10}$$

$\mathbb{S}^{0act}_{\sigma idsy}$ is the initial condition, $\mathbb{S}^{0}_{\sigma idsy}$, for the *first* occurrence of day type d in the *first* week of season s of the *first* year of yearly interval y. This is equal to the initial condition of the first occurrence of the previous day type plus the increase in inventory over that day type multiplied by the number of occurrences of that day type (equation 9.7). Similarly, equations (9.8) and (9.9), respectively define the initial conditions at the start of each season and the start of each yearly interval.

Since a series of repeated daily profiles is approximated by a single profile, the inventory levels will not be well represented if each single profile is the first of each series and there is a change in inventory over the profile. Using the fact that the inventory in any one profile is, to a good approximation, equal to the inventory in the first profile plus an integer multiple of the appropriate δ variable (the change in the inventory over one day, week or year), the average inventory levels over all profiles (of the same type) can be calculated. It can be shown that this is exactly equivalent to shifting the inventory levels by $(R_d - 1)\delta_{\sigma idsy}/2$ when averaging over the repeated days in each day type. Similar offsets are required to average over all weeks in a season and all years in a yearly interval. These offsets are used in equation (9.10) to define the initial conditions of the profiles directly modelled in the formulation.

Finally, the available storage capacity limits the amount of resource in storage. The inventory constraints must be applied to the first and last profile of each set of repetitions because it is not known *a priori* whether the inventory will rise or fall in a particular profile. The constraints must therefore involve shifting the inventory levels once more to obtain their values in the first and last occurrence of each day type. Similarly, the first and last weeks of each season and the first and last years of each yearly interval must be considered. The capacity constraints below use all possible combinations to ensure that the inventory does not exceed the available capacity in any scenario.

$$\mathbb{S}_{\sigma ihdsy} \pm \frac{R_d - 1}{2}\delta_{\sigma idsy} \pm \frac{W_s - 1}{2}\delta_{\sigma isy} \pm \frac{Y_y - 1}{2}\delta_{\sigma iy} \leq N^{S}_{\sigma iy}\mathbb{S}^{max}_{\sigma} \ \forall\, (\sigma, \bar{a})|\zeta^{put}_{\sigma\bar{a}st} = 1,$$

where $\mathbb{S}^{max}_{\sigma}$ is the capacity of a single storage technology σ, $N^{S}_{\sigma iy}$ are integer variables representing the number of storage technologies σ present in zone i during yearly interval y and $\bar{a}$ is an index representing all of the resources that can be stored. The domain specification for this constraint selects the storage technologies capable of storing each storable resource, $\bar{a}$.

As with conversion technologies, the number of storage facilities in each zone can change due to investments and retirements:

$$N^S_{\sigma iy} = N^S_{\sigma iy-1} + N^{S+}_{\sigma iy} - N^{S-}_{\sigma iy} \qquad \forall\, y > 1,$$

where $N^{S+}_{\sigma iy}$ is the number of technologies of type σ built in zone i in yearly interval y and $N^{S-}_{\sigma iy}$ is the number of technologies retired.

9.2.8 Performance metrics and objective function

The objective function is a weighted sum of a number of performance metrics, μ, which may include cost and various environmental impacts:

$$\min Z = \sum_{\mu} w_{\mu} \left(C^I_{\mu} + C^E_{\mu} + C^{\mathscr{P}}_{\mu} + C^{\mathscr{Q}}_{\mu} + C^{\mathscr{I}}_{\mu} + C^S_{\mu} \right),$$

where w_{μ} is the weight for metric μ and the terms inside the summation are the impacts due to, respectively: imports, exports, conversion technologies, transport technologies, infrastructure utilization and storage.

These impacts are defined in terms of the various rate variables in the model and a set of parameters for each impact: $c^I_{\mu\rho r}$, $c^E_{\mu\rho r}$ etc. The index ρ represents the dependence of the indicator: T indicates dependence only on the number of technologies present; R, the rate of a technology; D, the distance between two zones and RD indicates dependence on both the rate and the distance.

The metrics associated with imports and exports are:

$$C^I_{\mu} = \sum_{rihdsy} c^I_{\mu Rr} I_{rihdsy} \Delta h_h R_d W_s Y_y;$$

$$C^E_{\mu} = \sum_{rihdsy} c^E_{\mu Rr} E_{rihdsy} \Delta h_h R_d W_s Y_y.$$

These variables only depend on the rate of import and export, hence only the $c^I_{\mu Rr}$ and $c^E_{\mu Rr}$ parameters are needed. These are typically positive numbers, so that importing resources will increase the metrics (e.g. the cost and environmental impacts will increase due to imports). If a valuable resource is exported, then this would have the effect of reducing the total cost, due to revenues generated, so that $c^E_{\mu Rr}$ for $\mu =$ 'cost' would be negative.

The metrics associated with resource-conversion technologies are:

$$C^{\mathscr{P}}_{\mu} = \sum_{pihdsy} c^{\mathscr{P}}_{\mu Rp} \mathscr{P}_{pihdsy} \Delta h_h R_d W_s Y_y + \sum_{piy} c^{\mathscr{P}}_{\mu Tp} N^+_{piy}.$$

The first term represents the total operational impact of the conversion technologies, $c_{\mu Rp}^{\mathscr{P}}$ being the impact per unit rate of operation, and the second term is the impact due to the construction of the technology itself, for example, $c_{\text{cost}Tp}^{\mathscr{P}}$ would be the capital cost of a single unit and $c_{CO_2Tp}^{\mathscr{P}}$ would be the life-cycle CO_2 emissions due to the manufacture of a single unit.

The metrics for transport technologies are:

$$C_\mu^{\mathscr{Q}} = \sum_{tii'hdsy} c_{\mu RDt}^{\mathscr{Q}} l_{ii'} \mathscr{Q}_{tii'hdsy} \Delta h_h R_d W_s Y_y,$$

where $c_{\mu RDt}^{\mathscr{Q}}$ is the impact per unit rate of transport technology, per unit distance between the zones.

Whenever a transport task operates, it may also cause impacts due to the utilization of the supporting infrastructure (e.g. a maintenance cost). There will also be impacts due to the construction of the infrastructure. These are given by:

$$C_\mu^{\mathscr{I}} = \sum_{\varsigma ii'hdsy} c_{\mu RD\varsigma}^{\mathscr{I}} l_{ii'} \mathscr{I}_{\varsigma ii'hdsy} \Delta h_h R_d W_s Y_y + \sum_{\varsigma ii'} c_{\mu D\varsigma}^{\mathscr{I}} l_{ii'} X_{\varsigma ii'}^{\mathscr{I}}.$$

The impacts due to storage are given by:

$$C_\mu^S = \sum_{\sigma ihdsy} \left(c_{\mu R\sigma}^{\mathscr{S}\text{put}} \mathscr{S}_{\sigma ihdsy}^{\text{put}} + c_{\mu R\sigma}^{\mathscr{S}\text{hold}} \mathscr{S}_{\sigma ihdsy}^{\text{hold}} + c_{\mu R\sigma}^{\mathscr{S}\text{get}} \mathscr{S}_{\sigma ihdsy}^{\text{get}} \right) \Delta h_h R_d W_s Y_y$$

$$+ \sum_{\sigma iy} c_{\mu T\sigma}^S N_{\sigma iy}^{S+}.$$

The first three terms are due to the operation of the three different storage tasks; the last term is the impact due to investment into new storage facilities.

9.3 Planning case study

In this case study, the TURN framework is used to optimize the energy provision for a small eco-town. The site considered here, which just one part of the planned development, is an area of 87 hectares including homes for 6,500 people and other amenities. The proposed plan, shown in Figure 9.3, comprises 39 zones, most of which are dedicated to housing and a few are mixed purpose. For this study, these 39 zones have been aggregated into nine larger zones (as indicated by the heavy outlines and the numbers in the figure) in order to simplify the analysis.

The number and types of building in each zone are known and used to estimate the likely demands for services in each of the zones, assuming UK average figures for the types of building present. Water and space heating have been considered together and demand for natural gas has not been considered (this would be purely for cooking as the model must decide how to meet the demands for space and water heating).

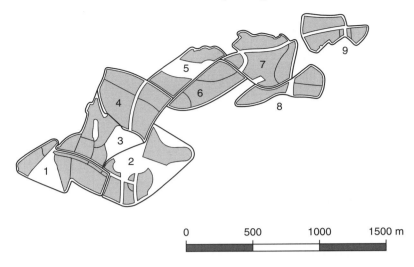

0   500   1000   1500 m

Figure 9.3 Eco-town layout with aggregated zones.

Table 9.3 Definition of hourly intervals

Time interval h	Duration (Δh_h)	Period of the day
1	2	07:00–09:00
2	3	09:00–12:00
3	1	12:00–13:00
4	5	16:00–18:00
5	4	18:00–22:00
6	9	22:00–07:00

The hourly demand profiles are determined for the intervals shown in Table 9.3. Three typical days are considered: one for summer, mid-season and winter. The summer season comprises 92 identical days; mid-season 153 days; and winter 120 days (Weber *et al.* 2010). The time horizon for a whole year is approximated by using the hourly time intervals (*h*) shown in Table 9.3 and by setting $\Delta d_1 = 7$, $W_1 = 13$, $W_2 = 22$, $W_1 = 17$ and $Y_1 = 1$: that is, each week comprises seven identical days; the summer season lasts 13 weeks, mid-season 22 and winter 17; and there is one annual interval of just 1 year.

The demands for the whole eco-town are shown in Figure 9.4. It can be seen that the demands are predominantly domestic and the total electricity demands are assumed to be independent of the season.

9.3.1 Resources and technologies

The problem is to satisfy as much of the demand for heat and electricity using two biomass resources, forestry residue and wood chips, with the surplus covered by grid electricity and natural gas. The forestry residues, which can be imported or collected

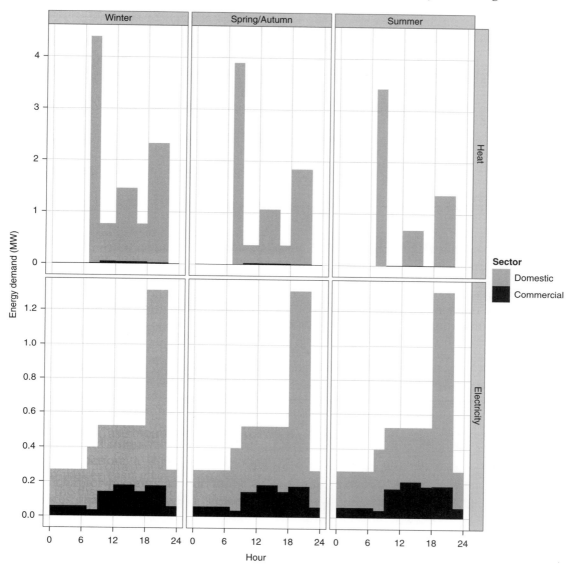

Figure 9.4 Total demands for heating and electricity in the eco-town.

locally, can be converted into wood chips. In turn, the chips, which may also be imported, are used to generate district heat and electricity in a variety of CHP plants. The CHP plants also produce a 'waste heat' resource. The district heat can then be transported throughout the eco-town and converted to heat in domestic and commercial heat exchangers. Heat can also be generated in domestic chip and natural-gas boilers. The resources and technologies modelled are shown in Figure 9.5 and their main features are given in Table 9.4.

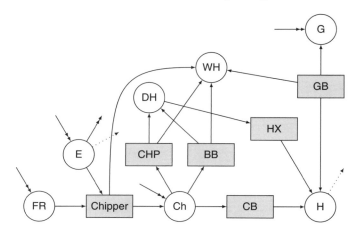

Figure 9.5 A simplified TURN for the planning case study. The resources (circles) are: forestry residue (FR), electricity (E), district heat (DH), wood chips (Ch), waste heat (WH), natural gas (G) and heating (H). The technologies (rectangles) are: Chipper, CHP (of various types and sizes), backup boiler (BB), domestic chip boiler (CB), heat exchanger (HX, commercial and domestic) and domestic natural gas boiler (GB). The double-headed arrows represent the possibility of importing or exporting a resource; the dotted lines represent demands for a resource.

Table 9.4 Costs and efficiencies of the biomass resource conversion technologies

Technology	Size	Efficiency		TKC (k£)	O&M (k£/year)
		η_{el}	η_{th}		
Chipping plant	5 t/h			250	37.5
Domestic boiler	25 kW$_{th}$	—	82%	6	0.5
ORC-small	0.5 MW$_{el}$	18%	78%	2,000	80
ORC-medium	1 MW$_{el}$	19.5%	78%	3,400	120
ORC-large	2 MW$_{el}$	20%	78%	6,400	220
ICE-small	0.5 MW$_{el}$	24%	50%	1,750	75
ICE-medium	1 MW$_{el}$	25%	50%	3,000	140
ICE-large	2 MW$_{el}$	26%	50%	6,000	260
Back up boiler	0.1–1 MW$_{th}$	—	85%	20–100	0*

Notes: ORC = organic Rankine cycle, ICE = internal combustion engine, TKC = turn-key cost, O&M = operating and maintenance costs,
*included in the O&M costs of the CHP.
Data source: Keirstead *et al.* (2012b).

Storage technologies

A closed wood chip storage facility is included in the case study. It has a capacity of 20 kt, an annualized capital cost estimated to be £210,000 and annual operating costs of £48,000. Since it is impractical to import biomass during all intervals, the upper bounds on the I_{rihdsy} variables for forestry residue and chips are set to zero for all but the first hourly interval to ensure wood chips are imported only in the first hourly interval of every day.

The second storage technology is a dummy technology needed to allow the 'waste heat' resource to accumulate.

Transport technologies

Both types of biomass can be transported by road, requiring diesel, which can also be transported by road. Dedicated networks are allowed for electricity, gas and district heat.

9.3.2 Objective function

The objective is to minimize the total annualized cost of satisfying the demands for heat and electricity. The cost of electricity production by the biomass CHPs also includes a reduction due to Renewable Obligation Certificates (ROCs), a UK policy incentive for renewable energy production. The costs are reduced by £108.74 for every megawatt-hour of electricity produced, which is based on a 2008–9 ROC price of £54.37/ROC and 2 ROC/MWh for 'dedicated biomass with CHP' technologies (Ofgem 2010).

9.3.3 Results

The results indicate that a bioenergy strategy for the eco-town is promising. Table 9.5 shows the technologies used to meet the heat demand in each zone, indicating that all of the demand is satisfied by one medium-sized ORC CHP and backup boiler. A domestic

Table 9.5 Technologies present in the optimal solution

Zone	Technology	Number
1	Domestic heat exchanger	308
	Commercial heat exchanger	1
2	Domestic heat exchanger	435
3	Domestic heat exchanger	448
	Commercial heat exchanger	1
4	1 MW ORC CHP	1
	0.5 MW backup boiler	1
	Covered wood chip storage	1
	Domestic heat exchanger	397
5	Domestic heat exchanger	175
6	Domestic heat exchanger	220
	Commercial heat exchanger	1
7	Domestic heat exchanger	474
8	Domestic heat exchanger	310
9	Domestic heat exchanger	375
	Commercial heat exchanger	1

heat exchanger is used to connect each domestic property to the district heating network. Since no gas-fired boilers are used, there is no import of natural gas into the eco-town.

The rate of heat generation is shown in Figure 9.6, where it can be seen that the backup boiler is only used during the winter peak to provide about 10 per cent of the heat. Electricity production and import is shown in Figure 9.7. The morning peak in the heat demand is the only time where the CHP can provide all of the electricity for the eco-town; at other times, the demands are predominantly satisfied by grid imports. Note that in this case study, the CHP is not permitted to over-produce heat in order to satisfy demand for electricity. Also, more electricity is generated than the actual demands because there are losses in the network and the storage and transport technologies require electricity to operate (e.g. pumps for district heating).

Wood chip storage is required because deliveries (imports) are only allowed in the first interval of each day and a continuous supply of wood chips is required for the CHP to operate throughout the day. The inventory of wood chips is shown in Figure 9.8, where the 07:00–09:00 interval represents the initial inventory for each day ($\mathbb{S}^0_{\sigma idsy}$), and the amounts[2] imported and consumed in each interval are shown in Figure 9.9. These figures also show the effect of the constraint on $\delta_{\sigma idsy}$: the total consumption is equal to the import for each day and the inventory returns to its initial value (zero) at the end of each day.

Figure 9.10 shows the operation of the district heat and electricity networks for intervals 1 and 5 in the winter season, corresponding to the peaks in heat and electricity demands. The arrows indicate the direction and magnitude (width of the arrow) of

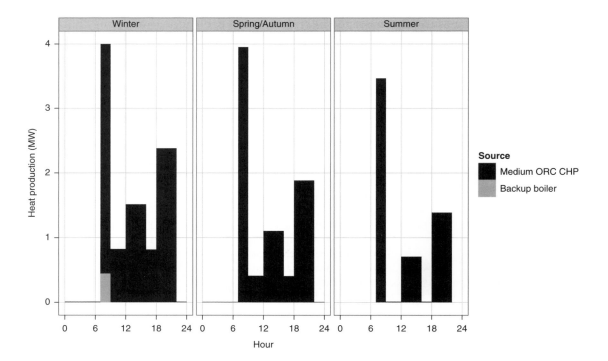

Figure 9.6 Total rate of heat generated.

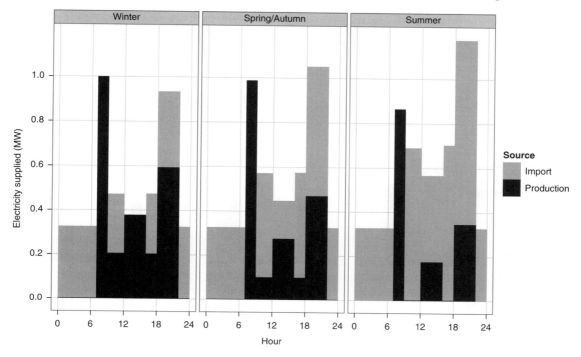

Figure 9.7 Total electricity supplied.

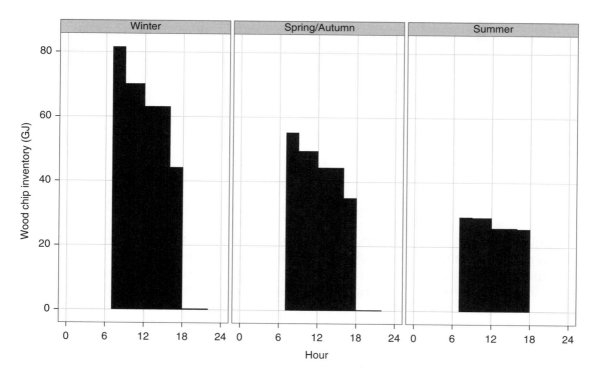

Figure 9.8 Inventory levels of wood chips at the end of each time interval.

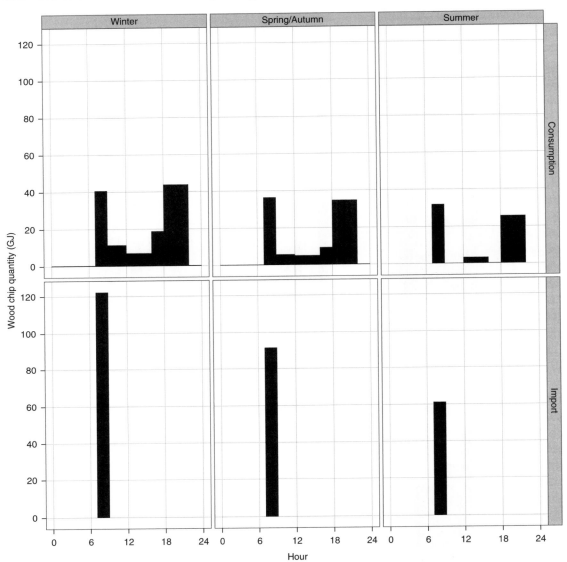

Figure 9.9 Amount of wood chips imported and consumed in each time interval.

the flows. Figure 9.10a corresponds to interval 1, where the heat demands are high and this is reflected in the heat network, on the left, where the flows are large; the electricty demand is lower in this interval and so the flows are quite small in the network on the right. There is no import of electricity as the heat demands are high enough for the CHP (in zone 4) to generate all of the required electricity.

Figure 9.10b shows the same network in operation during interval 5. The district heat network looks similar to interval 1 but with lower flows. The lower heat demands result in the CHP only being able to supply electricity for zones 1–4, with the remaining demand being fulfilled by imports, shown by the short bold arrow entering zone 8.

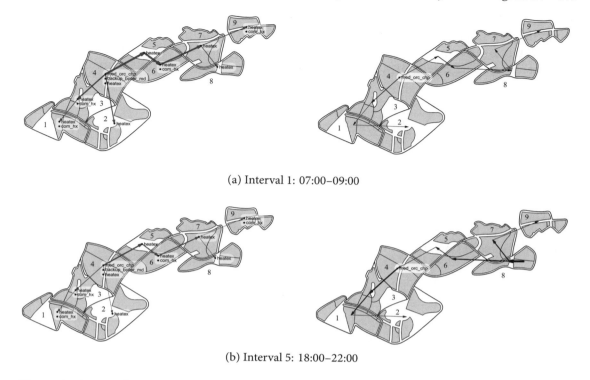

(a) Interval 1: 07:00–09:00

(b) Interval 5: 18:00–22:00

Figure 9.10 Distribution networks for district heat (left) and electricity (right) during the winter season.

9.4 Retrofit case study

The TURN formulation is suitable for both green-field studies, where a new urban energy system needs to be designed, and brown-field or retrofit studies where there is an existing urban energy system that needs to be augmented with new technology. Retrofit studies can be performed simply by fixing the initial state of the urban energy system: that is, the numbers of technologies present in the first yearly interval are set to correspond to the existing urban energy system and the optimization will be free at any time to add further technologies or remove existing ones.

In this case study, a more complicated situation is considered, where the retrofits include building refurbishments. Since retrofits of this type will change the demands for energy (e.g. adding loft insulation will reduce demand for heat) and the TURN formulation presented above assumes a fixed set of demands, it is necessary to extend the formulation to account more explicitly for this type of retrofit.

9.4.1 Modelling retrofits

The central constraint in the TURN framework is the resource balance (9.1). In order to model the effect of retrofitting building stock, the demands, D_{rihdsy}, are now considered as state variables that can be reduced by investment in retrofit technologies, indicated by the index j. The effects of these technologies on the demands are then expressed using the parameter ΔD_{rj}, which is the amount of additional demand for resource r generated by the presence of technology j in tenure type e. ΔD_{rj} will be negative in most cases. The effect of all retrofit interventions on the existing demands is then given by:

$$D_{rihdsy} = D^0_{rihdsy} + \sum_j \sum_e N_{jeiy} \Delta D_{rj},$$

where D^0_{rihdsy} are the predicted demands in the absence of any refurbishment interventions, and N_{jeiy} are the number of retrofit technologies in place in yearly interval y.

It may also be desirable to limit the rate of retrofits to a given value. This can be achieved by including the following constraint:

$$\sum_j \sum_e N^+_{jeiy} \leq H_{iy} \widehat{N}^+ Y_y,$$

where H_{iy} is the number of buildings in zone i, and $\widehat{N}^+$ is the maximum annual rate of building stock retrofit. A similar inequality constraint can be used for the minimum annual rate, if one wants to drive the solution towards an increasing deployment of refurbishments. Examples of empirical rates are provided in Table 4.2.

A final constraint of note is that of the available finance. The index e is included to indicate the building tenure for a particular zone i. Essentially, a tenure is the right to occupy a building: the two main types are tenancy and owner occupancy (shared ownership is a less common option). The owner of the building may be the occupier, a private landlord or a public landlord. For example, in Great Britain residential buildings are mostly owner-occupied (68 per cent), with the rest being rented from a social landlord/local authority (18 per cent) or privately rented (15 per cent). In this model, each building tenure will have a separate budget which limits the total costs, C^{ref}_{eiy}, in terms of technological expenditure, as below:

$$C^{ref}_{eiy} = \sum_j c^{ref}_{costTje} N^+_{jeiy}$$

$$C^{ref}_{eiy} \leq \widehat{C}^{ref}_{eiy},$$

where $\widehat{C}^{ref}_{eiy}$ is the maximum budget. A similar inequality constraint may be enforced for the minimum budget. A simpler method to include the building tenure is to aggregate tenures to particular zones i of building type. For instance, all the owner-occupiers who live in terraced/row housing will be included in one zone. This simplification will lose some of the spatial benefits of the TURN framework, however, unless the tenures do fit into the disaggregate zonal representation.

9.4.2 *Analysis and results*

The case study is based on a simple research question: 'Given an existing building stock of over 200 residences, composed of different tenures, what is the optimal ordering of investments that should be made for efficient refurbishment?'. The answer may appear straightforward if ignoring the sociological barriers discussed in section 4.3.2. Yet the question is made slightly trickier when the majority of buildings were built before cavity walls became a mainstream feature of construction in the UK. In the absence of cavity wall insulation as an option, what demand-side measures should be prioritized?

There are 205 residential buildings in this case study, of which about 50 per cent are owner-occupied, 30 per cent are rented from private landlords and the rest are rented from public landlords. The area in question has set a target of 40 per cent greenhouse gas emissions reductions by 2020 relative to a 2005 baseline. Most of the building stock is in relatively poor condition, with the majority being built before the 1930s. The current annual rate of refurbishment lies between 0.8 per cent and 1.4 per cent per year with various retrofit measures being installed, but is thought to be too low to achieve the 2020 target. Parameters from national and local databases on building demand and size have been used to develop a model in the TURN framework (Jennings *et al.* 2011a,b). As with the previous case study, there are six hourly periods per day and three seasons per year. In addition, the investments may take place during any of 96 monthly intervals from 2013 to 2020 inclusive. The objective function is to minimize the gap between the actual emissions and the 2020 target. The constraints on finance are given by maximum budgets for each type of tenure. The worst case can similarly be modelled by inverting the objective function, that is maximizing the gap between actual emissions and the targets.

In this case study, there are two main ways in which the building stock's emissions may be reduced. Existing inefficient supply-side technologies may be replaced by more efficient technologies, and existing building thermal envelopes may be improved by installation of demand-side options. A small set of demand-side interventions has been used in this case study: external wall insulation, loft insulation and double glazing. Note that large-scale heating supply has not been included here, as the seasonal nature of the residential heat demand is assumed to impact upon the business case for combined heat and power, and there are no large swimming pools/industrial centres in this case study to provide a steady baseload demand for heat.

There are two objective functions considered for this study, those of minimization and maximization of the gap between actual and targeted emissions. Figure 9.11 shows that the uptake of external wall insulation and loft insulation is greatest, as expected, when emissions are being reduced. The allowable number of interventions (set by the number of buildings without interventions at initialization) is fulfilled by 2015. When optimizing for a maximization of emissions, some uptake of loft insulation by private landlords is induced mainly on account of fulfilment of the constraints on typical rates of refurbishment.

Figure 9.12 illustrates the temporal property of GHG emissions due to space and water heating in residential buildings. Reflecting the residential heating demand profiles illustrated in Table 4.4, aggregated emissions peak and trough at particular periods

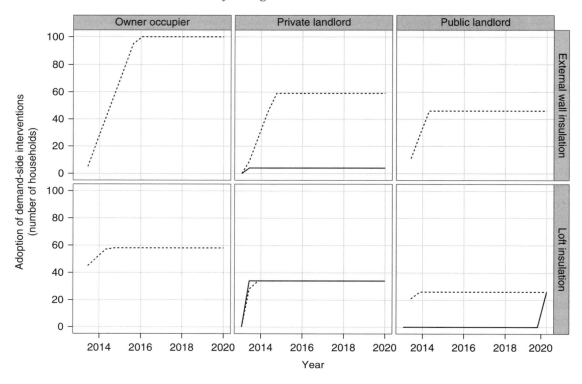

Figure 9.11 Number and timing of retrofit interventions for the two objective functions, by building tenure. Objective functions indicated by line-type: solid = maximize emissions gap, dashed = minimize emissions gap.

of the year. The emissions reductions for the minimization of emissions can be seen in Figure 9.12 as reaching their lowest point by 2015, whereas emissions under the maximization objective do not change noticeably.

These results can be informative for building owners and public/private landlords. It is a strength of this model that large-scale building retrofits, based on local data (i.e. a bottom-up framework), can be modelled for expected changes in years to come. Nonetheless, there remains work to be done in improving the fidelity of the model: two examples being to assess the learning rates of contractors at the large-scale and to improve the modelling accuracy of the financing of the types of refurbishment contract in each tenure.

9.5 Conclusion

There have been many models developed to optimize urban energy systems. A large proportion of these focus on conventional energy pathways and some more recent models have begun to consider renewables, such as wind and biomass. However, few models have been developed that are completely generic: even those that consider many

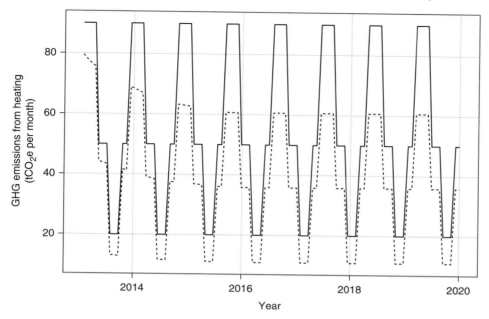

Figure 9.12 Greenhouse gas (GHG) emissions from residential heating demand, by objective function. Objective functions indicated by line-type: solid = maximize emissions gap, dashed = minimize emissions gap.

forms of energy contain specific equations for each type of energy-generation technology, which limits their extensibility.

In this chapter we have presented an MILP model that we believe is truly generic. It is based on resources that can represent energy or materials required or generated by a city, and technologies that can convert one set of resources to another or that transport or store resources. These fundamental building blocks allow the user to model any urban resource system desired, which can be done simply by defining any new resources and the technologies that interact with them, without the need to modify the model equations. Another feature of the model is its use of an efficient hierarchical non-uniform time discretization. This allows large time horizons to be considered without the need for a large number of time intervals, by exploiting whatever periodicity there is in the demand data.

The model was illustrated using a small eco-town in the UK, focusing on biomass as the primary energy source. It was shown that biomass CHPs can provide all of the heating needed for the eco-town but that electricity needs to be imported from the grid during times of high demand (especially if the heating demand is low). This solution illustrates the model's ability to assess complex energy pathways with spatial and temporal variation.

The TURN model was also applied to a large-scale retrofit project. It could be seen that uptake of low carbon interventions such as external wall insulation was directly related to the objectives, and the building owner's budget. The rate of retrofitting labour available is an important consideration for such schemes, as is the effect of building tenure on finance available for such retrofits.

There are a number of future directions for the model. First, more case studies shall be undertaken with a view to establishing the size of problem that can be solved. Second, the current model considers domestic technologies in a slightly limited way. There are two avenues currently under consideration: the first involves modifying the equations specifically to include domestic conversion and storage technologies in a more general and realistic manner; the second is to use the flexibility of the resource-technology network to segregate domestic and commercial activities without the need to modify the formulation. An example of the latter approach would be to define two resources for electricity: domestic electricity and grid electricity. Domestic users could install either a technology that converts grid electricity to domestic electricity (with no costs whatsoever) or install domestic technologies that generate domestic electricity. Similarly, domestic storage technologies would only store domestic electricity. In this way, domestic technologies are prevented from behaving unrealistically. This latter approach would, however, require more resources and technologies, hence a larger model, but would be a useful test of the flexibility of this framework.

Finally, future work is planned on the arrangement of realistic financing by tenure, with a particular interest on optimizing available finance by type of development and tenancy arrangements. For example, a retrofit bond could be of interest to public landlords, and the temporally dynamic provisions could be incorporated into the existing TURN framework. The optimal levels of finance available for private landlords and property developers, subject to the fees of creditors and investors, should be analysed and stochastics applied to appropriate financial parameters. Capital management for retrofits, building on the powerful representation of urban energy systems provided by TURN, has the potential as a robust financial decision-making tool in the future.

Notes

1 Here, $\bar{\bar{s}}$ is used to represent the cardinality of the set whose index is s. This avoids the need to define an extra symbol to represent the set itself and confusion between the usual notation for set cardinality and the notation for absolute value, both of which are $|.|$.

2 The amounts are calculated by multiplying the appropriate rate variables, I_{rihdsy} and $\mathscr{S}^{get}_{\sigma ihdsy}$, by the duration of each interval, Δh_h.

10 Ecologically-inspired optimization modelling of urban energy systems

Nicole C. Papaioannou

Imperial College London

10.1 Introduction

As discussed in Chapter 1, over half of the world's population now lives in cities and this fraction is expected to rise in the coming decades. The trend raises questions about the sustainability of urban settlements and the extent to which they will be able to 'meet the needs of the present without compromising the ability of future generations to meet their own needs' (WCED 1987). To address this challenge, urban governments are looking for ways to measure the state of their city and to evaluate progress against desired targets, such as reducing the environmental impact of an urban area.

Naturally, the energy sector plays a pivotal role in attempts to achieve sustainable urban development. It has been estimated that 75 per cent of global energy consumption occurs in cities and 80 per cent of greenhouse gas emissions also come from them (UN Habitat 2007). Approximately half of these emissions result from the burning of fossil fuels for transport and the other half arises from energy use in buildings and appliances – both these practices being necessary to maintain the human quality of life in urban systems. Climate change, sustainable development and urbanization therefore go hand-in-hand and a better understanding of urban energy systems and their impact on the environment is consequently at the forefront of urban government agendas.

This chapter explores how modelling frameworks inspired by natural ecology can help us understand the sustainability of urban systems, and energy systems in particular. After first discussing the similarities between cities and ecological systems, a review of relevant environmental impact assessment methods is presented, before demonstrating how these methods can be combined with optimization techniques to assess the life-cycle impacts of an urban energy system.

10.2 Cities as eco-systems: the concept of urban metabolism

Cities can be thought of as having a metabolism, like plants or animals (Chapter 2). The first urban metabolism study was performed by Wolman (1965), triggered by the rapid expansion of urban America at the time and concerns about the difficulties of providing an adequate water supply to an average US mega-city. His article analysed the metabolism of a hypothetical American city, quantifying the overall fluxes of resources and wastes into and out of an urban region of 1 million people. According to Wolman (1965: 179), 'the metabolic requirements of a city can be defined as the materials and commodities needed to sustain the city's inhabitants at home, at work and at play... The metabolic cycle is not completed until the wastes and residues of daily life have been removed and disposed of with a minimum of nuisance and hazard'. Although he focused largely on water, as the input required in the greatest quantities, estimates for food and fossil energy inputs were also calculated, as well as those fluxes associated with chosen outputs such as refuse and air pollutants.

Subsequent to Wolman's work, more studies were conducted around the world and, over several decades, a body of literature has emerged highlighting the significance of urban metabolism. One of the earliest attempts was that of ecologists Duvigneaud and Denaeyer-De Smet (1977) who produced a study of Brussels, Belgium. They took into consideration the detailed quantification of urban biomass, including even organic discharges from cats and dogs. Newcombe *et al.* (1978) studied the city of Hong Kong, determining the flows in and out for construction materials and finished goods. This particular study was further elaborated upon more recently by Warren-Rhodes and Koenig (2001), demonstrating that the per capita consumption of food, water and materials had risen, respectively, by 20, 40 and 149 per cent in the period between 1971 and 1997. Newman (1999) also portrayed growing trends in per capita resource inputs and waste outputs while observing the city of Sydney, Australia. More recently, Sahely *et al.* (2003) reported that while studying the city of Toronto, some per capita outputs (namely, residential solid waste) had decreased between 1987 and 1999, even if most inputs to this North American city were constant or increasing. Additional metabolism studies of cities include those for Tokyo (Hanya and Ambe 1976), Miami (Zucchetto 1975), nineteenth-century Paris (Stanhill 1977), Greater London (Girardet 1995, CIWM 2002), Vienna (Hendriks *et al.* 2002) and part of the Swiss Lowlands (Baccini 1997).

Together, these studies provide a quantitative approach to describing the metabolism of a diverse set of global cities, and their respective changes over time. Generally, the metabolism of cities is analysed in terms of the four fundamental flows: those of water, materials, energy and nutrients. Variations in the flows can be expected from city to city because of age, the level of economic development, the availability of technologies, cultural factors and, in the case of energy flows in particular, factors such as climate or urban population density (Kennedy *et al.* 2008). Such review studies show that cities generally exhibit increasing per capita metabolism over time with respect to water, wastewater, energy and materials. More importantly, urban metabolism has shown itself to be a valuable tool for pin-pointing the metabolic processes that potentially threaten the sustainability of cities. These include changes in ground water levels, the depletion of local materials, the regular and irregular accumulation of toxic materials and nutrients and the urban heat island effect. By understanding urban metabolism, urban policy-makers are able to better comprehend the extent to which local resources are approaching exhaustion and hence, devise relevant strategies to delay exploitation.

10.3 Environmental impact assessment methods

When seeking to understand the detailed impacts of an urban energy system, descriptive techniques such as urban metabolism are often inappropriate. While the technique provides an excellent overview of the energy flows needed to sustain a city, it does not allow the researcher or policy-maker to understand the potential environmental impacts of alternative energy system configurations, e.g. the difference between a district heating system based on biomass sources and one based on natural gas. Although a vast number of techniques exist to measure the environmental 'friendliness' of a system, three environmental impact assessment (EIA) methods were chosen for the purposes of this chapter:

- Material Flow Analysis (MFA)
- Ecological Footprinting (EF)
- Life-Cycle Assessment (LCA).

These methodologies can be applied to urban energy systems individually or in combination, in order to indicate what is environmentally acceptable for different technologies and scenarios. Ultimately, the aim is to give a recommendation for the design of a specific urban energy system configuration, one which is not economics-oriented but rather puts an emphasis on environmental impact. In this review, we hope to highlight some of the practical issues associated with these methods, such as the types of data required, the feasibility of obtaining this information, the ease of applying the method and the technique's general relevance to the study of urban energy systems.

10.3.1 *Material Flow Analysis*

Material Flow Analysis (MFA) is defined by the holistic analysis of the inputs and outputs of process sequences, including material extraction or harvest, chemical transformation, manufacturing, consumption, recycling and disposal of materials. The method relies upon accounts, measured in physical units (typically tonnes), that quantify the throughput of such processes. Chemical substances, e.g. carbon or carbon dioxide, can be accounted for in this way, as well as natural and technical compounds, such as coal and wood. MFA carries a clear resemblance to financial accounting and to traditional economic principles. Although there are subtle differences in approach depending on the research questions at hand, the concepts of an industrial metabolism and mass balancing (i.e. ensuring that the net mass in and out of a system is zero) provide a common basis for MFA studies.

Policy-makers are becoming more and more familiar with the benefits of MFA and related techniques (see Box 10.1) in their decision-making processes. The aspiration in such analyses is to identify resource-efficient system designs, comprised of consistent and minimized physical exchanges between the system and its environment, with the internal material cycles being driven by renewable energy flows (Richards *et al.* 1994).

Box 10.1 MFA by any other name

There are several different 'flavours' of material flow analysis and, while the underlying principles are similar, the focus of the study varies, as described below.

Substance Flow Analysis (SFA) Studies of the material flow of a specific substance are often known as substance flow analyses. The substance of interest is typically an environmental pollutant, like lead or, in the case of fossil-fuelled energy systems, carbon dioxide. By determining the main entrance routes of such substances to the environment, understanding the processes associated with their emissions, the flows within the systems in which they exist, and taking into consideration biochemical and physical changes in the environment, one can quantitatively assess the risks linked to such substance-specific outcomes.

Life-Cycle Assessment (LCA) If the main concern is the environmental impact of a certain product or service, then MFA may be used as a vital step within a wider LCA. The system boundary for an LCA is broad, encompassing all aspects of the anthroposphere, environment, technosphere or physical economy related to the product in question. It should be noted that shifting from single substances to products containing multiple substances causes an increase in the complexity of the analysis due to additional potential impacts.

Urban metabolism A very significant and increasingly popular domain of MFA is depicting and analysing the metabolism of cities, regions and national or international economies. As described above, the distinguishing feature of the urban metabolism approach is its use of a system boundary of direct relevance to a city (rather than a discrete process as in MFA), although recent research is beginning to expand beyond this strict geographic interpretation (Kennedy *et al.* 2009). This type of study is becoming an essential tool for the evaluation of the sustainability status and trends within cities.

To achieve these goals, the strategies of 'detoxification' and 'dematerialization' of the industrial and urban metabolism are typically adopted. The former refers primarily to pollution reduction, wherein steps are taken to mitigate the releases of particular substances to the environment. These have been primarily controlled using governmental regulations banning certain substances, and additionally by introducing cleaner technologies. Solutions are typically provided to address the needs of specific regions and relatively short-term periods (i.e. years). However, in the case of climate change and greenhouse gas emissions, the drive for long-term and global solutions creates the need to analyse flows of critical substances, materials or products in a systems-wide approach, incorporating cradle-to-grave techniques.

The second strategy, dematerialization, relies on the principles of resource efficiency. By controlling the quantity of primary resources 'ingested' by the urban and industrial systems, the method aims to tackle the problem at its root. The desire for eco-efficiency includes not only major inputs (e.g. materials, energy, water) but, at the same time, it takes into consideration major outputs to the environment (e.g. emissions to air, water, waste) and associates them to the products or services produced (EEA 1999, OECD 1998).

MFA highlights the importance of acquiring data about industrial and urban processes. In order to conduct an MFA and validate any subsequent improvements in the environment, the human-induced material flows through the urban system must be monitored. These data collection systems help researchers evaluate the balance of anthropogenic and natural flows in the economy and environment, but also help policy-makers to demonstrate the cost efficiency and effectiveness of their policies. Consequently, a recent review study (Bringezu 2000) concluded that the results and findings of MFA-related studies have been and currently are used in environmental protection policies:

- to support public debate on environmental goals and targets, particularly with respect to resource and eco-efficiency matters, and the meshing of environmental and economic policies
- to compile economy-wide material flow accounts for use in official statistics databases
- to progress monitoring through the derivation of sustainability indicators.

10.3.2 *Ecological footprinting*

There is strong agreement between natural and social scientists that sustainability is directly related to the maintenance of natural capital (Wackernagel *et al.* 1999). This includes species, ecosystems and other biophysical entities. Indicators such as the ecological footprint attempt to provide a simple framework for the accounting of natural capital and the formal definition of the ecological footprint is: 'the total area of productive land and water required continuously to produce all the resources consumed and to assimilate all the wastes produced by the defined population, wherever on Earth that land is located' (Rees and Wackernagel 1996: 228).

The concept of ecological footprints is based on material and energy flow accounting. If a city is thought of as having an 'industrial metabolism', then it can be compared with a large animal grazing its pasture. Even though the city uses up resources, all the energy and matter is returned to the environment. So, as Rees and Wackernagel (1996: 228) state, the question becomes: 'How large a pasture is required to support the city indefinitely – to produce all its "feed" and to assimilate all its wastes sustainably?'.

In the case of energy systems, a key consideration is the linkage between the emissions of carbon dioxide from fossil fuel combustion and its absorption by the biosphere. In other words, how much land is needed to soak up energy-related greenhouse gas emissions? Although this narrow perspective of the ecological footprint of energy systems is strongly critiqued in Ayres and Ayres (2002), calculations such as the example shown in Box 10.2 are common (e.g. Barrett *et al.* 2002).

Table 10.1 CO_2 emissions created by the direct combustion of different fossil fuels (i.e. Scope 1 emissions)

Fuel	Conversion factor (kgCO₂ per kWh)
Natural gas	0.184
Coal	0.325
Oil	0.267
Petrol	0.241

Data source: Defra (2011).

Box 10.2 Calculating the ecological footprint of an energy system

Here is a brief example of how to calculate the ecological footprint of an energy system, or at least the footprint associated with greenhouse gas emissions (see text).

1 *Calculate the greenhouse gas emissions associated with fossil fuel combustion.* Fossil fuels emit different amounts of carbon dioxide for the same energy value. Reference tables, such as Table 10.1, provide these conversion factors though one must be careful to ensure comparisons are made on a like-for-like basis. Life-cycle analyses of fuel production, for example, can change the final emissions level by 30–50 per cent (Barrett *et al.* 2002).

2 *Calculate the land equivalent needed to absorb the emissions.* Assuming that we burned 1 GWh of natural gas, 184 tonnes of CO_2 would need to be absorbed. This number is then converted into the amount of land required to absorb the CO_2, using reference tables to find that 5.2 tonnes of CO_2 is absorbed by 1 hectare of forest: $184/5.2 = 35.4$. Since the ecological footprint works on the concept of a globally average hectare of land, a further conversion factor must be used to express the relative productivity of the land type in question to the world average. In this case, it is assumed that forest is 1.78 times more productive than the global average hectare and so the ecological footprint can be calculated as: $35.4 \times 1.78 = 63.0$ global hectares (gha).

By taking into consideration the total amounts of different energy forms used in an urban energy system and multiplying by the appropriate conversion factors, one can calculate the ecological footprint of energy consumed in a city. This has been done for cities such as London (CIWM 2002), York (Barrett *et al.* 2002), Santiago de Chile (Wackernagel 1998), Macao (Lei and Wang 2003), and for countries such as Australia (Lenzen and Murray 2001), Sweden (Wackernagel *et al.* 1999) and New Zealand (McDonald and Patterson 2004).

The advantage of the ecological footprint is that it allows the analyst to compare a system's impact to a useful benchmark: the total amount of land on Earth. At this global scale, the 2006 Living Planet Report (WWF 2006) highlighted that, since the late

1980s, the world has been in overshoot. That is, the ecological footprint has exceeded the Earth's biocapacity by approximately 25 per cent from 2003 onwards. This means that the demands upon the planet's land – for resource provision and waste disposal – is growing at such a rate that the Earth's regenerative capacity cannot keep up. People are turning resources into waste faster than nature can turn waste back into resources (WWF 2006). In absolute terms, in 2003, the world's average ecological footprint per capita was at 2.2 global hectares per person, whereas it is estimated that only 1.7 hectares per person are available (Global Footprint Network 2007).

10.3.3 *Life-Cycle Assessment*

Life-Cycle assessment (LCA) is a methodology developed to evaluate the environmental performance of a process, product or activity. It encompasses all stages in a product's life-cycle, from natural resource extraction to final product use and disposal and is therefore often known as a 'cradle-to-grave' analysis. The main stages of the LCA, as defined by the Society of Environmental Toxicology and Chemistry (Consoli *et al.* 1994) are:

1 Define the goal and scope of the analysis, including the system boundaries
2 Compile an inventory of materials and energy use and of emissions and solid waste, known collectively as 'burdens', associated with the product or service
3 Assess the impact of the burdens, with the following steps:

   • Classification, in which the burdens are aggregated into a smaller number of environmental impacts
   • Characterization, where the potential environmental impacts are quantified
   • Valuation, by which the environmental impacts are reduced to a single measure of environmental performance

4 Identify steps to reduce the product's environmental impacts.

One of the more recently developed methods of life-cycle analysis is the Eco-Indicator 99 methodology published by PRé Consultants (2001). It has been used to calculate the environmental impact of a large number of commonly used materials and processes. The technique attempts to model the potential damages of a product or service in three damage categories or 'endpoints'. These are:

Human health: All human beings, in present and future, should be free from environmentally transmitted illnesses, disabilities or premature deaths
Ecosystem quality: Non-human species should not suffer from disruptive changes of their populations and geographical distribution
Resources: The nature's supply of non-living goods, which are essential to human society, should be available also for future generations.

Four steps are taken to arrive at a single Eco-Indicator value for a particular process (Hugo *et al.* 2004) and the method is summarized graphically in Figure 10.1.

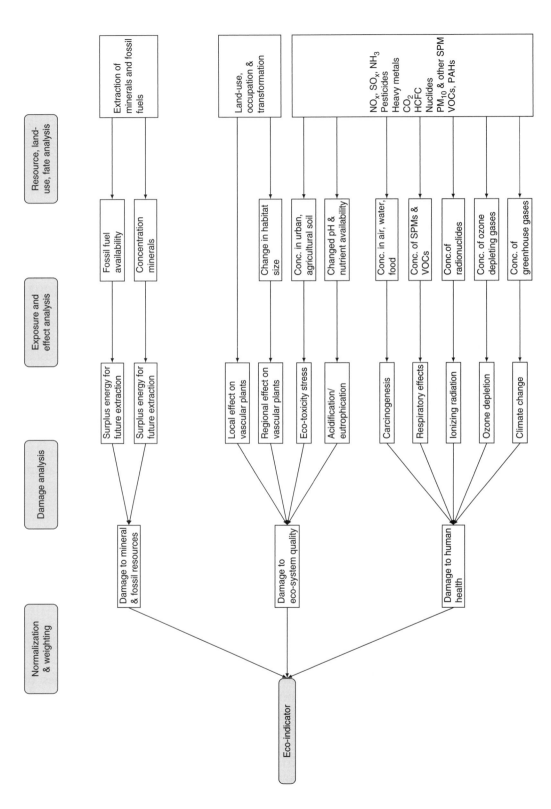

Figure 10.1 The Eco-Indicator 99 methodology (PRé Consultants 2001). SPM = suspended particulate matter, VOC = volatile organic compound, PAH = polycyclic aromatic hydrocarbons, HCFC = hydrochlorofluorocarbons.

1   The *Resource, Land-use and Fate Analysis* step, in which the fate of emissions and the extraction of resources is analysed based on emissions inventories and feedstock requirements.

2   The *Exposure and Effect Analysis* step, to determine how much of a substance released into the environment is actually taken in by living life forms, such as plants, animals and humans. The effect that resource extraction has on future generations is also analysed at this stage. The result is the classification of the impacts according to 12 indicators such as climate change, ozone layer depletion, acidification and eco-toxicity.

3   The *Damage Analysis* step quantifies the impact of each category by using a damage factor and the 12 indicators are then aggregated to form the main three damage categories mentioned earlier.

4   The *Normalization and Weighting* step, in which a single aggregated Eco-Indicator 99 value is calculated through normalization and by assigning scores and relative weighting factors to each impact category. The assignment of relative importance is a subjective exercise, hence the methodology provides three different 'perspectives' based upon the principles of the cultural theory of risk (see Box 10.3).

Box 10.3 Cultural theory of risk and LCA

The cultural theory of risk (see Thompson *et al.* 1990) is based on the type of relations that people have within a group and how an individual's life has been moulded by external events, otherwise known as their 'grid'. The assumption is that the position of each individual in this group-grid setup has a large influence on the value system of individuals and their groups. Hence, these value systems are used frequently to tackle the problem of modelling subjectivity. The five most important archetypes that have been identified to generally explain people's attitudes are:

1   *Individualists*, who are relatively free of control by others, yet are often engaged in controlling others.

2   *Egalitarians*, who portray no character differentiation, and relations between group members tend to be ambiguous, causing conflicts.

3   *Hierarchists*, who are both controlling others, but are also subject to control by them, bringing a sense of stability to the group.

4   *Fatalists*, who act alone, lack opinion and are typically controlled by others.

5   *Autonomists*, who are usually the smallest group that manages to escape any manipulative forces from the rest and thinks completely independently.

The representatives of the first three archetypes are often the most useful in decision-making, since they have such different opinions between them, whereas the last two categories represent a smaller range of perspectives. Based on these three personality profiles, Hofstedder (1988) stated that three versions of an LCA could be produced:

1   The individualist version, where only proven cause-effect relations are included and short-term perspectives are chosen. For example, in human health issues, age-weighting is used, due to the individualist's typically higher value of ages 20–40.
2   The hierarchical version, where facts from recognized scientific and political bodies are included. A classic example is the accordance with the IPCC climate change guidelines.
3   The egalitarian version, where the largest amount of data is included, with little omissions and long-term perspectives, since this version is of a precautionary nature and does not ignore possible future problems.

Consequently, following the completion of an LCA, three possible scores can be obtained, depending on the cultural theory perspective. The hierarchical version is usually the default method since its ideology is most common in the scientific community and in political organizations. The other two methods can provide basis for sensitivity analyses if desirable. Therefore, the final weighting is heavily determined by the basic value system a person is using and the concepts of cultural theory of risk play a large role in its derivation.

The Eco-Indicator 99 Methodology is considered advantageous over other techniques due to its systematic ideology, which produces a single value that is representative of all the different environmental concerns. Equally important is the fact that the calculations can be 'interrupted' at any stage according to the particular needs of the impact assessment (Hugo *et al.* 2004).

10.4   Combining EIA and process optimization techniques

Various authors have attempted to incorporate LCA within a formal process systems' optimization framework, in order to help design products with a minimized environmental impact. An early example of this approach is the methodology for environmental impact minimization (MEIM) performed by Stefanis *et al.* (1995). The authors highlight the need to account for the waste associated with inputs to a process (e.g. upstream wastes associated with raw materials, energy generation, etc.) and not solely the outputs. The paper illustrates the MEIM method using a case study of vinyl chloride production and demonstrates the three key steps: (1) definition of the system boundary (expanded to include raw material extraction and energy generation); (2) environmental impact assessment where, for each pollutant released from the process, a vector denoting its corresponding environmental impact (air and water pollution, solid wastes, global warming, photochemical oxidation and stratospheric ozone depletion) can be obtained leading to an aggregated impact vector and (3) design of environmentally benign processes, where the incorporation of the environmental impact criteria in process optimization assists in finding the best plant design that achieves both minimum global environmental impact and minimum economic cost. A significant result from this study is the fact that when only local emissions are minimized, the minimum global impact is not necessarily achieved. Through the vinyl chloride case study, it is shown that the zero avoidable pollution (ZAP) approach is not always the best environmental choice. Indeed, 'a minimum site-specific discharge limit exists beyond which the global environmental impact increases due to trade-offs in waste generation over the whole life cycle' (Stefanis *et al.* 1995).

Azapagic and Clift (1995) brought LCA and linear programming (LP) together in order to environmentally optimize a product system. The problem lies in how to allocate environmental burdens (such as resource depletion, emissions, solid waste, etc.) in a system where more than one product is being produced. By representing the whole productive system with an LP model, the environmental performance of the product system can be analysed and managed. The authors first identify the environmental burdens and then aggregate them into different impacts such as greenhouse effect, ozone depletion, and so on. As with methods like Eco-Indicator 99, weights are then assigned to the impacts according to their importance and based on environmental and socio-economic objectives that are of interest to the decision-makers. The additional contribution of the optimization model is that the shadow values[1] of the LP model can be used to identify products and activities in the system which make the greatest contribution to the overall environmental impact. Furthermore, the LP model accounts for the different states of the system and multi-objective LP models can help to resolve conflicting demands on system performance.

Chapter 9 and Appendix A provide an introduction to the use of optimization models in energy systems analysis, highlighting in particular the distinction between models where all of the decision variables are real-valued (i.e. decimal numbers) and models where the decision variables can take integer values. Distinctions can also be made between linear and non-linear models. These extensions of the basic linear programming framework are vital to the application of optimization and LCA methods to urban energy systems, as presented in the example model below.

10.5 A hypothetical case study

In order to illustrate the way in which LCA and urban energy systems modelling can fuse, a hypothetical case study was devised for an area of 980 hectares (divided into 49 cells with 20 ha area each). The purpose is to show potential energy technology combinations and reductions in environmental impact achieved through scenario planning and integrated design.

The model for the hypothetical case study was built using the mixed-integer programming technique. It includes various technologies, resources and products and was run to select the least environmentally harmful combination, by minimizing one damage category at a time. The model is presented in this section, stating the main

Table 10.2 Energy technologies and their approximate capacities

Technology	Capacity (kW)	Efficiency (%)	Capital cost (£)	O&M cost (£)
Biomass CHP	45,000	80	35 million	3 million
Biomass boiler	2	90	600	50
Coal CHP	45,000	70	15 million	0.5 million
Electricity grid	45,000	0.033 per km	80,000 per km	900 per km
Natural gas grid	100	Negligible	150,000 per km	2,500 per km
Natural gas CHP	1,500	70	1,000,000	35,000
Natural gas boiler	1	90	850	50

Note: O&M = operating and maintenance costs.
Data source: author estimates.

assumptions and the mathematical formulation. As the equations are linear and include binary and continuous variables, the problem is classed as an MILP and can be solved using GAMS/CPLEX.

10.5.1 *Input data and assumptions*

The model begins by specifying the final energy service requirements of the city. The demands for heat and electricity for this fictional area were estimated on a per-capita basis to suit the needs of this case study. The algorithm generates random values for each cell by assuming high demands in the middle cells, or otherwise 'city-centre', and decreasing demands as one proceeds to the outskirts. For each cell, it is assumed that the demand for heat and electricity can be generated locally, otherwise resource flows take place between cells, as long as a feasible route exists.

A number of energy conversion technologies are then used to convert between raw input fuels and final energy services. Each technology has a series of parameters giving their capital cost, operating costs and resource efficiency, as shown in Table 10.2.

Since minimizing the cost of the energy system configuration is not of primary interest here, the values shown in the table are estimates. Indeed, costs are completely omitted from the objective function for the present case study so that only environmental impact affects the results. However, if economic considerations become of interest at a later stage, the model can be modified, for example, using the constraint method to pursue a multi-objective optimization (e.g. minimizing costs subject to limits on environmental impact, or minimizing environmental impact subject to cost constraints). The constraint value can be modified and the model re-run to generate multiple solutions and thereby explore the interaction between the economic and environmental objectives.

10.5.2 *Model formulation*

The model is formulated as follows: (definitions of the symbols can be found in section 10.8).

First, the set of resources included in the model is given by $r \in R = \{\text{Resource}_1,$ $\text{Resource}_2, \ldots\}$. Some of the resources are used as feeds to the technologies (e.g. natural gas), whereas others are also considered as products (e.g. electricity). Environmental impacts (e.g. global warming potential) were also seen as exiting 'resources' from these technologies, in order to facilitate the formulation of the model. The resources were further sub-divided into accumulating (i.e. all waste and environmental impacts) and non-accumulating.

The set of technologies is similarly given by $t \in T = \{\text{Technology}_1, \text{Technology}_2, \ldots\}$. These technologies were sub-divided into technologies that import resources (e.g. electricity grid) and those that consume resources (e.g. a natural gas boiler).

A resource balance is defined for each cell, stating that local production, plus imports from outside the side, plus net transfers within the city, minus demand must equal to accumulation:

$$P_{r,i} + I_{r,i} + \sum_{i'} Q_{r,i,i'} - \sum Q_{r,i,i'} + -D_{r,i} = A_{r,i}$$

Accumulation must be zero, unless the resource is an environmental impact or waste, in which case:

$$A_{r,i} \leq 0$$

The production rate of resource r, in cell i, is given by:

$$P_{r,i} = \sum_t \tau_{t,i} \text{ConvFact}_{t,r}$$

where $\text{ConvFact}_{t,r}$ is a conversion factor for a given technology and resource, and τ is the production rate of technology t in cell i, given by:

$$\tau_{t,i} \leq N_{t,i} \text{TechCap}_t$$
$$\tau_{t,i} \geq 0$$

The flow of resources between the cells can be described by the following equations:

$$Q_{r,i,i'} \leq X_{r,i,i'} \text{MaxTrans}_r$$
$$Q_{r,i,i'} \geq 0$$

The capital cost can be defined in terms of the technology capital costs and the transportation capital costs as shown below. Similar terms could be introduced to capture the operating and maintenance costs but are not shown here for clarity.

$$\text{Capital Cost} = \sum_{r,i} \text{TechCC}_t N_{t,i} + \sum_{r,i,i'} \text{TransCC}_r X_{r,i,i'} d_{i,i'}$$

The environmental impact can therefore be defined as the cost of each resource multiplied by the resource accumulation in a cell and is given by:

$$\text{Environmental impact} = \sum_{r,i} \text{RCost}_r A_{r,i}$$

Recalling that we are only interested in the environmental impact for this case study, the objective function Z can be defined as shown below. Total financial costs could also be incorporated into this objective function if desired.

$$\min Z = \text{Environmental impact}$$

10.5.3 Results

The model was run under three different scenarios for minimizing different kinds of environmental impact: respectively, global warming potential (GWP); resource depletion (RD); and air quality (AQ). For each scenario, the results were calculated for all three environmental impact metrics and these are shown in Table 10.3.

It can be observed that when the objective is to minimize air quality, the values of GWP and RD increase by seven times and approximately 12 times against their optimized values, respectively. Similarly, when resource depletion is chosen as the objective to be minimized, a 10-fold increase is seen for GWP and a six-fold increase in the value of AQ. However, when global warming potential is chosen to be minimized, the AQ value is only around 30 per cent away from the minimum value which can be achieved when air quality is the objective. Similarly, for resource depletion, the obtained value when GWP is minimized is less than 20 per cent from the equivalent achieved when minimization of resource depletion is the objective. It therefore seems that out of the three environmental impacts, minimization of GWP affects the optimal values of the other two categories the least.

For each scenario, the model also provides the choice of technologies which led to least environmental impact according to the damage category of interest. The values in Table 10.4 represent the relative contribution of each selected technology to total urban energy demand.

Table 10.3 Changes in environmental impact values when minimizing one objective at a time

Objective function (minimize …)	Output metric		
	Global warming potential (kg CO_{2eq}/h)	Air quality (10^{-9} DALY/J)	Resource depletion (kW)
Global warming potential	489	119	221
Air quality	3,460	81	2,150
Resource depletion	4,940	498	182

Note: DALY = Disability Adjusted Life Years, and therefore DALY/J = DALY/kg (of harmful substance exposed to) × kg (of harmful substance released)/TJ (of natural gas burnt) = 10^{-9} DALY/J.

Table 10.4 The model's choice of technologies according to environmental impact objective. Values indicate the percentage of total urban energy demand satisfied by a given technology

Technology	Objective function (minimize …)		
	Global warming potential	Resource depletion	Air quality
Biomass CHP	48	–	–
Biomass boiler	52	52	–
Coal CHP	–	48	–
Gas CHP	–	–	48
Gas boiler	–	–	52

Note: CHP = combined heat and power.

To achieve the best results in terms of GWP, the model chose only the biomass technologies, whereas for RD, it found that a combination of biomass and coal is better to use as resources. On the other hand, when AQ was the objective, all natural gas technologies were preferred. The results agree with what was perhaps to be expected, i.e. biomass technologies tend to have low carbon dioxide emissions, which is considered a significant contributor to global warming. Furthermore, coal is still abundant in comparison to natural gas reserves and so is biomass, so if resource depletion minimization is the objective, then natural gas technologies would not constitute a satisfying solution. Finally, in terms of air quality, all natural gas technologies were chosen by the model to satisfy the electricity and heat demands, since very little particulate emissions occur from them.

These results point back to the issue of how different stakeholders might define 'environmental impact'. Each country, government, community and industry will have its own priorities with respect to which environmental impact is to be minimized. But as can be seen, the choice of one objective tends to hinder the others and some view will need to be taken on whether, e.g. air quality or global warming potential is more important.

However, one can say that in order to achieve minimum environmental impact, different system designs need to be considered and incorporated in an urban area. An optimum combination of energy technologies based on demand and geographical considerations could possibly satisfy the minimum environmental impact objective. Of course, the issue becomes even more interesting and complex when cost is introduced as a more significant component of the objective function. LCA may prove to overcome the hurdles of other environmental assessment methods by providing a link between the environmental impacts and the economics of an urban energy system.

10.6 Selecting EIA methodologies for UES design

Generally, optimization techniques are applied to a problem or case study as a final step in the process of its analysis and understanding. Optimization results typically provide quantitative improvement with respect to a system's initial state. However, what about

Table 10.5 Material needed to run the UK eco-town based on five different scenarios

Scenarios		Fuel consumption (tonnes/year)		
		Natural gas	*Coal*	*Biomass*
A	Natural gas and electricity grid (base case)	12,800	7,470	0
B	Natural gas only	17,900	0	0
C	Biomass (combination of electricity grid and biomass CHP)	12,800	4,980	6,110
D	Waste to energy (combination of anaerobic digestion, electricity grid and natural gas)	10,300	5,980	0
E	Mixed renewables (combination of PV cells, wind turbines, heat pumps, electricity grid and natural gas)	11,800	6,640	0

taking one step back, and analysing the system in question in a rather qualitative manner, to see what can possibly be revealed prior to any quantitative improvements?

To demonstrate this ideology, a new case study is introduced and put through environmental impact assessment methodologies. By performing this step, one can obtain more qualitative information. Specifically, this can show which EIA methodology (or combination of them) is most relevant and appropriate to the design of a particular urban energy system and how readily it can be applied.

Three different methods of measuring environmental 'friendliness' have been analysed and applied to a UK eco-town. These methodologies can be applied to an urban energy system individually or in combinations to indicate what is environmentally acceptable for different UES solutions and scenarios.

The case study is a proposed eco-town in England with an estimate of 15,440 dwellings, a population of 32,000 and an overall energy demand of 273 GWh/year, split into 188 GWh of heat demand and 75 GWh of electricity demand. Five different energy system scenarios were taken into consideration and compared with respect to the three EIA methodologies: MFA, LCA and EF.

Table 10.5 shows the results of the material flow analysis, indicating how much of each fuel (natural gas, coal and biomass) is required for the smooth operation of each energy system scenario. All materials are reported in tonnes and give an idea of how sustainable (or not) each technology is. The major component of each scenario involves natural gas, with the base case (scenario A) and the mixed renewables option (scenario E) requiring the largest amounts of coal. Even though the biomass option (scenario C) achieves lower amounts of coal and natural gas, an environmental impact can be associated with the introduction of just over a million tonnes of wood chips to fulfil the end-use energy demand of this eco-town.

Based on the results of section 10.5, it has been concluded that out of the three environmental impacts, namely, air quality, resource depletion and global warming potential (GWP), the minimization of GWP affects the optimal values of the remaining two categories the least. GWP is the factor with the greatest impact on a system and therefore an LCA is conducted on a greenhouse gas (GHG) emissions basis in this case. Subsequently, the energy systems were compared with respect to their GHG emissions per capita as shown in Figure 10.2. It appears that the waste-to-energy option

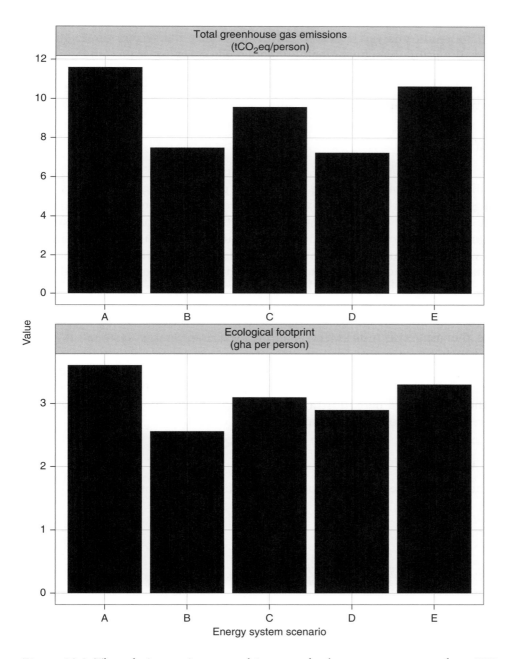

Figure 10.2 The relative environmental impact of urban energy systems for a UK eco-town based on greenhouse gas emissions per capita (LCA) and ecological footprint per capita.

(scenario D) achieves the lowest GHG emissions per year, whereas the base case, i.e. using the combination of natural gas for heating purposes and the grid for electricity demands seems to score the highest.

An ecological footprint analysis was also performed. Since the system under study is an energy system, the primary environmental impact is the greenhouse gas emissions and therefore, the comparison of energy systems based on ecological footprint follows closely the pattern of the LCA study. However, the lowest ecological footprint is achieved by scenario B this time, the natural gas only set-up.

10.7 Conclusion

Until recently, the optimization of urban energy systems has taken a relatively limited view of environmental impact, focusing largely on greenhouse gas emissions. This chapter has sought to widen that perspective, acknowledging the complex characteristics of urban energy systems that make them stand out with respect to other energy systems and seeking to provide a better understanding of how such systems function and interact with respect to various measures of environmental performance. By evaluating the three different environmental impact assessment methodologies, it was shown how qualitative and quantitative analysis techniques can be used to evaluate alternative energy system configurations at various stages in the design process.

By applying a combination of these methodologies, a more thorough insight is provided to the dynamic and multi-faceted nature of urban energy systems. It is not easy to decide which single metric is most important over others; that is the task of policy-makers and key stakeholders worldwide. Yet, while there is little doubt that global warming is still a main concern for cities everywhere, the environmental impact assessment methods presented here illustrate how a range of impacts on health, resource scarcity and the local and global environments can be evaluated.

10.8 Symbols used

The following notation is used for the model presented in this chapter.

Sets

R	Set of resources, $r \in R$
T	Set of technologies, $t \in T$

Variables

$Q_{r,i,i'}$	Flow of r from i to i'
$P_{r,i}$	Production of r in cell i
$I_{r,i}$	Import of r to cell i
$\tau_{t,i}$	Production rate of t in i

| $A_{r,i}$ | Accumulation of r in i |
| Z | Objective function |

Parameters

TechCC_t	Technology capital cost (k£)
TechCap_t	Technology capacity in kW
RCost_r	Cost (environmental impact) of each resource
$D_{r,i}$	Demand of each resource in each cell
TransCC_r	Transportation capital cost of r per m
MaxTrans_r	Maximum rate of transport for resource r
$d_{i,i'}$	Distance from i to i'
$\text{ConvFact}_{t,r}$	Rate of r produced per rate of operation of t

Binary variables

| $X_{r,i,i'}$ | If network exists for r between i and i' |

Integer variables

| $N_{t,i}$ | Number of technology t at cell i |

Note

1 In a maximization optimization problem, the shadow value (often called the 'shadow price' in economic models) represents the amount by which the objective function would increase if the constraint were relaxed by one unit.

11 Activity-based modelling for urban energy systems

Aruna Sivakumar

Imperial College London

In this chapter, we present activity-based modelling techniques as a means of enhancing the behavioural modelling of demand within integrated urban energy model systems. The activity-based modelling paradigm is commonly used in state-of-the-art land use-transport model systems to develop bottom-up and policy-sensitive predictions of the demand for travel. This chapter describes the activity-based modelling paradigm and explores the role of activity-based models within urban energy systems.

11.1 The potential for activity-based models

Urban energy systems (UES) models can be conceptualized as broadly comprising a supply component and a demand component. The supply models integrate the various urban infrastructure networks (such as electricity, water, heat, etc.); and aim to analyse and understand the current state of these systems on the one hand and to design optimal and reliable systems on the other. *Integrated models of urban infrastructures* are therefore typically optimization-based resource management models (e.g. Geidl 2007, Schenk 2006). Most such UES model systems use aggregate energy demand predictions (aggregate both in time and space) that are typically made using static accounting and rule-based techniques, and focus more on the technology and network-level modelling of the supply systems. Chapter 9 discusses such a model in greater detail.

Another class of UES models, generally identified as *urban dynamics models*, focus on the geographical and temporal patterns of resource demands and produce ecological and environmental simulations of water and waste flows, building heating, and so on. Simulation models, such as climate risk models, fall under this category. Specific examples of urban dynamics models include the tool developed by Robinson *et al.* (2007) for the assessment of energy, water and waste consumption at a building or small neighbourhood level; the GIS tool developed by Girardin *et al.* (2010) for estimating the spatial pattern of energy requirements in an urban area or the model by Mori *et al.* (2007) assessing the interactions of heat demand and locally available heat sources, e.g. lakes or incinerators. However, such urban dynamics models simply simulate observed resource demands without attempting to predict or necessarily understand the behaviour underlying these demands.

In fact, the demand component in most UES model systems simply plays the role of providing the inputs to the integrated and detailed supply models. This is typically achieved by using regression models that are calibrated to capture the correlations in observed aggregate energy consumption data (see, e.g. Brownsword *et al.* (2005) on modelling urban energy supply and demand).

When embedded within state-of-the-art land use-transport (LUT) models, the activity-based modelling paradigm (referred to as ABM in this chapter and not be confused with agent-based modelling, see Box 11.1) holds the potential to produce detailed energy and resource demands that are not only spatially and temporally disaggregate but also sensitive to a wide variety of policies and scenarios. In this chapter, we first take the reader through a history and description of ABMs within the LUT modelling context, followed by a critical analysis of the value of this approach for UES modelling and a more detailed examiniation of a specific activity-based model of resource and energy demands. We end the chapter with an analysis of the challenges in developing such models and the emerging opportunities for these models within the UES context.

Box 11.1 Activity-based models and agent-based models

The term 'agent-based modelling' is far more commonly known and used, and must not be confused with the term 'activity-based modelling'. Agent-based modelling is a computational technique used to simulate the actions of agents as a means of understanding the properties and state of the entire system, where the agent could be an individual or a group of individuals. Activity-based modelling, on the other hand, is a conceptual framework or a modelling paradigm with the objective of developing a behavioural and individual-level model of demand. A model of resource demands could be aggregate and statistical at one end of the range or activity-based and behavioural at the other end. Agent-based modelling techniques lend themselves easily to implementing activity-based models of demand; however, not all agent-based models are activity-based.

11.2 History of LUT models and ABMs

Integrated land use-transport model systems form a body of research that is generally overlooked by urban energy system modellers, though this is slowly changing (Ghauche 2010, Sivakumar *et al.* 2010b, Keirstead and Sivakumar 2012). Integrated LUT models were traditionally developed as a means of estimating travel demand in response to land use changes, and over the years, they have evolved to be rich descriptors of the activity and travel patterns of all the agents in the study area including households and individuals, businesses, real-estate developers and others. Figure 11.1 illustrates the complex linkages in urban systems that LUT models address.

The land use components of such integrated model systems describe medium to long-term urban processes such as household (re)location, work (re)location, real-estate

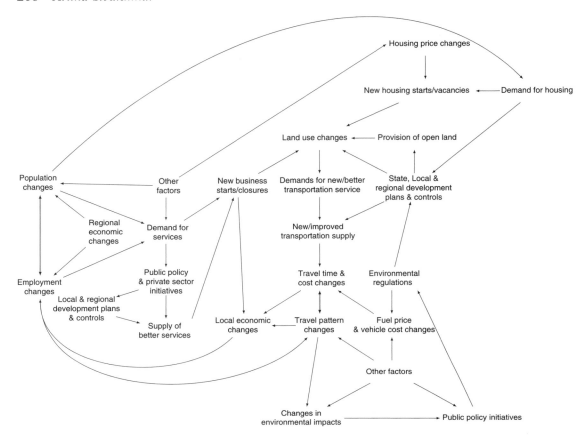

Figure 11.1 Complexity of functional linkages in urban system dynamics.
Based on Southworth (1995).

development, business (re)location and automobile ownership, thus providing urban planners with a tool to forecast future land use layouts of urban areas. This is integrated with model components that predict the travel patterns generated by the agents within the urban landscape; the transport flows created through these processes, in turn, feed back to the land use models guiding further real estate development, business and household relocation in response to the conditions on the transport networks. Such integrated land use-transport model systems thus attempt to produce reliable and policy-sensitive travel demand by capturing the complex relationships between transport and land use in a system of descriptive models (for detailed reviews, see Wegener 1994, 2004, Hunt *et al.* 2005).

However, it is not until recently that operational LUT models have actually attempted to capture a number of these interactions in a behaviourally realistic manner. Figure 11.2 presents the development pathways of these various LUT model systems over the last five decades. At the left end of this figure are the aggregate models and at the right end are the disaggregate models, with a range of models in between.

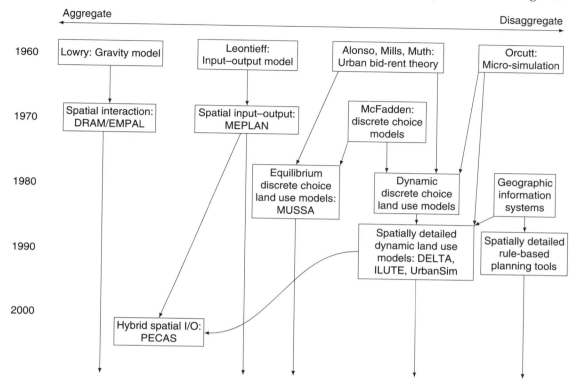

Figure 11.2 Integrated land use-transport modelling frameworks.
Citations for specific models can be found in the main text. Based on Waddell (2005).

The earliest land use-transport models were essentially static models, driven typically by gravity, entropy-based or input–output formulations (Lowry 1964). These models are linked to a four-stage transport model[1] through iterative feedback of network flows in the form of an accessibility index to estimate equilibrium patterns of land-use and transport (see, for instance, LILT, the Leeds Integrated Transport package (Mackett 1983), DRAM/EMPAL (Putman 1995), IMREL (Anderstig and Mattsson 1991), MUSSA (Martinez 1996)). These static models do not model market processes behaviourally and cannot realistically capture urban spatial processes and are therefore not very responsive to policy and scenario analyses.

The next generation of LUT models were the general spatial equilibrium models such as MEPLAN (Hunt and Simmonds 1993) and TRANUS (de la Barra 1989), which are typically also spatially aggregate models like the static models but with more closely integrated land-use and transport elements. The interactions between these elements are determined by input–output analysis or discrete choice models, and these interactions are used to derive the demand for transport. General spatial equilibrium model systems are based on random utility theory and theories of competitive markets. These models treat land-use and transport systems endogenously and therefore capture the interactions between these systems more accurately. However, these models, in being spatially aggregate, are not entirely behaviourally realistic. Despite these limitations, TRANUS was one of the first LUT models that was used to analyse urban energy demand.

The third generation of LUT models, i.e. the agent-based micro-simulation models, combine the strengths of micro-simulation and the disaggregate modelling of behaviour and land-use processes. Notable examples include DELTA (Simmonds 1999), ILUTE (Miller *et al.* 2004), UrbanSim (Waddell 2002), and PRISM (Alberti and Waddell 2000). These are agent and activity-based models with the individual (person, household, firm or any other agent in the urban system) as the unit of analysis. Hence, these models are intuitive in their formulation, and capture the interactions between land-use and transport systems to the greatest extent possible. For instance, in DELTA, households follow a utility maximizing formulation with the market adjusting prices within a time period. In UrbanSim, on the other hand, households maximize a consumer surplus measure and use a bid-choice function. The key tool used in these models is micro-simulation, which provides a practical methodology to apply probabilistic models at the level of the individual. In turn, a key element in such micro-simulation models is the concept of a synthetic population of decision-makers and the development of improved methods for the generation of such synthetic populations (Muller and Axhausen 2011).

11.3 So what are activity-based models?

Activity-based models were essentially designed to enhance the behavioural realism and policy sensitivity of travel demand model systems at a time when transport planning and policy needs moved from the 'predict and provide' phase to the 'demand management' phase. These models operate at the level of the individual (and household) as the decision-maker(s) and are based on the economic principle of travel as a derived demand. In other words, they are based on the principle that travel demand is derived from the need for individuals and businesses to participate in activities that are distributed over space and time – a principle that could apply equally well to the demand for energy and other resources.

Figure 11.3 illustrates an example of a string of activities undertaken by a working person over a 24-hour period, beginning with an early morning home-stay duration until the person leaves home to pick up some milk and fill up the car with petrol, which forms a 'before-work tour' with two stops (S_1 and S_2). After another short home-stay duration, presumably getting ready and eating breakfast, the person leaves for work as represented by the 'home-work commute'. The work duration defines temporal fixities in the individual's activity-travel pattern, as illustrated by the time of arrival at work and the time of departure from work. There is a short work-based tour when the person goes out to pick up lunch (S_3). After the work-home commute in the evening, the person stays at home for a bit (S_4), followed by one last tour for the day, the 'after-work tour', which involves two stops, one at the pub to meet friends and the other at the cinema (S_5 and S_6).

This is just an example of the many activity-travel patterns undertaken by individuals, and the most advanced ABMs attempt to predict the details of these patterns including the timing, the location, the mode of travel and other information for the entire

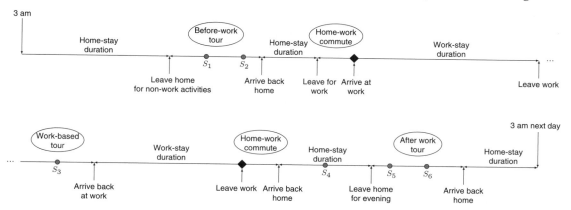

Figure 11.3 Activity-travel pattern of a worker for a 24-hour period.
The diamonds represent temporal fixities in the schedule and the shaded circles represent activity locations or travel destinations. Based on Bhat *et al.* (2004). See acknowledgements for clip-art sources.

population of the study area. The travel demand derived from such models is therefore sensitive to a range of policies and future scenarios.

For instance, let's examine the staggered work hours programme (also called 'gliding time' in Germany and flex-time in the USA), whose intended objective is to ease peak hour congestion and consequently reduce fuel use and improve air quality. Figure 11.4 illustrates an individual who typically leaves work at 5:00 pm, drives to a supermarket that is located 15 minutes away, spends 25 minutes shopping and is back home by 6:00 pm. The expected impact of a staggered work programme that releases this individual from work at 4:00 pm, instead of 5:00 pm, is a 1-hour shift in his pattern that would see him back home by 5:00 pm instead of 6:00 pm, thus spreading out the traffic flow and reducing congestion. This is what most aggregate, statistical models of travel demand would predict. However, a model that is sensitive to the activities that drive the individual's travel patterns, and a model that maintains the links between these activities, is better placed to capture a variety of different behaviours. For instance, given the additional free time, the person may choose to drive to a larger supermarket that is

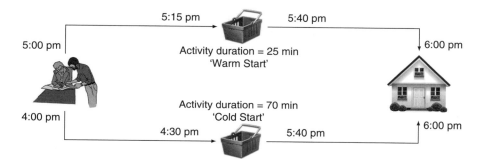

Figure 11.4 Illustration of the potential impacts of a staggered work programme.

further away (30-minute drive instead of 15 minutes previously) and shop for longer, leading to a vehicle cold start, which is in fact worse from an air quality perspective. Heterogeneous behaviours such as these are better predicted by the state-of-the-art activity-based travel demand models that understand weekly and longer-term rhythms in the activity-travel patterns of individuals and households, and derive the travel demand as a function of the preferences in activity patterns.

11.4 The state-of-the-art in ABMs

From the urban energy modelling perspective, state-of-the-art activity-based integrated land use-transport models form a behaviourally realistic means of simulating consumption, and therefore resource demands. For instance, a model capable of predicting detailed activity-travel patterns, as in Figure 11.3, is a short step away from an energy and resource demand model that can make policy-sensitive, temporally and spatially disaggregate demand predictions. Before we proceed to examine ABMs from this perspective, we will first take a little time to understand the state-of-the-art in these models.

LUT model systems are typically a suite of interconnected descriptive and normative models that can jointly predict urban processes and activities. The embedded models are typically probabilistic, micro-econometric, heuristic or rule-based, or combinations thereof. For instance, CEMDAP (Bhat *et al.* 2004) contains probabilistic and micro-econometric models; ALBATROSS (Arentze and Timmermans 2000) is primarily based on decision heuristics; TASHA (Miller and Roorda 2003) combines heuristic rules with econometric models; and AMOS (Pendyala *et al.* 1995) contains elements of each of these approaches.

The state-of-the-art LUT models, regardless of the underlying model types, are typically implemented as agent-based micro-simulation systems with the activities of all the agents in the study area being simulated. In this sense, they are distinctly different from the older school of LUT models, which were predominantly based on aggregate flows of people, goods and resources. The state-of-the-art models are also highly disaggregate with respect to time and space, and some of the models are capable of operating on a second-by-second time scale with parcel-level[2] spatial detail, effectively producing predictions of activity and travel demand over continuous time and space.[3] The flip side of such descriptively rich models is the quantity of data and computational time (effort) required to implement them. On the other hand, once operational, such models are excellent test beds for a variety of policy scenarios, engineering and technological solutions. Being integrated models, they effectively capture both direct and indirect effects of the scenarios of interest.

A key feature of the state-of-the-art LUT models is the underlying models of individual and group behaviour. These models, unlike typical engineering models which are employed in the supply components of UES model systems, acknowledge the stochasticity of human behaviour and the intrinsic heterogeneity in this behaviour which not only results in observationally identical individuals making very different choices but also in situations where the same individual makes very different (seemingly irrational) choices at different times. Similar ideas have been explored in some engineering models such as those of Richardson *et al.* (2010) and Widén and Wäckelgård (2010).

The earliest models of agent-level behaviour focused principally on predicting the choice of specific facets of individual trips (such as transport mode, route or household location) and tended to be deterministic in nature (typically assuming that behaviour was driven solely by considerations of cost or travel time minimization). From the 1970s onwards, these approaches were gradually replaced by models which, at the conceptual level, consider travel decisions explicitly as part of the broader context of an individual's programme of activity participation (Hägerstrand 1970, Lenntorp 1976, Jones *et al.* 1983), and at a methodological level, treat decision-making as a stochastic (rather than deterministic) process (McFadden 1978, Train 2003).

The current state-of-the-art is predominantly represented by techniques based on discrete choice modelling methods (see Box 11.2), which can accommodate a wide variety of decision-making contexts, including both individual and group decisions, decisions regarding both discrete and continuous outcomes, static and dynamic decisions, decisions with single or multiple expressed outcomes, decisions made under uncertainty and those influenced by qualitative as well as quantitative factors. These methods can also be used as a means of integrating data both from real market outcomes ('revealed preference data') and data from hypothetical market studies ('stated preference data').

Box 11.2 Econometric techniques for demand modelling

Demand modelling can be achieved using a variety of different techniques, including decision rules, neural networks, probabilistic modelling, econometrics and more. The choice of modelling methodology is usually motivated by the conceptual framework, the data availability and the computing capability. Econometric techniques are commonly used to calibrate demand models that are embedded within activity-based model systems.

The basic hypothesis of the econometric modelling techniques used in ABMs is that there exists a 'population model', which is a perfect predictor of the quantity of interest. For instance, say we wish to predict the activity duration for an individual (such as length of time spent shopping in Figure 11.4). However, it is only possible to get data for the quantity of interest from a sample of the population, and the 'sample model' is clearly subject to sampling errors. This is the basis for a wide range of econometric and micro-econometric models, such as regression models, hazard duration models, discrete choice models and so on. These models are calibrated using real data that is collected through surveys such as travel surveys and time use surveys, from a (ideally representative) sample of the population.

Let us take discrete choice models (DCMs) as an example, as this is a widely used technique in travel-demand modelling and consumer behaviour modelling. DCMs can be used to predict an individual's choice from a set of discrete alternatives, e.g. modes (car, bus, train, cycle, walk, etc.) or destinations (neighbourhood pub, pub near work, favourite pub, friend's favourite pub, famous pub, etc.). Such a disaggregate model therefore opens up the possibility of developing behaviourally realistic and policy-sensitive demand models. In fact, the derivation of DCMs, based on the theory of random utility maximization, by McFadden in the late 1970s, provided a big impetus towards the modelling of individual and firm choice behaviour. The discrete choice paradigm not only brings the problem of understanding choice behaviour down to the individual decision-maker but also acknowledges the discrete nature of many of these choices.

We now consider a brief presentation of a very basic DCM specification used in travel-demand models, the multinomial logit model (MNL). According to random utility theory, individuals associate a utility with each alternative they consider and the alternative with the highest utility is chosen. The utility that individual i attributes to alternative j is then given as:

$$U_{ij} = V_{ij} + \varepsilon_{ij}$$

where V_{ij} represents the measurable component of the utility and ε_{ij} represents the random error component, from the analyst's perspective. Based on the theory of utility maximization and conditional on assumptions placed on the random error component, several formulations such as the multinomial probit or logit models may be derived. The MNL model, for instance, is based on the assumption of an extreme value (EV1) distribution of the error term and is given by:

$$p_{ij} = \frac{e^{V_{ij}}}{\sum_{j \in C} e^{V_{ij}}}$$

where p_{ij} is the probability that individual i selects alternative j, and C is the choice set of alternatives available to the individual. Over the last few decades there have been several advances in discrete choice modelling that support the development of flexible and behaviourally realistic models of behaviour, and therefore demand. For more on econometric modelling and DCMs, the reader is referred to textbooks such as Wooldridge (2004) and Ben-Akiva and Lerman (1985).

11.5 Energy in land use-transport models

In the last few decades, several LUT models have been extended to include air quality, energy consumption and sustainability indicators. For instance, we now have PROPOLIS (Lautso *et al.* 2004), which is an LUT model with energy and sustainability indicators; CEMUS (Bhat and Waller 2008) which is a very detailed model of individual travel behaviour and the energy and environmental implications thereof; ILUMASS (Strauch *et al.* 2005) which combines land use, transport and the environment; I-PLACE3S or Internet-PLAnning for Community Energy, Economic and Environmental Sustainability

(Czachorski *et al.* 2008) which is not a full-fledged integrated urban model but rather a GIS-based land-use mapping/scenario building platform. This is just a subset of the models available; for more detailed reviews of other currently operational urban models, please see Wegener (1994, 2004) and Kazuaki (2006).

Another new urban energy model system, being developed by a collaborative team led by the Universidade de Coimbra under the Massachusetts Institute of Technology (MIT) Portugal programme, is iTEAM or Integrated Transport and Energy Activity-based Model (Ghauche 2010, Almeida *et al.* 2009). As the name suggests, iTEAM is focused on urban form, transport and energy demands working up from behaviour at the household/individual level and the firm/organizational level. SynCity, being developed at Imperial College London (see Chapter 8 and Sivakumar *et al.* 2010b, Keirstead and Sivakumar 2012, Sivakumar *et al.* 2010a) is one of the very few urban energy model systems that integrates full-fledged and detailed supply and demand model components, and in fact, implements an activity-based model of resource demands as described further in the next section.

In addition to these, most operational LUT model systems are loosely integrated with transport air quality and energy assessment models that translate the predicted transport flows into pollutant and fuel consumption estimates (see, for instance, Wagner and Wegener 2007). In fact, as Wegener (1994) claimed, these models can now be referred to as integrated land use-transport and environment (LTE) model systems.

11.6 An example implementation: AMMUA

The BP Urban Energy Systems project at Imperial College London has produced an integrated urban systems model, entitled SynCity, which incorporates both demand and supply elements. The demand component of SynCity is an activity-based model system called AMMUA or the Agent-based Micro-simulation Model of Urban Activities. AMMUA is designed to predict the spatial and temporal distribution of demand for resources generated by the activity and travel patterns of the agents in the study area (Sivakumar *et al.* 2010a, Keirstead and Sivakumar 2012, Sivakumar *et al.* 2010b, Keirstead *et al.* 2012a).

AMMUA effectively takes a bottom-up approach to modelling demand, based on the theory that cities use energy as a result of human activity – economic, social, recreational, etc. – and that to understand and model energy use in cities, we must model this human activity. Human activity (of individuals, households, firms and organizations) is spatially and temporally distributed and the transport, energy and communication networks facilitate, constrain and modulate all these activities. The underlying principle of AMMUA thus draws on the activity-based modelling paradigm. Figure 11.5 illustrates this principle; the activities undertaken by individuals and households drive the demand for resources, including transport and energy, and generate economic activity to meet their consumption requirements. This economic activity in turn drives the behaviour of firms and organizations and their demand for resources including transport and energy.

Figure 11.6 presents an overview of the AMMUA model system. Along the lines of a typical operational activity-based land use-transport model, it comprises a medium-term model system of land use processes, residential and work location choices, and a

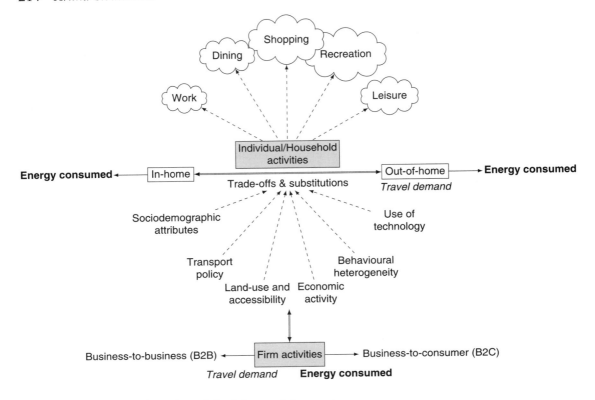

Figure 11.5 Activity-based model of demand for resources.

short-term (daily) model system of agent activities distributed over time and space. The activity scheduling model within the daily model system draws on the TASHA model from the University of Toronto (Miller and Roorda 2003). The medium-term models in AMMUA include a new component that predicts the technology holdings of households and individuals, in addition to the vehicle ownership models that exist in most LUT systems.

In addition to these components, AMMUA also includes an urban goods and services model system (UGSM) that models the business-to-business (B2B) and business-to-consumer (B2C) activities of firms and organizations in response to the consumption behaviours in the urban area. The UGSM therefore predicts the flow of light and heavy goods vehicles and commercial vehicle fleets within the urban system. Finally, the resource demand models take the outputs of the short-term models and the UGSM to convert them into spatial and temporal demand for resources, including heating, electricity, transport fuel, water and wastes.

Furthermore, as acknowledged by many other researchers (ILUMASS, CEMUS, ILUTE, etc.) such a disaggregate, agent-based and bottom-up model system is best implemented as a micro-simulation model. AMMUA is accordingly designed to take a micro-simulation approach with a combination of econometric (based on random utility maximization) and heuristic agent behaviour models. Therefore, one of the initial

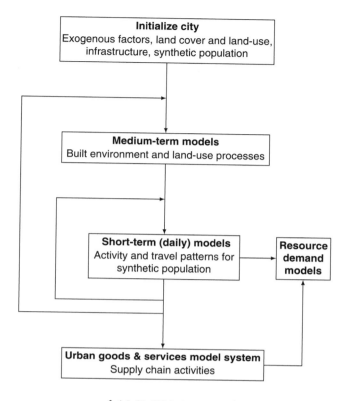

Figure 11.6 Overview of the Agent-based Micro-simulation Model for Urban Activities (AMMUA).

components of AMMUA is a synthetic population generator, which generates the entire population of the study area to be simulated.

11.7 Critical analysis of activity-based modelling methods

Activity-based resource demand models such as AMMUA presented in the previous section, and land use-transport and environment (LTE) models in general, are essentially extended versions of LUT models and are inherently complex. Consequently, there are several challenges involved in developing an operational model of this kind. In this section, we discuss a few of these challenges; for a more detailed and critical analysis, please see Keirstead *et al.* (2012a).

11.7.1 Data

The primary challenge with these models is one of data. LUT models are essentially descriptive models and need sizeable real-life datasets in order to be calibrated prior to implementation. As the models get more complex (and closer to real life) – in terms of integrating land use, transport and the environment; operating at the level of the decision-making agent (individuals, households, businesses, etc.); and operating at detailed spatial and temporal scales – the corresponding data needs to grow very quickly. However, there are potential avenues to overcome this challenge such as the use of data pooling techniques to draw together a variety of datasets for econometric modelling (Sivakumar and Polak 2009). Moreover, in an increasingly digital world, we have many new potential sources of data such as GSM and GPS traces that mitigate the challenge posed by the data requirements; just as the development of parallel computing technologies and cloud storage serve to mitigate the potential computational challenge.

11.7.2 Model system complexity

The second major challenge is associated with the complexity of LUT and LTE model systems. They are essentially systems of models predicting different quantities that are linked together (e.g. type of activity an individual undertakes in a day, time of day of activity participation, choice of technology in undertaking activity). There are several challenges involved in such a system of models, such as validity of the sequentiality in implementing the sub-models and the econometric issues associated with micro-simulation, quantification of the flow of uncertainty through the model system and so on. Current good practice is to use the micro-simulation model system as a test bed to undertake rigorous sensitivity analyses (see Chapter 12); however, this is an area that clearly needs further research as it is a relatively small body of research compared to the proliferation of such complex, model systems.

11.7.3 Different modelling paradigms

The development of urban energy system models with detailed LUT-based demand components involves the integration of descriptive models of human behaviour with normative models of urban supply systems. This is inherently a challenging task, which questions the concept of an equilibrium state. The juxtaposition of normative and descriptive models within a single modelling framework has very little precedent and needs to be explored further.

11.7.4 Complexity of application context

Most urban models have been applied mainly to a narrow set of planning problems such as evaluating transport pricing policies, estimating the impacts of new low-energy technologies, examining the environmental impacts of demographic evolution, or evaluating the social equity of a specific policy. This is limiting as the narrow perspective often fails to account for indirect effects of policies on the urban system. It also fails to account for the conflicting effects of different policies. In the face of the complex problems faced by urban areas and in light of the tightly integrated sub-systems in urban areas, probably the most difficult challenge faced by UES models is to produce reliably policy-sensitive and meaningful answers.

11.8 Conclusion

While there are numerous challenges to overcome in adopting activity-based demand models within urban energy systems, there are clearly opportunities to be exploited as well. The strengths of the ABM approach are broadly two-fold. First, the activity-based paradigm ensures a behavioural approach to predicting resource demands that acknowledges their derived nature. As a result, the models produce reliable and policy-sensitive forecasts that retain the links between different demand vectors. Second, the agent-based disaggregate approach provides high resolution detail, which serves as an effective input to integrated supply models.

These attributes of activity-based models are owed to the significant developments in micro-econometric and simulation methods. An agent-based micro-simulation model of urban energy systems supports not only the simulation of the infrastructure networks but also the behaviour, activities and consequent resource demands of every entity in the study area potentially by the second. Moreover, econometric advances in behavioural modelling that have occurred over the last few decades have enabled the development of disaggregate behavioural models that predict highly heterogeneous behaviours of individuals such as loyalty, variety-seeking, learning, spatial cognition, socially constrained preferences and norms, and more.

Underpinning all the state-of-the-art models of land use-transport interaction mentioned in this chapter, are models describing the decision-making behaviour of individual agents (e.g. travellers, developers, employers, etc.). Developments in the technologies available for modelling these agent-level behaviours serve as a strong driving and enabling force in the development of behaviourally-realistic and policy-sensitive resource demand prediction models.

A strong future direction for UES is to explore some of these advanced demand modelling techniques, not only with the objective of producing policy-sensitive and disaggregate predictions of resource demand but also to explore the nuances of resource demand and the underlying behaviour including lifestyle choices, impacts of social networks, new technologies and information. Understanding and being able to model the resource consumption behaviour is as important to the future to urban energy systems modelling as the ability to model urban infrastructures in an integrated manner.

Notes

1 A four-stage transport model represents transport activities with distinct models for trip generation (how many workers live in a particular urban zone?); trip distribution (where do those workers need to travel to within the city?); modal split (by which mode do they travel?) and traffic assignment (which routes do they follow?).

2 A land parcel is an area of land that is uniquely defined for ownership or land use purposes. A parcel is therefore a fundamental cadastral unit: a piece of land which can be owned, sold and developed.

3 In reality, these models are operated in 15–30 minute intervals and over zonal configurations that range from land parcels of the size of a building to zones that are several square kilometres in size.

12 Uncertainty and sensitivity analysis for urban energy system models

James Keirstead

Imperial College London

12.1 Introduction

As has been discussed in previous chapters, computer models are invaluable tools when trying to assess the performance of an urban energy system. Sometimes these assessments are retrospective, in that the calculations are aimed at determining the impacts of current or past performance. An example might be an assessment of the environmental impacts of observed fuel consumption. More often though, we use models in a predictive sense, in order to make a forecast about the likely performance of a system at some point in the future. However, as Niels Bohr famously said, 'prediction is very difficult, especially about the future' and for modellers of any system, the vagaries of prediction are something of a professional hazard. Yet decision-makers in the private and public sector rely upon these tools to make decisions about the future and therefore the predictive accuracy of these tools, or lack thereof, can have serious consequences.

There is one well-known model that illustrates this problem particularly well. In 1972, a group of scientists, industrialists and diplomats commissioned researchers from the Massachusetts Institute of Technology to explore the 'predicament of mankind' and the result was *The Limits to Growth* (Meadows *et al.* 1972). Using a system dynamics model called World3, the report showed that, for almost any reasonable set of alternative model inputs, the dominant mode of behaviour in the world system was one of overshoot-and-collapse. That is, if current trends in population growth, capital formation, resource consumption and pollution continued, mankind would encounter a series of a 'limits to growth' that would lead to severe declines in population, food supply and industrial output, well before 2100.

The report caused an international sensation. Newspaper headlines warned of impending doom, the book became a worldwide best-seller, and a vigorous debate about the report's findings began (for a critical view of the *Limits to Growth* models, see Nordhaus 1992). However, in the process, some of the subtlety of the original report was lost. It explicitly warned that the modelling results 'are *not* exact predictions of the values of the variables at any particular year in the future. They are indications of the system's behavioural tendencies only'. (Meadows *et al.* 1972: 92). When assessed on these terms, the report stands up well: recent research indicates that observed data in the 30 years since the report was published compares favourably with the standard business-as-usual run of the World3 model (Turner 2008).

The authors later conceded that they could have framed their message slightly differently to avoid overstating the timing of the collapse (Meadows *et al.* 1992). But it could be said that they had taken reasonable precautions to show the uncertainty of their model's output. In addition to explicitly stating that the results were not precise predictions, they also ran a series of alternative scenarios in which the input parameters were varied. These included measures such as doubling the amount of non-renewable resources available on the planet, increasing agricultural yields and improving the efficacy of birth control. Each new scenario was plotted so that taken together, readers could see the range of possible outcomes in the system (see Meadows *et al.* 1992: 11). Indeed, later on, they note that thousands of model runs were performed with overshoot and collapse emerging as the most frequent outcome (Meadows *et al.* 1992: 136).

This case illustrates some of the perils of modelling complex systems, but also some of the strategies for coping with that complexity. Although one cannot expect to remove all bones of contention from a piece of analysis (otherwise, the work was probably not worth doing in the first place) or to avoid any misinterpretation of the findings, it is clear that running a model multiple times under different scenarios and assumptions helps both the analyst and the reader to understand the behaviour of complex models and to draw appropriate conclusions.

This chapter will therefore explore a series of techniques, collectively known as *uncertainty and sensitivity analysis*, that can be used for such problems. Although they are not widely applied in the analysis of urban energy systems, we believe that there is a strong argument for their use. As we shall see, there can be widespread disagreements over the values of key assumptions and the structure of an efficient energy system can change dramatically depending on the values chosen. This is particularly true with optimization models, which may switch from one solution to another on very small changes in assumptions. Only by applying these techniques can we gain a robust understanding of the behaviour of such systems.

12.2 Post-normal science and urban energy systems

Before getting into the details of the techniques, it is worth considering why these analyses are important. What is it about modelling complex systems like urban energy systems that necessitates these additional analyses? After all, extra model runs take time and effort: why can't we just run the model once with the best-known parameter values and take that result?

The answer is that it depends on the type of analysis that is being done. Suppose, for example, that we want to throw a ball and hit a target. If by hitting the target we could win an apple, then we might decide that a basic approximation of the problem is fine. We simply need to know the mass of the ball, the distance and height of the target, and from that, we can calculate at what angle and how hard to throw the ball. In this case, the uncertainty in the calculation and the consequences of missing are relatively low: the fairground is a relatively controlled environment, we can estimate the parameters we need with reasonable accuracy and, even if we miss the target, we've only missed out on a small prize. This type of problem might be described as 'applied science'.

Now let us change the problem slightly. In this case, we find ourselves not at a fairground but in the woods, trying to hunt for our next meal with a bow and arrow. The basic structure of the problem is the same, calculating the flight of a projectile, but there are a few key differences. For a start, the problem of uncertainty has increased. The arrow is very lightweight and our scales may not be able to measure its mass accurately. Let us also assume that there is a breeze blowing from the side, creating uncertainty about the direction and effect of air resistance on the arrow's flight. More importantly, we are quite hungry and so the consequences of missing the shot are much greater than before. So in this situation, if we were not seasoned hunters ourselves, we might call in a more experienced hunter to take the shot for us. The application of experience and science in systems of increased uncertainty and greater consequences of failure might therefore be called 'professional consultancy'.

As a final example, let us consider the proverbially difficult problem of rocket science. Again, the rough structure of the problem is the same. However, in trying to fire a rocket to the moon, the system boundaries expand significantly incorporating a range of new uncertainties. Generating and controlling the necessary thrust, for example, requires the cooperation of thousands of components and the tolerances for failure shrink significantly. The distance between the launch point and the target has changed from about 10 to 375,000,000 m. And, as several accidents have shown, the consequences of failure are very serious, involving the loss of life and equipment.

These scenarios show a clear progression in two directions: the uncertainty in our understanding of the system and the consequences of a decision. We can extend the figure even further to incorporate some of the most significant problems facing mankind, such as anthropogenic climate change. In these extreme cases, the complexity of the system and the consequences of decisions are so great that we have arguably moved beyond the basic application of scientific understanding and professional judgement found in applied science and professional consultancy, into a domain known as 'post-normal science' (Ravetz 1999) (see Figure 12.1). The increased uncertainty in this domain demands a different strategy for problem-solving, one that calls upon both professional expertise but also the opinions of 'non-experts', others potentially affected by the decision or with unexpectedly relevant experience.

So where, then, can we place the analysis of urban energy systems in this space? Certainly, aspects of the design of urban energy systems resemble applied science, such as the calculation of heat loss in a building or the sizing of transformers and other components in an electrical network. For these analyses, the uncertainty in the system is relatively low and the consequences of failure, within the narrow system under design, are comparatively small. As these components accumulate into wider systems however, the complexity grows towards the domain of professional consultancy. The design of a district heating system is a good illustration. Designers will often need to estimate the heat requirements of multiple buildings, calculate pumping losses through lengthy pipe networks with multiple connections and fittings, and programme control systems that must respond to the variations in individual behaviour and equipment performance. In this case, the consequences of poor decisions can be larger, as seen in poor performance of district heating systems in Romania (Leca 2008).

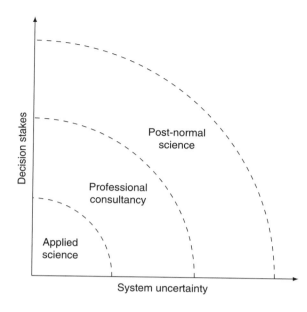

Figure 12.1 The impact of system uncertainty and decision stakes on problem-solving strategies. Based on Ravetz (1999).

Ultimately, urban energy systems contribute to questions of post-normal science. As discussed in previous chapters, cities are globally significant consumers of energy and contributors to climate change. The decisions we make in the design of urban energy systems can therefore have very significant consequences, and yet the complexity of cities at this scale makes it difficult to know exactly what the consequence of any particular design decision might be. Similarly, 'design' decisions are more than just questions of technical specifications. The way that consumers interact with energy system technologies can create unexpected results and therefore, they need to be engaged with in the creation of efficient urban energy systems.

These scales of practice, from applied science to professional consultancy to post-normal science, therefore create different demands on modelling practice and the handling of uncertainty within these tools. In applied science, uncertainty may be negligible, particularly in cases where design parameters are well-known (e.g. the performance of a pump, efficiency of a particular engine, and so on). However, once we start assembling these components into systems, and particularly once we start to expand our system boundary beyond technologies to include human behaviour and the drivers of energy service requirements, we get into domains where uncertainty in model parameters and structure can have a significant impact on the decisions we make and their possible consequences. It is for these applications that the techniques of uncertainty and sensitivity analysis are necessary.

12.3 Uncertainty and sensitivity analysis

In this section, we will review the definitions of uncertainty and sensitivity analysis, provide a basic outline of the relevant methods, and look at the types of situations in which their use is appropriate. The reader is referred to Saltelli *et al.* (2008) for a more detailed treatment of this material.

12.3.1 *Uncertainty analysis*

Uncertainty analysis is a technique for quantifying the uncertainty, or variability, of a model's output. In other words, if the true values of a model's input parameters are unknown, we might like to try some different assumptions and see how the model's output changes. The result will then give us some indication of how variable the model's result is under a particular set of assumptions. If the model's results are practically equivalent in the different model runs, then we can say that there is low variability in the model's output and high certainty that the result is robust. Conversely, if there is a wide range of model results, then the result is highly uncertain and we need to be careful about interpreting the results in decision-making.

Often, when people say that they have performed a 'sensitivity analysis' of their model, they really mean an uncertainty analysis. For example, Lozano *et al.* (2009) present an optimization model that chooses CHP systems and their operation schedules so as to minimize the total annual cost. This is a typical urban energy system modelling application and, as noted by the authors, the optimal system structure could change for different values of total energy demands, technology performance and financial conditions. To address these questions, the model was run five times with different prices of natural gas, ranging from €0.015 to 0.035 per kilowatt-hour (kWh), at €0.005 per kWh intervals; a summary table is then presented to show the variations in the technologies chosen by the model and overall key performance indicators. A similar analysis is also presented for changes to a key financial term encapsulating maintenance and capital recovery costs. The confusion in terminology arises because, in the above application, the analyst wishes to understand how 'sensitive' the model result is to a change in parameters. However, as we will see below, 'sensitivity analysis' applies to a more specific case where the goal is more complex than simply describing the variability of the output.

There are four steps to performing an uncertainty analysis, illustrated in Figure 12.2. First, we must identify the variable input parameters and describe their variation. For example, we might assume that the capital cost of a key piece of equipment can be described as a normal distribution with a mean cost μ and a standard deviation σ. In the case above, the authors have assumed a uniform distribution of the natural gas price ranging from €0.015–0.035 per kWh. Second, we draw n random values from these distributions. Third, we compute the model n times using each sampled value drawn in the previous step. Finally, some measure of the variability of the model's output is calculated. This can include a histogram of model outputs, as well as formal metrics of variability such as variance.

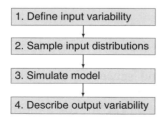

Figure 12.2 The steps of an uncertainty analysis.

Let us turn to each of these steps and describe them in more detail. Throughout we will provide the relevant code to see how the analysis can be performed in R (R Development Core Team 2011). For the sake of argument, let us assume that we have a simple model $y = x_1 + x_2 + x_1 x_2$.

Define input variability

Identify the variable input parameters and describe their variability. For this case, let us assume that the true value of neither x_1 nor x_2 is known. Drawing on our knowledge of the situation, we can only describe the variability as a probability distribution. As Table 12.1 shows, there are many different types of probability distribution. In this case, we might assume that x_1 is normally distributed with a mean of 10 and a standard deviation of 2; we can then say that x_1 is drawn from a normal distribution with the notation, $x_1 \sim N(10, 2)$. Similarly, let us assume that x_2 is uniformly distributed between 0 and 1; that is, $x_2 \sim U(0, 1)$.

Sample input distributions

Ideally, we are trying to describe the output of the model as a probability distribution of some kind, which summarizes the uncertainty in the model's output. For all but the simplest models, this is very difficult to do analytically and therefore the preferred technique is Monte Carlo simulation. This means running the model many times and

Table 12.1 Some common statistical distributions

Distribution	Notation	R command	Probability distribution
Uniform	$U(a, b)$	`runif`	
Normal	$N(\mu, \sigma)$	`rnorm`	
Triangular	$T(a, b, c)$	`rtriangle`	
Weibull	$W(k, \lambda)$	`rweibull`	

To find out more about each R command, type `?command` at the command prompt, e.g. `?rnorm`. Note that the triangular distribution requires the `triangle` add-on package (Carnell 2009).

using the simulation results to approximate the true probability distribution. Typical analyses might involve 500 to 1,000 runs, depending on the complexity of the model, the time it takes to solve, and the precision requirements for the result.

Once we have decided on how many simulation runs to perform, we need to create a sample of parameter values drawn from the distributions specified in step 1. Most software systems have a means of creating random numbers with a given distribution (at least for common distributions like the normal or uniform); however, this naïve method creates a problem particularly when calculating large models. As described in Box 12.1, naïve sampling might result in the model running very similar parameter combinations. If our model takes a long time to run, we would therefore prefer to *quasi-random* numbers that cover the parameter space evenly.

Box 12.1   Quasi-random sampling

One of the drawbacks of the Monte Carlo simulation is that it can potentially take long periods of time to complete the necessary model runs. If a model needs to be run hundreds or thousands of times in order to generate a reliable distribution of outputs, this can take hours or days for even fairly simple models. Efficient sampling of the input parameter space is therefore a key consideration.

To illustrate this problem, let us consider a basic Monte Carlo problem: estimating π. To do this in a Monte Carlo framework, we will draw random x and y values assuming a uniform distribution $U(0, 1)$ for each. For each simulated point, p, we will calculate its distance d from the origin $(0, 0)$. If d is less than or equal to 1, then we can say that point is within a quarter circle; if not, it lies on the unit square. We can then calculate π by assuming that the ratio of the point counts is equal to the ratio of the true areas, namely:

$$\frac{N_{\text{circle}}}{N_{\text{square}}} = \frac{A_{\text{circle}}}{A_{\text{square}}}$$

$$= \frac{\pi r^2 / 4}{r^2}$$

$$= \pi / 4$$

Figure 12.3 compares the two sampling methods. Notice that in Figure 12.3a, which was created using R's standard random number generator, we get locations within the parameter space that are under- or over-represented. In other words, there are locations with no points at all and others with two or more in nearly the same location. Figure 12.3b on the other hand uses Sobol' low-discrepancy quasi-random sampling to cover the space much more evenly.

We can compare these methods by simulating π using each approach in R and Figure 12.4 shows the resulting efficiency of the quasi-random sampling.

Suppose we needed to know π within a 1 per cent error. With Sobol' quasi-random sampling, this would take approximately 100 simulations, whereas with a naïve random sample we would require approximately 10,000 simulations. As expected, with a large enough sample, both methods converge to the same answer. However, the example clearly demonstrates the efficiency gains to be had by using intelligent sampling techniques to randomly, but evenly, cover the parameter space.

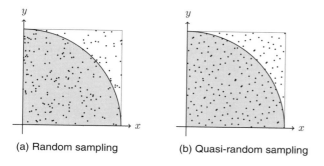

(a) Random sampling   (b) Quasi-random sampling

Figure 12.3 Comparison of naïve random and quasi-random sampling for the Monte Carlo estimation of π. Notice the much more uniform coverage in the quasi-random case, which was generated using a Sobol' sequence.

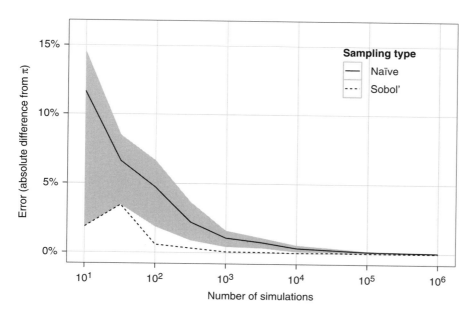

Figure 12.4 Comparison of error rates between naïve random sampling and Sobol' sequence quasi-random sampling in a Monte Carlo estimation of π. The naïve sampling results include an inner quartile confidence interval based on 50 repetitions of the experiment at each simulation value, whereas the Sobol' method generates the same sequence each time.

This can easily be done in R. The code shown in Listing 12.1 below demonstrates how we can create a sample of 1,000 values for x_1 and x_2 from their respective distributions. The method works by first generating a series of quasi-random values uniformly distributed between 0 and 1 for p parameters. In this case, we have two parameters so the `sob` variable in the code below will be a $1,000 \times 2$ matrix of 0–1 values. These values are then treated as cumulative probabilities from their respective distributions and converted back to the original values using R's quantile functions (e.g. `qnorm(0.5,0,1)` returns a value of 0, the 50 per cent quantile value for a normal distribution $N(0, 1)$). There are many types of quasi-random sequence; the following code uses the Sobol' sequence.

Listing 12.1 R-code for creating quasi-random parameter samples. An online version of the sample generation code can be found at `https://gist.github.com/1730440`.

```
# Define a function to generate a Monte Carlo sample
generateSample <- function(n, vals) {

  # Packages to generate quasi-random sequences
  # and rearrange the data
  require(randtoolbox)
  require(plyr)

  # Generate a Sobol' sequence
  sob <- sobol(n, length(vals))

  # Fill a matrix with the values inverted from
  # uniform values to distributions of choice
  samp <- matrix(rep(0,n*(length(vals)+1)), nrow=n)
  samp[,1] <- 1:n
  for (i in 1:length(vals)) {
    l <- vals[[i]]
    dist <- l$dist
    params <- l$params
    fname <- paste("q",dist,sep="")
    samp[,i+1] <- do.call(fname,c(list(p=sob[,i]),params))
  }

  # Convert matrix to data frame and label
  samp <- as.data.frame(samp)
  names(samp) <- c("n",laply(vals, function(l) l$var))
  return(samp)
}

# The number of sample we wish to draw
n <- 1000
```

```
# A list describing each variable and
# their respective distributions
vals <- list(list(var="x1",
                dist="norm",
                params=list(mean=10,sd=2)),
           list(var="x2",
                dist="unif",
                params=list(min=0,max=1)))

# Generate the sample
samp <- generateSample(n,vals)

# Access the input variable samples
x1 <- samp$x1
x2 <- samp$x2
```

Simulate model

Once we have generated the sample, these values can be run through the model. If the model is a simple one, like we have here, the calculation can be done directly in R:

```
y <- x1 + x2 + x1*x2
```

However, it may be that your model is much more complicated and needs to be run on specialist software. In that case, the best strategy is to export the samp data frame to a comma-separate text file and then import it with appropriate software. For example:

```
write.csv(samp,"sample.csv",row.names=F)
```

This will give a file with the following structure:

```
"n","x1","x2"
1,10,0.5
2,11.3489795003922,0.25
3,8.65102049960784,0.75
4,9.36272127207125,0.375
5,12.300698760752,0.875
6,10.6372787279288,0.125
```

Describe output variability

Once the model has been run, you will need to store the results in a file and then import it back into R (or other software for further analysis). Assuming your data is stored in a file called `results.csv`, you can run a command like:

```
results <- read.csv("results.csv")
```

This can then be analysed in R in many different ways. In the case below, we have made a simple histogram of the result (Figure 12.5) and provide a few key statistics.

```
> summary(y)
   Min. 1st Qu.  Median    Mean 3rd Qu.    Max.
  5.866  12.220  15.120  15.500  18.430  30.550
```

These, then, are the steps of an uncertainty analysis. The aim is to describe the variability of a model's output in response to the variability of model inputs. In many ways, it can be thought of as an exploratory statistical technique in which we do not try to draw specific inferences, but merely attempt to gain an understanding of the resulting data. Uncertainty analysis is a vital tool in the assessment of any model's uncertainty and even if you are planning to perform a more detailed sensitivity analysis, as described below, you should also include an uncertainty analysis, particularly when

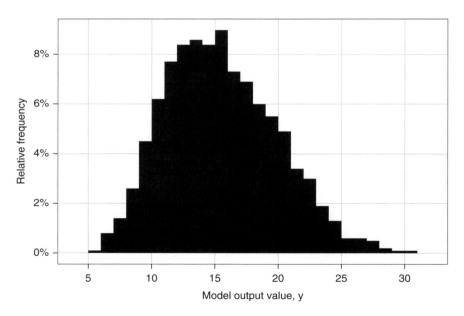

Figure 12.5 Histogram for the results of the model $y = x_1 + x_2 + x_1 x_2$, where $x_1 \sim N(10, 2)$ and $x_2 \sim U(0, 1)$.

trying to communicate model results to non-experts. In our experience, plots of trends with confidence intervals showing the uncertainty of key parameters are often more effective and more intuitive to understand than specific sensitivity indices.

12.3.2 *Sensitivity analysis*

Describing the variability of a model's output is a useful technique as it prevents us from drawing our modelling conclusions too narrowly. But having identified this uncertainty, wouldn't it be nice to figure out its source? In other words, how is the variability of a model's inputs feeding through the model to lead to output variability and can we somehow use this information to reduce the variability of the model's output?

This type of analysis is known as *sensitivity analysis*, 'the study of how uncertainty in the output of a model (numerical or otherwise) can be apportioned to different sources of uncertainty in the model input' (Saltelli *et al.* 2008: 1). Performing a sensitivity analysis is more complicated than an uncertainty analysis, but the general process is the same. First, we design a sample of model inputs to reflect our knowledge about the uncertainty of each input parameter. Quasi-random low-discrepancy sequences can be used as described in Box 12.1 to ensure that we cover the sample space efficiently. We then run the model multiple times using these values. However, rather than simply presenting the resulting variability of the model's output, e.g. as a histogram or summary statistics, sensitivity analysis involves the calculation of indices that capture the effects of each input on the output.

Consider a simple model, $y = x_1 + x_2 + x_1 x_2$, and suppose that variable x_1 is drawn from a normal distribution with a known mean and variance. In a sensitivity analysis, we would like to know how much of the variability in the output metric y can be attributed to the variability of x_1. Notice however, that x_1 can effect the value of y in two ways: on its own, as the standalone term x_1, and in conjunction with other variables, such as $x_1 x_2$ here. In sensitivity analysis, we often wish to distinguish between these effects and so the term *first-order sensitivity* is used to refer to the effect of the variable on its own and *interaction* or *total sensitivity* effects to capture linkages with other variables. These concepts can be formalized into measures such as Sobol' first-order and total-sensitivity indices as discussed below.

When planning a sensitivity analysis, it is important to be clear of the purpose of the analysis. Table 12.2 shows some of the common applications of sensitivity analysis. All of these methods may be applicable in the analysis of urban energy systems. Variance cutting, for example, may be particularly useful if the model is simulating the reliability of a generating facility, storage device or other component, and you wish to identify those inputs which need to be better understood in order to ensure reliable performance. Factor mapping is very useful in the context of optimization modelling where certain parameter combinations can result in infeasible problems. By sorting the results into feasible and infeasible solutions, and comparing the distributions of input parameters on each set (e.g. via box-plots or χ^2-tests), it can be possible to determine the feasible and infeasible input parameter space.

In general, we often wish to calculate a list of first-order and total sensitivity indices for each of the uncertain input parameters in our model in order to identify those

Table 12.2 Settings for sensitivity analysis

Setting	Description
Factor prioritization	Identify variables which, when fixed to their 'true' values, would lead to the greatest reduction in output variance
Factor fixing	Identify variables which have very little effect on model output and therefore can be fixed at any value within their uncertain range
Variance cutting	To reduce the variance of the output below a certain threshold, e.g. for risk analysis
Factor mapping	Which input values lead to desired or undesired outputs?

See Saltelli *et al.* (2008: 156) for more details.

parameters whose variability has a significant impact on output variability and those with an insignificant impact. Saltelli *et al.* (2008) and de Rocquigny *et al.* (2008) provide detailed descriptions of the relevant methods and their application. The key issue is to be aware of the computational cost of the various sensitivity analysis techniques. For example, variance-based global sensitivity indices represent best practice but they run at a cost of $n(p+2)$ model runs where n is the number of simulation (typically ~ 500) and p is the number of uncertain input parameters. Large optimization models, particularly those that contain network routing or other combinatoric features, may take an hour to reach a satisfactory optimization gap. That means that for a model with 10 uncertain inputs, one might be looking at $500(10+2) = 6,000$ runs taking over 8 months to complete! For these cases, elementary effects techniques can be very useful for initial screening of variables. The Morris elementary effects method, for example, calculates proxies for first and total sensitivity indices at a cost of $r(p+1)$ runs, where r is a value typically between 4 and 10.

Most of the common sensitivity analysis methods are implemented in the R package `sensitivity` (Pujol 2008). However, as this discussion shows, it is important that the specific method is carefully chosen depending on the analysis at hand. For models with hundreds or thousands of input parameters, it simply is not feasible to conduct variance-based global sensitivity analysis in most cases[1] and elementary effects or other filtering methods will be needed to whittle the parameter list down to a few key variables of interest.

12.4 An example: meeting future carbon targets in Newcastle-upon-Tyne

To illustrate these principles, we will now present an example based on the city of Newcastle-upon-Tyne.[2] Like many urban areas, Newcastle is trying to come to grips with the policy challenge of climate change and has identified a number of possible options for increasing the efficiency of its buildings and energy systems. These include retrofitting domestic buildings, installing large amounts of renewable energy (including small- and large-scale wind turbines, geothermal, waste-to-energy and biomass options), and creating a large district heat network. The overall goal of these measures is to reduce carbon emissions by 80 per cent by the year 2050, but also to achieve a series of interim targets as a way of ensuring that progress is on track.

This problem can be framed as an optimization problem, similar to the one described in Chapter 9. As input, we have known demands for energy services such as electric power and space heating for the year 2008. We also have a list of available interventions, each characterized by their costs, performance and resource requirements. The overall goal is to meet each of the carbon budgets, including the final 80 per cent savings target, at the minimum total cost.

If this problem involved a single period in time, e.g. designing what to do to Newcastle's energy system next year, then we might be able to get away with a single optimization run. That is, we might be able to pull together reliable data for all of the model's parameters so that when we run the model and get an optimal solution, we have a degree of confidence that this is the correct solution. In the post-normal science model above, this corresponds to the applied science or professional consultancy domains, depending on the difficulty of the case study. It is a large and complicated problem, but with some degree of research and judgement, the uncertainties with the problem can be reduced to an acceptable level.

However, our goal is to estimate the energy system's evolving structure over a 40-year period. This significantly increases the level of uncertainty in the problem, as the final solution might include technologies that do not even exist yet. For those technologies that do exist, there may also be questions about when investments should be made. For example, as more and more photovoltaic panels are installed, the cost per kilowatt decreases through learning effects and economies of scale. Does this mean that we should install the panels in 2020, or hold off until 2030? Similarly, we cannot be certain how demands for energy services will change over such a long period of time, or whether the national electricity grid might be substantially decarbonized in future. At best, we can only hope to summarize these parameters as a set of scenarios, or distribution of parameter values.

This example is therefore arguably a case of 'post-normal' science. Newcastle-upon-Tyne faces a high degree of uncertainty in choosing its energy strategy over the next 40 years. Even if we accept that the process should be iterative, with an initial strategy revisited year after year and amended as necessary, there are significant costs to an incorrect strategy: expensive assets may be constructed only to languish in disuse as cheaper alternatives come along shortly thereafter. Uncertainty and sensitivity analysis allow us to assess these issues in a systematic way, to evaluate a range of scenarios and identify the common solutions, and to determine which parameters need greater certainty in order to better know the future. The rest of this section will go through the analysis step-by-step, illustrating how these methods can be put into practice using R and GAMS.[3]

12.4.1 Problem definition

We begin with a formal definition of the problem. To keep things simple, we will consider Newcastle as a single entity, rather than breaking it up into its constituent districts and neighbourhoods. As discussed in Chapter 9, this has the disadvantage that the spatial characteristics of technologies such as district heating are not properly considered. However, this simplification enables us to demonstrate the techniques of uncertainty and sensitivity analysis more clearly and with a faster model.

Next, we will define the temporal periods of the model. We will start in 2008, the last year of available observed demand data from the UK Department of Energy and Climate Change. We will then model the energy system, going forward to 2050 at decadal intervals (i.e. 2008, 2010, 2020,..., 2050). At each period, a carbon savings target is defined based on national policy proposals, carrying the implicit assumption that all UK cities will have to contribute equally to the national targets.

We then describe the energy services demands. For this example, we will concentrate on the demands for two energy services: electric power and space heating. Other energy demands, e.g. transport fuel, will be ignored.

To meet these demands, we must declare a number of possible technologies. These include domestic gas boilers, heat pumps and storage heaters, as well as retrofit measures such as double-glazing, loft insulation or wall insulation, all of which help to satisfy effective demands. With these characteristics defined, the question is: what is the lowest cost energy system that meets the carbon budgets?

12.4.2 Selecting the parameters and analysis setting

To begin the analysis, the uncertain model input parameters of interest must be identified. Table 12.3 shows seven parameters for the Newcastle problem. The list is not exhaustive but represents the types of uncertainty that a decision-maker might face when planning an urban energy system. For example, project finance uncertainty is captured by the discount rate, inflation rate and estimated cost of achieving a behavioural change in demand; changes in the physical constraints on the problem are represented by the growth rate of electricity demand and the potential build rates in the domestic and commercial sectors; and the maximum CHP limit represents a policy constraint

Table 12.3 Description of parameter distributions for Monte Carlo uncertainty and sensitivity analyses

Parameter	Distribution	Notes
Discount rate	$U(0.035, 0.10)$	Treasury Green Book rate for large investments, through to a commercial threshold.
Inflation	$\Gamma(3.09, 1.13)$	Γ distribution fitted to historic UK data (1989–2010)
Electricity demand growth	$U(0, 0.05)$	Based on Hong *et al.* (2009) estimated rebound effect.
Build rate (domestic)	$T(2, 10, 20)$	Dwellings per day that can be retrofitted, based on Newcastle City Council estimates.
Build rate (commercial)	$T(1, 10, 20)$	Installations of commercial technologies per year, based on Newcastle City Council estimates.
Behavioural change cost	$U(0, 200)$	Cost of achieving a behavioural change in demand, e.g. via an energy display monitor.
Max. CHP output	$U(0, 134)$	MW, compared to average 2008 electricity demand of 134 MW.

Notes: $U(min, max)$ = uniform distribution, $\Gamma(s, r)$ = Gamma distribution with given (s)hape and (r)ate parameters, $T(min, mode, max)$ = triangular distribution.

that might be affected by local decisions. In all cases, these are variables whose true value is unknown but the corresponding uncertainty can be expressed as a probability distribution using observed or hypothesized data.

Next, we must decide what type of analysis to perform. Certainly a descriptive uncertainty analysis will be beneficial, helping us to understand the range of possible outcomes. More specifically however, we can refer to the sensitivity analysis settings listed in Table 12.2. In this case, we wish to know which parameters are most important in determining the policy direction for Newcastle, suggesting the use of the factor prioritization and factor fixing settings. This means that Sobol' first-order and total sensitivity indices should be calculated as they reflect the amount of variance in a given model output in response to variation in the input parameters, both on their own and through interactions with other variables. The analysis requires $n(p+2)$ model runs, with a suggested n value of 500 and the number of parameters $p = 7$ as described above (Saltelli *et al.* 2008). Therefore, 4,500 model runs were performed and sampling was done with a Sobol' quasi-random sequence to ensure efficient coverage of the parameter space. We can use the same model simulations for our uncertainty analysis exploring the variability of key model outputs.

Using the distributions described in Table 12.3 and the R code shown in Listing 12.1, we then create quasi-random samples for the seven input parameters. The resulting list of values is saved to a comma-separated file, which can be read into GAMS. As Listing 12.2 below shows, the general idea is to set up the model as normal and then put the main SOLVE statement within a loop. Each time through the loop, the relevant values in the model are updated and the model solved again. Appropriate results are also stored with each iteration and then written to an output file for later processing. More information on the syntax for these expressions can be found in the GAMS manual (Rosenthal 2012), but note the variable names used below are specific to a particular model and will need to be changed as appropriate.

Listing 12.2 Example GAMS code for reading in parameter values, running a Monte Carlo simulation and storing the results for later analysis.

```
* Previous model code describing variables, parameters, equations, etc
* ...

* Define model as normal
Model Newcastle /ALL/ ;

* Load input data from file
SET k /1*4500/;
SET input / "INP1","INP2","INP3","INP4","INP5","INP6","INP7" /;
TABLE inputs(k,input)
    INP1 INP2 INP3 INP4 INP5 INP6 INP7
$ONDELIM
$include inputs.csv
```

```
$OFFDELIM

* Define overall output metrics
SET labTotal /"COST","CARBON"/;
PARAMETER result(k,labTotal);

* Define year-by-year output metrics
SET labYear / "LOFT","WINDOWS","WALLS","CAVITY" /;
PARAMETER yearresult(k,y,i,labYear);

* The main solve loop
LOOP(k,

  * Load the parameter values for iteration k
  discount = inputs(k, 'INP1');
  infl = inputs(k, 'INP2')/100;
  demInf('elec') = inputs(k, 'INP3');
  maxRetrofit = inputs(k, 'INP4')*365;
  maxComm = inputs(k , 'INP5');
  PCC('beh_chng_heat','y0') = inputs(k, 'INP6');
  maxCHP = inputs(k, 'INP7')*3600;

  * Update the relevant uses of these values
  * For example, the following lines change capital cost
  * values to account for the new inflation value
  PCC(p, y) = PCC(p, 'y0')*((1 + infl)**elapsed(y));
  demand('elec',i,t,y) = demand('elec',i,t,'y0')*((1+demInf('elec'))**
      elapsed(y));

  * Solve the model
  SOLVE Newcastle USING MIP MINIMIZING SS_Z ;

  * Write the results
  * If the model solves successfully, save the results
  * If not, save placeholder values
  IF(Newcastle.modelstat EQ 8,
    * The overall results for each model run
    result(k,'COST') = SS_Z.L;
    result(k,'CARBON') = totcarbon.L;

    * The detailed data for each year
    yearresult(k,y,i,'LOFT') = installed('loft',i) + NPO.L('loft',i,y);
    yearresult(k,y,i,'WINDOWS') = installed('windows',i) + NPO.L('
      windows',i,y);
    yearresult(k,y,i,'WALLS') = installed('walls',i) + NPO.L('walls',i
      ,y);
```

```
      yearresult(k,y,i,'CAVITY') = installed('cavity',i) + NPO.L('cavity'
          ,i,y);
   ELSE
     result(k,labTotal) = -99;
     yearresult(k,y,i,labYear) = -99;
   );

);


* Write the overall results to a file
FILE f /"overall-results.txt"/;
f.nd = 10;
f.lw = 0;


* Write the file header
PUT f;
PUT "k";
LOOP(labTotal, PUT "," labTotal.tl; );
PUT /;


* Write the values
LOOP(k,
   PUT ord(k):0:0;
   LOOP(labTotal, PUT "," result(k,labTotal); );
   PUT /;
);
PUTCLOSE f;


* Write the annual results to a file
FILE fy /"annual-results.txt"/;
fy.nd = 8;
fy.lw = 0;
fy.nw = 0;


* Write the file header
PUT fy;
PUT "k,y,i";
LOOP(labYear, PUT "," labYear.tl; );
PUT /;


* Write the data
LOOP(k,
   LOOP(y,
     LOOP(i,
       PUT ord(k):0:0,",";
       PUT year(y):0:0,",";
```

```
    PUT ord(i):0:0;
    LOOP(labYear, PUT "," yearresult(k,y,i,labYear); );
    PUT /;
  );
 );
);
PUTCLOSE fy;
```

A significant challenge with Monte Carlo simulation of optimization models is that some parameter combinations result in infeasible solutions. The presence of these values means that the sensitivity indices cannot be correctly calculated and so a preliminary analysis should be performed to check how many model results were infeasible. This is an example of the factor mapping setting as we wish to determine which inputs lead to feasible and infeasible results. In the present case, it was found that 12.1 per cent of the parameter combinations resulted in an infeasible model solution. By examining the distribution of the input parameters in the feasible and infeasible result sets, it was determined that growth in electricity demand was the key variable. When demand growth rates were above 4 per cent, there was a 99.6 per cent chance of an infeasible solution versus a 90.8 per cent chance of a feasible result when demand grew at less than 4 per cent per annum. The corresponding distribution was therefore changed to $U(0, 0.04)$ and the analysis run again; this led to only 0.20 per cent infeasible solutions. These remaining infeasible solutions were then replaced with boot-strapped values (i.e. randomly selected with replacement from feasible model results) before calculating the sensitivity indices.

12.4.3 *Results and interpretation*

Four output variables of interest were identified: the cumulative energy system cost (i.e. the objective function value), the cumulative carbon emissions over the analysis period, the percentage of heat demand met by combined heat and power, and the percentage of total energy demands met by renewable energy generated within the city. As Figure 12.6 shows, the most important input variables differ depending on the output metric of interest. For example, the total system cost is affected primarily by the variability of growth in electricity demand and the discount rate, whereas for carbon emissions, the contribution of each variable is similar. The amount of heat provided by CHP within the optimal energy system is most sensitive to the maximum CHP output constraint (as expected) but also the growth in electricity demand. The share of total energy demand met by renewables is sensitive primarily to electricity demand growth. In the context of a factor prioritization setting, these results indicate that the practical next step is to identify the 'true' value of the electricity demand growth and maximum allowable output from CHP parameters, for example, by conducting further research or asking policy makers to make a decision. The model can then be re-run and we would expect a smaller range of outcomes, providing more better decision support.

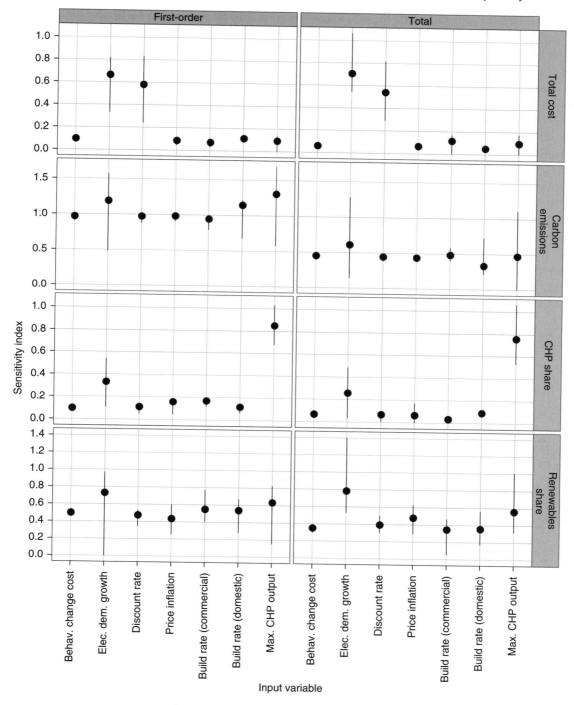

Figure 12.6 Sensitivity analysis results. First-order index values refer to the effect of variability of a single input (x-axis) on the output of interest (y-axis), whereas the total index captures the effects of that parameter's variability on its own and through interactions with other variables. Error bars give 95 per cent confidence intervals.

Similarly for the factor fixing setting, we can use these results to fix the values of the insignificant variables (e.g. the discount rate, price inflation, maximum build rates and cost of achieving a behavioural change). One would typically choose the expected value of the distribution and so in the case of the cost of achieving a behaviour change, which is uniformly distributed between £0 and £200, a value of £100 is sensible. Finally, we note that the infeasibility analysis discussed above is itself a type of sensitivity analysis, specifically an application of Monte Carlo filtering (factor mapping). This is primarily useful for the modeller during the preparation of the analysis, but it can have policy implications. For example, if a model is found to be infeasible with a certain range of parameter values, this may indicate that a policy goal such as a carbon constraint cannot be satisfied.

In some cases, it will not be possible to completely fix the value of all input parameters and for these situations an uncertainty analysis is more appropriate. Figure 12.7 uses the results of the above simulation to show such an analysis. The plot gives the range of penetration levels for key domestic retrofit measures, as well as the maximum potential installation rate of each technology based on the Home Energy Efficiency Database (HEED) (EST 2011). The results indicate full penetrations of loft insulation and cavity wall insulation by 2030, and solid wall insulation by 2040. While the

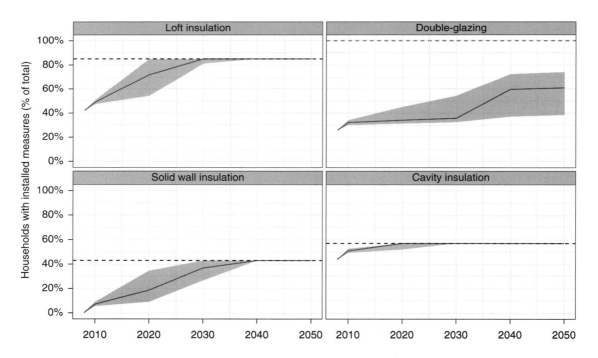

Figure 12.7 Penetration of domestic energy-efficiency measures as part of Newcastle's overall energy strategy. The shaded area represents the interquartile range and the solid line gives the median value. The dotted line indicates the maximum potential installation rate, based on information about building types available in HEED (EST 2011).

rate of double-glazing installation does increase over the analysis period, the full 100 per cent penetration level is not achieved. This reflects the relative cost of double-glazing against other retrofit and low carbon generation technologies. These rates of penetration are also highly dependent on the HEED data, which describes both the current levels of efficiency measures and their maximum potential. The key feature is that we can explicitly illustrate the uncertainty associated with each retrofit option. The results make it clear that the exact levels of penetration will vary depending on the assumptions being used. But they also show consistent trends, meaning that we can be more confident in our conclusions about which measures to pursue first and at roughly what rate.

12.5 Conclusion

Computational models are a vital tool in the analysis of urban energy systems. However, one needs to be careful not to put too much confidence in the results of such tools. We must remember that assumptions and uncertain data are part of the modelling process and affect the validity of the conclusions drawn from the results. In the case of urban energy systems, which are problems with large uncertainties and potentially large consequences of a poor decision, we can refer to the idea of post-normal science as a way of informing our inference processes. That is, we need to recognize the uncertainty inherent in the modelling, describe it appropriately, and discuss it with decision-makers and other stakeholders.

To help this process, this chapter has focused on two relevant techniques. The first is uncertainty analysis, where we describe the variability of our model inputs (e.g. using probability distributions) and then simulate the model many times, plotting the resulting distribution of model outputs. The resulting uncertainty can be shown as summaries of key statistics, histograms or error bars. In the Newcastle case, this helped to demonstrate the consistent importance of key energy efficiency measures such as loft or cavity wall insulation and help decision-makers to better understand the potential impact of their choices.

In contrast, sensitivity analysis is more of a modeller's tool. It helps the model developer to understand which inputs lead to variation in the model outputs. As shown here, a method like Monte Carlo filtering might be used to identify which model inputs lead to feasible or infeasible model results. Or, as shown in the Sobol' indices, a sensitivity analysis can illustrate which parameters are relatively unimportant (and can be fixed to an appropriate value within their range of uncertainty) and which parameters have significant impacts on output variability, either alone or through interactions with other variables.

Both of these methods rely upon Monte Carlo simulation. By running our model multiple times with different assumptions, a better understanding of its behaviour can be gained. However, the use of low discrepancy quasi-random sequences is vital if the input parameter space is to be covered efficiently. We have illustrated how such sequences can be generated with the R language and integrated within a GAMS optimization model.

Notes

1 The exception is, of course, very fast models (<1 second running time) or where parallel processing can be used to speed up the computation.
2 The work in this section was done in collaboration with Carlos Calderon at Newcastle University and has been published as Keirstead and Calderon (2012).
3 GAMS (or the General Algebraic Modeling System) is a popular language for mathematical programming and optimization problems. More information can be found at: http://www.gams.com/

Part IV

Implementing solutions

13 Managing transitions in urban energy systems

James Keirstead

Imperial College London

13.1 Introduction

The previous chapters have presented a range of technologies and modelling techniques to improve the efficiency of urban energy systems. Whether the intervention occurs at the master planning stage, in technology design and operation or in systems integration, the analysis has suggested that significant efficiency gains are possible. However, these chapters also discussed the current situation in many cities, noting high levels of energy consumption and carbon emissions that are linked to slow changing infrastructures. While some new eco-cities will be built in the coming decades, most of the world's population will live in existing cities, meaning that retrofit and other techniques for a more gradual energy transition will be needed if urban energy efficiency is to be improved.

This is a daunting challenge. Buildings, roads and pipelines are very capital intensive and the resulting infrastructures are designed to last for decades, if not centuries. Fortunately, as a recent UK Foresight report noted, '[t]here is extensive evidence that major changes in technologies, institutions and policy approaches are indeed possible over timescales as short as a few decades.' (Foresight 2008: 44). The goal of this chapter is therefore to examine the literature on infrastructure transitions in order to highlight the policies needed to achieve urban energy efficiency goals. The first section provides the theoretical background, before turning to the case of Copenhagen as an example of how a city has transformed its energy system over a relatively short period of time. Finally, we discuss the case of Nakuru, Kenya, as urban energy transitions in developing countries are quite different, with a greater emphasis on the provision of basic commercial energy services.

13.2 Transitions in theory

'Transitions' is a somewhat vague term. It implies the gradual change of a system from one state to the next, but there are questions about what defines a 'state' and what aspects of a system need to change between these states. For example, in an electricity system, we might talk about a generation transition which is concerned with changes in the mix of generating technologies. The beginning and end states might simply be defined by a point in time, but more likely, it will be characterized by the performance of the system, for example, a transition to a low-carbon electricity system with a given carbon intensity factor. In this context, shifts in other system technologies that are not directly related to the generation mix, such as changing domestic meters to smart meters, may be of less interest.

The question of what is or is not a transition therefore depends on the nature of the system and the problem at hand. For the purpose of this chapter, we will distinguish between two types of transition: the 'energy ladder' fuel transition relevant to cities in developing countries, and socio-technical systems transitions, a general theoretical approach applicable in all locations but discussed in the literature mainly from the perspective of developed countries.

13.2.1 *Fuel transitions in developing countries*

The 'energy ladder' refers to a hypothesized transition in energy fuels in developing countries (Leach 1992). In its ideal form, it represents a 'smoothly sequenced evolution' in fuels 'from firewood, to charcoal and kerosene, and ultimately to LPG and electricity consumption' (Barnes *et al.* 2005: 6). This transition is of interest to policy-makers, largely because of the importance of access to high-quality energy sources in promoting economic and social development goals.

At the first stage of the transition, biomass and animal wastes are used as the primary energy resource. These are often burned in inefficient stoves, creating substantial local health impacts such as respiratory illness (Bruce *et al.* 2000). Furthermore, the low efficiencies of these stoves put significant pressure on local biomass resources. As cities grow, the demand for biomass increases and locally available resources decline, greatly increasing the price of these fuels through both scarcity and increased transportation costs. This encourages consumers to seek out alternative fuel sources, particularly as their incomes rise and other options become affordable.

The second stage is therefore marked by the use of alternative fuels such as charcoal and kerosene. For urban residents who can afford these fuels, they offer improved convenience, efficiency and cleanliness. Kerosene stoves, for example, can be started on demand and do not require the lengthy kindling time necessary for wood fires. The fuel also burns more cleanly with evidence from rural India showing that lung function is less impaired in kerosene-using households, compared with biomass households (Beherc *et al.* 1994). However, the urban growth that encourages existing residents to make this switch is driven by new residents arriving from rural areas. These households may maintain a preference for traditional biomass fuels, meaning that the city does not transition wholesale from one fuel type to the other.

With rising incomes, modern fuels such as LPG (liquefied petroleum gas) and electricity become increasingly affordable marking the third stage. A key question in many developing countries is the capacity of local markets to deliver these fuels. Are transportation and distribution networks sufficient to meet demand? Private supplies, for example, diesel generators for the provision of electricity rather than grid connection, may be the most convenient form of energy infrastructure. It should be noted that these transitions are also accompanied by shifts in technology (e.g. from an open fire, to a kerosene stove, to an electric range) as well as increased consumption, as the availability of electricity, for example, enables consumers to diversify their consumption beyond basic heating and cooking to include additional services like lighting, refrigeration and leisure appliances.

Overall then, the 'energy ladder' suggests a fairly linear shift in fuel types, from biomass to charcoal and finally to commercial fuels like electricity (see, e.g. Figure 13.1). However, the literature now recognizes that this model is overly simplistic. Factors such as the rate of urbanization, per capita incomes, the availability of local energy resources and cultural preferences are all known to affect consumer preferences for specific fuel types. As a result, neither cities nor individual urban households move exclusively from one fuel category to the next, and multiple fuel types can be used by a single consumer for different purposes. This will be demonstrated in the Nakuru case study below.

13.2.2 Socio-technical systems transitions

In developed countries, the commercial fuels that mark the third stage of the 'energy ladder' are the dominant forms of consumption. In the UK, for example, biomass accounts for only 1.1 per cent of total primary energy consumption (DECC 2011a) and in urban areas, the dominance of electricity and natural gas is even more complete: only 0.02 per cent of London's total energy demands are met by biomass (DECC 2008).

However, these mature urban energy systems have their own challenges. National energy policy issues such as affordability, energy security, ageing infrastructures and

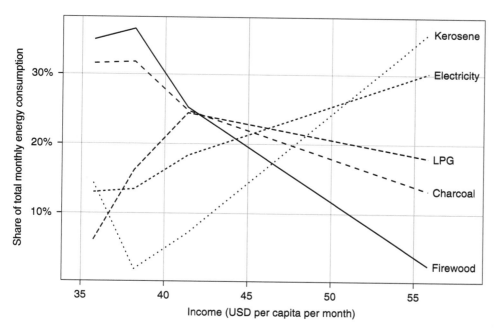

Figure 13.1 Share of total monthly energy consumption by fuel type. 1988 data from 45 developing country cities.
Data source: Barnes *et al.* (2005).

climate change are driving talk of energy system transitions toward visions of 'low carbon' futures (DTI 2007). But, whereas in the developing country context, the dominant causes of energy transitions are rising levels of income and urbanization, changes in mature infrastructures are more complex. The electricity system in the UK has existed in its present form arguably since the creation of the National Grid in 1933. Transitions in generating mix have happened, such as the 'dash for gas' in the 1990s, but as the result of changes in national and international markets for technology and fuels, and shifts in the domestic policy environment. For city dwellers, these changes have made very little difference to their energy consumption as the dominant mode of the grid – centralized supply with distribution to local consumers – has remained unchanged. In this sense, we can talk of transitions on the supply side but very few on the demand side, particularly in cities.

That may change however. Recent research, such as the UK Government's Foresight report on Sustainable Energy Management and the Built Environment (Foresight 2008), has highlighted the need for transitions in our existing energy systems and noted that these transitions are likely to affect our cities. Of the four scenarios discussed by the report's authors, local environments play a significant role in all of the future energy system configurations through the provision of new distributed generation capacity, informed and actively engaged consumers, and radical shifts in the efficiency of buildings and patterns of consumption.

How then can we analyse and understand such transitions? One of the most popular techniques, widely used in the analysis of energy system transitions, is the notion of *technological transitions*, defined by Geels (2002: 1257) as 'major, long-term technological changes in the way societal functions are fulfilled'. In addition to the obvious changes in technology, this framework also considers the other factors that must also change to ensure that a new innovation is widely adopted: user preferences, regulation and laws, economic structures, cultural norms, and other complementary infrastructures.

Central to the analysis of technological transitions is the division of the technological and social environment into three levels: the landscape, regime and niches (see Figure 13.2). At the top level, we have the *landscape*. This represents the dominant trends in the current socio-technical system and therefore includes things such as the underlying level of economic development, cultural norms, environmental concerns, and so on; in other words, the landscape is the context in which technologies and actors function. The next level is the *regime*. There can be more than one regime, each representing a consistent combination of technologies, institutions and users and use patterns. For example, in their analysis of the Dutch electricity system, Verbong and Geels (2007) identify three regimes: a shift from regional to national energy policy between 1960–1973, a period of direct government intervention from 1974–1989, and a period of economic reform that restructured the electricity industry to essentially its present form (1990–2004). Each regime period is demarcated by relatively stable arrangements of technologies and institutions, such as a liberalized market with competition in upstream centralized electricity generation or an activist national energy policy focused on nuclear power and natural gas systems.

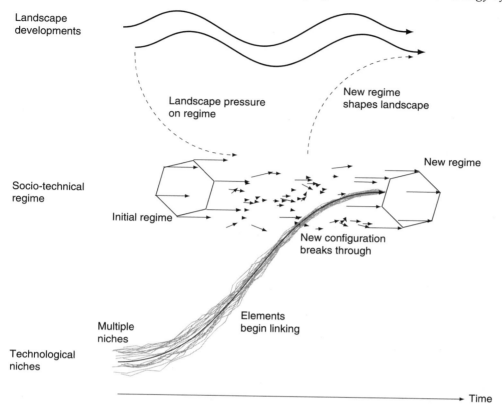

Landscape
developments

Landscape pressure
on regime

New regime
shapes landscape

New regime

Socio-technical
regime

Initial regime

New configuration
breaks through

Elements
begin linking

Multiple
niches

Technological
niches

Time

Figure 13.2 A multi-scale view of technological transitions.
Based on Geels (2004: 915).

These regimes can co-exist at a single time, provided they serve complementary functions within society, but it takes a long period of time for regimes to change. For example, in the current situation, there is increased concern about climate change at the landscape level. The electricity generation regime is dominated by fossil-fuelled power stations, and so the question arises: how can low carbon alternatives enter the system and become the dominant regime themselves? This is achieved through *niches*. Niches represent a protected space in which technologies and the people that use them can experiment, learning both about the performance of the technology itself but also building a constituency of users, manufacturers, policy-makers and others to support the innovation. Returning to the example of the Dutch electricity system, Verbong and Geels (2007: 1036) note that, at the time of writing, the regime did not provide a sufficient window of opportunity for new technologies like renewable energy to enter the system. 'But if such a window emerges in the future, for instance because of climate change shocks or shifts in public opinion, then radical niche-innovations should be sufficiently developed to take advantage of it. That is why it is important to keep them alive, and facilitate learning processes, experimentation and network building'.

To promote these niches, the concept of *strategic niche management* has been developed (Kemp *et al.* 1998). Although the idea of these niches is that the technology and supporting systems evolve through the process of experimentation, there is some scope for government to promote niche development. Five steps are necessary:

Choosing a candidate technology Technologies for niche development should be outside the existing technological regime and offer significant benefits (e.g. to mitigate greenhouse gas emissions in the case of climate change and energy systems). The technologies should be sufficiently developed that a reasonable assessment of their potential can be performed, but at the same time, offering significant scope for learning and improvement (e.g. cost reduction, performance improvement, business model development). Furthermore, it should already be attractive for certain applications, outside of the mainstream but sufficiently attractive to facilitate a real trial. Readers may wish to refer to NASA's technology readiness levels (TRLs) as a guideline for distinguishing between the different stages of technological development (NASA 2010).

Selecting an experiment This means choosing an application in which to trial the technology. For solar photovoltaics, off-grid communities would be a good example as the dominant regime (grid electricity) is not applicable.

Setting up the experiment The goal here is to balance protection of the technology (i.e. giving it space for trial-and-error improvements) with sufficient pressure to result in a viable mainstream technological system. Variables may include market design, support incentives and the use of complementary technologies.

Scaling up the experiment If a technological system proves its potential in a protected niche, the next step is to ease its introduction into the market by the provision of some support relative to existing technologies. A good example, in the case of renewable energy, is a feed-in tariff, whereby electricity from such technologies receives a price premium in the market.

Removing niche protection Finally, if the technology appears to be settling in well, or if it appears to be unsuccessful, the above elements of niche protection are removed and the system competes with the dominant regime directly.

These techniques have been applied in a number of relevant contexts, although they are sometimes labelled as using a different theoretical approach. Examples include the development of Jatropha biofuels (Eijck and Romijn 2008), solar hot water heaters (Ornetzeder 2001) and green electricity tariffs (Wüstenhagen *et al.* 2003). On an urban energy scale, Evans *et al.* (1999) analysed the urban energy strategy for combined heat and power in the city of Newcastle-upon-Tyne, concluding that although the technology was ready, the necessary social networks could not be constructed to support the widespread adoption of the technology. Specifically, by focusing purely on the price of the new innovation, other motivations for adopting CHP (such as the demonstration value of creating a partnership between the university and industry) were neglected and a coalition around the technology could not be established. The key conclusion is that '[w]ithout the presence of a niche, system builders would get nowhere.' (Kemp *et al.* 1998: 184).

Technological transitions are therefore about much more than the technology alone. In the case of the energy ladder transition, these factors still apply but the structure of the landscape and regimes are arguably less complex. Modern energy systems, representing the accumulated infrastructure and social relations of decades, can be much harder to change and strategic niche management therefore offers significant opportunities to change the system structure. A key observation is that there exists more than one possible niche. There are many possibilities for achieving the sustainability of cities and their energy systems (Guy and Marvin 2001) and, as a result, multiple experiments should be conducted to see which technology or technologies offer the best solution for local circumstances (Moss *et al.* 2000).

13.3 Technology transitions: the case of district heating in Copenhagen

Although for some, the city of Copenhagen is inextricably associated with the ignominious end to the 2009 COP15 climate conference, the reality is that Denmark and its capital have been very successful in reducing their greenhouse gas emissions. The Danish economy has the lowest energy intensity in the European Union and, while the economy has grown by 45 per cent between 1990 and 2007, CO_2 emissions fell by 13 per cent. These results have been achieved 'through persistent and active energy policy focus on enhanced energy efficiency', through measures such as high-efficiency building and appliance standards, public awareness campaigns and taxes on energy consumption. Renewable feedstocks are also important with 19 per cent of final energy consumption coming from renewable sources like biomass, waste and wind (Kemin 2009).

However, perhaps the most important factor is the widespread cogeneration of heat and electricity. As discussed in Chapter 5, cogeneration technologies allow more of the fuel's energy to be used, e.g. by capturing the waste heat of combustion and providing it to a district heating system. Since 1980, the share of Danish electricity cogenerated with heat has risen from 18 per cent to 53 per cent (Kemin 2009) and nowhere is this transition more apparent than in Copenhagen, where 61 per cent of local residents in the greater Copenhagen area (250,000 households) currently receive their space heating and hot water from a city-wide district heating system (CTR 2009, Denmark 2009) (Figure 13.3).[1] Nor is the system static: a recent assessment explored the options for increasing the share of renewable energy in the system from 35 per cent in 2008 to at least 70 per cent by 2025, and expanding the network to households currently using gas heating (Madsen *et al.* 2009).

In this section, we will examine how Copenhagen's district heating system evolved using the technological transitions concepts discussed above. What were the factors that led to these technical innovations and what does it tell us about transitions in urban energy systems generally?

13.3.1 Origins

The story of the Danish adoption of district heating begins with the 1973 oil crisis. When the Organization of Arab Petroleum Exporting Countries announced an oil embargo in response to the Yom Kippur war, the price of oil quickly doubled and, buoyed by

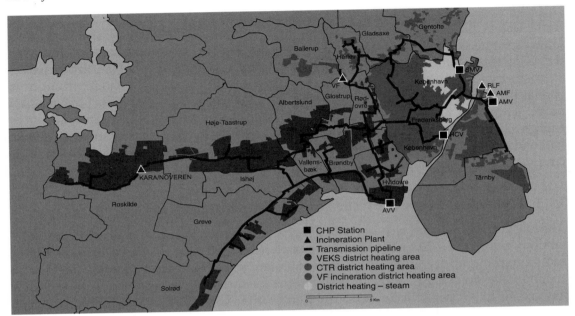

Figure 13.3 The district heating system of the Greater Copenhagen area.
Data source: Madsen *et al.* (2009).

subsequent events, remained high until the mid-1980s. This period of high oil prices led to significant changes in energy policy in the oil-consuming OECD nations, for example, creating fuel efficiency standards in US automobiles and the establishment of the International Energy Agency. It also spurred a new wave of oil exploration and development leading to new supplies such as the North Sea fields.

In their survey of Danish distributed generation, Vleuten and Raven (2006) highlight three distinct phases of Danish energy policy. Between 1900 and 1950, the Danish electricity system was established and characterized by the co-existence of multiple isolated independent systems. Municipalities in particular were active in this phase, promoting electricity supply as a vital public service but also as an opportunity to increase revenue. When efforts were made to centralize electricity production, these municipalities resisted by arguing that Denmark's geography lacked the resources necessary for large centralized power plants and until the 1950s, their arguments were largely supported by financial analyses from the national electricity council. Combined heat and power was an integral part of these local systems, with the first Danish CHP system being built in 1903 at Frederiksberg Hospital (DEA 2011). By the 1950s, CHP had spread with approximately 30 of 70 urban power generation facilities using the technology.

However, throughout the 1950s and 1960s, a transition to a regime of more central-ized electricity production occurred. The main driver for this shift was a changing view of the economics of centralized versus decentralized generation. With rapidly growing demand for electricity in the post-war period, and owing to 'temporary' interconnections created during the war, a centralized model seemed like the most promising option for future electricity provision. As pressure grew to unify the Danish system, many municipalities joined into local groups, either operating their own larger power plants or working in partnership with electric utilities. In this context, CHP systems were maintained either as separate heat utilities or small decentralized generators. Therefore unlike many other countries, Denmark's cities and utilities maintained a level of familiarity with distributed generation and CHP technologies which facilitated the energy system's present strengths.

From the 1970s however, the system evolved into its present hybrid regime. The key factor here was the 1973 oil crisis. This led to significant policy intervention at the national level to reduce Denmark's dependence on imported oil. Through a series of Energy Plans, the proposed solution to this problem morphed from one based on natural gas and nuclear power, to a greater emphasis on renewable energy and CHP. These goals have been supported since then, securing Denmark's present status as a world leader in both wind energy and combined heat and power technologies.

13.3.2 Contrasting policy environments

This national context is important, but obviously our concern is at the urban scale. Here the 1979 heat planning law was a key innovation, as it required local authorities to report on local heat requirements and technologies, as well as to assess future needs and technological options. This local information was fed into the creation of regional heat plans which identified areas where heat supply should be prioritized and possible locations for heat generation and pipework. Equally important was a 1982 executive order that gave local authorities the power to compel buildings, both new and existing, to connect to public supplies of gas or district heating. Similarly, the use of electric heating in new buildings was banned in 1988, and later extended to all buildings in areas with alternative public heating supplies. Crucially, these measures ensure that the economics of district heating are favourable for utilities, who can be assured of a certain customer base (DEA 2011).

A recent assessment of urban energy capabilities globally suggests that these powers are somewhat exceptional (Arup 2011). Only four of 14 C40 cities own or operate a district heating system and five of 46 have implemented actions to promote CHP in district heating. The case of London provides a good counter example. Schulz (2010) studied the combined heat and power situation in London and found multiple factors contributing to its relatively limited use (175 MW installed in London, compared with 3,576 MW in Copenhagen, CTR (2004)). These include:

Geographic factors Although the climates of London and Copenhagen are similar, the fact that the UK had, and has, greater indigenous energy resources than Denmark meant that national energy policy has not focused on efficiency to such a large extent.

Low settlement density Much of London is occupied by residential buildings at a density of approximately 4,000 people per square kilometre. This pattern lacks the concentration and diversity of heat loads necessary for the efficient operation of large CHP systems. In contrast, the average density of Copenhagen is approximately 6,200 people per square kilometre.

Image problems Existing CHP installations in London are primarily located in social housing and are often poorly configured and maintained. This has led to a degree of stigmatization of the technology, with many people reluctant to 'depend on the government' for their hot water and heating.

DH requires long-term coordinated investment Perhaps the largest obstacle is that London is made up of 32 quasi-independent boroughs with a mobile population. This makes it difficult to encourage individual home-owners to invest in district heating or for boroughs to collaborate across parts of London.

Limited local governance The Greater London Authority has very limited ability to raise funds and promote large infrastructure projects, and arguably has less 'urban power' than the much smaller city of Copenhagen (EU 2007). Furthermore, frequent changes in London's governance (not just the elected officials, but the actual structure of the city's governing bodies and institutions) mean that the policy continuity seen in Copenhagen simply does not exist in London (Pimlott and Rao 2002).

Toke and Fragaki (2008) also raise a number of national-level energy policy issues that help explain the observed differences in CHP and district heating use between the UK and Denmark. Chief among these is the UK's liberalized electricity sector, the structure of which discriminates against the sale of electricity from small-scale generation facilities. However, the authors conclude that innovations like a feed-in tariff for CHP, the use of thermal stores to improve the load factors on individual plants, or the aggregation of multiple facilities into a single 'virtual' power plant could all help CHP and DH to thrive within a liberalized setting.

Both Copenhagen and London have ambitious future plans for the use of CHP. However, these policy differences mean that, on the basis of accumulated experience and track record to date, Schulz's (2010: 20) conclusion seems robust; it is indeed 'questionable if the new-founded institutions [in London] have the funding and power to actually deliver those improvements in the ambitious timeframe necessary'.

13.3.3 Lessons learned

After this brief review of district heating and CHP use in Copenhagen, what can we say about energy transitions in mature developed cities? The first conclusion is that relatively rapid transitions are possible. It has been a little over 30 years since the first Danish heat law created the foundation for the expansion of district heating and nearly two-thirds of Copenhagen households are now provided with district heating, fired by biomass, waste and other fuels, leading to significant carbon savings. However, the second observation is that such a transition requires a consistent policy framework, leading to the accumulation of capabilities and experience within government, citizens and utilities that support such highly integrated systems. A notable feature, particularly compared with the relatively *laissez-faire* approach to energy markets in the UK, is the use of active market intervention, such as compulsory connection and the banning of electric heating, to ensure the success of the network. This approach may be unfamiliar in liberalized markets but the results are hard to dispute. Most households in Copenhagen merely have to call up the local district heat utility and they will be connected in a week (Danish Energy Saving Trust 2011), something that is unthinkable to residents of London.

Using the language of technological systems discussed above, we might therefore say that district heating represents a *regime* in Copenhagen and just about registers as a *niche* in London. Fortunately, the Copenhagen experience offers a few lessons for how this niche might be expanded. The early history of the Danish system had its roots in local experimentation with individual municipalities and industries providing their own cogenerated heat and power. Small schemes in the UK, for example, in office blocks and social housing, are a current example. For this niche to grow though, there will need to be appropriate policy support. In the Danish case, the co-benefits of district heating systems – namely the ability to raise revenue for local municipalities, the environmental benefits of these technologies and the overlap with strategic national energy priorities – all helped to support the growth of district heating. These synergistic effects are widely recognized as a key part of urban action on climate change (Bulkeley and Betsill 2003, Hammer *et al.* 2011), and so UK authorities should look at the opportunities to combine district heating with other policy goals, such as reducing fuel poverty.

Unfortunately, *landscape* factors, namely the 1970s oil crises, were critical in the development of the Danish solution and these are not always controllable or predictable. It may take a similar situation – but with a different policy response – to create support the growth of district heating in the UK, but equally, there may be other opportunities that are more appropriate for British cities. As Jollands *et al.* (2010) note, '[e]nergy saving potentials vary between cities, reflecting their particular geographic setting and biophysical resource and infrastructure endowments (including industrial structure), their social environment in terms of policy and institutional framing conditions. Of equal importance is the availability of a skilled workforce, public-private partnerships and networks with other cities.' In other words, there is no such thing as one-size-fits-all solution.

13.4 Transitions in developing cities: the case of Nakuru, Kenya

As a contrast with Copenhagen's relatively mature energy system, we will now look at the case of Nakuru, Kenya (Figure 13.4). Located in the Rift Valley, Nakuru is the fourth largest city in Kenya with a population of approximately 420,000 people (KNBS 2005). An estimated 46 per cent of its population lives below the national poverty line (World Bank 2005) and, as we shall see, its energy system is comprised of a mix of commercial and non-commercial fuels.

Nakuru makes an excellent case study for several reasons. First, UN population forecasts suggest that it will be in developing countries like Kenya where the biggest shift towards urban living will come in future decades (UN 2011). In Europe, for example, the UN expects the urban population to rise from 531 million in 2009 to 582 million in 2050, meaning that the overall rate of urbanization will grow slowly from 72.5 per cent of the total population to 84.3 per cent. Kenya, on the other hand, is expected to see the urban population rise from 8.7 million in 2009 (21.9 per cent of the total population) to 41.1 million (48.1 per cent) by 2050. The second related issue is that this general

Figure 13.4 View of Nakuru, Kenya.
© Lucas Chancel. Reprinted with permission.

urbanization trend is likely to occur primarily in smaller cities. Mega-cities may attract many headlines but the bulk of the new urban population will be living in secondary cities like Nakuru.

This population growth will increase the demand for urban energy services and the IEA anticipates that 81 per cent of the projected growth in global energy demand over the next 20 years will come in developing countries (IEA 2008b). The challenge is therefore to provide this energy from clean and sustainable sources so that poverty alleviation and environmental protection goals can be met.

The final reason to focus on Nakuru is that it has been the focus of two studies on urban energy consumption. The first, Milukas (1993), was originally conducted in 1981–2 and provides valuable information on fuel consumption by residents of Nakuru at different income levels. A second study updated this information and also assessed the impact of energy consumption on the local environment (Chancel 2010). By comparing these two sets of results, a clear picture of the actual energy transition in Nakuru can be described and compared with the theoretical models discussed above.

13.4.1 The relationship between income and fuel consumption

At the core of both studies was an investigation into the link between household income and the consumption of different energy resources. Households in Nakuru have access to many alternatives including charcoal, firewood, kerosene, liquefied petroleum gas (LPG) and electricity but biomass is the dominant supply source, with firewood and charcoal representing 74 per cent of Kenya's total primary energy supply in 2009 (IEA 2009).

As discussed above, the idealized 'energy ladder' model suggests that households would smoothly progress from one fuel to the next, progressing from the traditional fuels like charcoal and other biomass fuels, to paraffin as a transitional fuel and finally to LPG and electricity as modern commercial fuels. However, it was noted that several authors have critiqued the 'energy ladder' as being too simple a transition model. What does the evidence tell us about Nakuru's transition?

Milukas (1993) found that, in the residential sector, 539,000 GJ of charcoal and wood were consumed representing 75.7 per cent of final energy consumption. In contrast, electricity provided only 3.3 per cent of final energy consumption and LPG 1.5 per cent. These fuels were used primarily for cooking, as heating, cooling, lighting and appliance demands were comparatively low. By focusing on cooking, the fuel split can be readily explained. It makes sense to compare the relative costs based on the delivered energy service since each fuel requires a different cooking technology. On this basis, Milukas found the following costs, all in 1981 KSh/GJ: 179.2 charcoal, 342.1 kerosene, 355.6 LPG, 198.4 electricity. There was no equivalent price for firewood, as it was typically not sold in the market but collected informally (though of course this activity takes significant time). These costs, however, do not include the capital necessary for the cooking appliance itself, or indeed the grid connection necessary for electricity.

To explore the results in more detail, the paper lists the amount of energy consumed by different income categories. The summaries, shown in Table 13.1, indicate that low income households use primarily charcoal, medium income families LPG and electricity and high income families mainly electricity. However, the distinction is not clear cut, with high income families also consuming large amounts of wood and charcoal.

In the 2010 update (Chancel 2010), many of these income effects were repeated (Table 13.2). For example, 91 per cent of households used charcoal for cooking in 2010 compared with 90 per cent in 1980. Similarly, higher income groups again show greater consumption of electricity and transport fuel. Note however, the decline in the total level of charcoal and wood consumption in low and middle income households; this will be discussed in the next section.

These two studies offer support for an alternative energy transition, not energy ladder, hypothesis. As Chancel notes, rates of wood, kerosene and charcoal use show no significant connection with household income. One can discern a slight decrease in the use of these fuels at higher incomes but the key point is that a range of fuels are still used (in particular, charcoal for cooking purposes as dictated by local culinary preferences). In contrast, access to commercial fuels such as LPG, electricity and petrol show very strong trends. This suggests that there is an income barrier which households

Table 13.1 Annual household energy consumption and fuel access rates in Nakuru by monthly income category

	Income group		
	Low	*Medium*	*High*
Fuel use (GJ/household)			
Wood	3.1	18	14
Charcoal	29	22	9.4
Kerosene	4.1	1.5	0.27
LPG	1.4	6.0	4.9
Electricity	0.43	4.3	15.2
Total	37.7	51.4	44.1
Fuel access (% of households)			
Wood	9.6	14.7	18.2
Charcoal	93.4	79.4	45.5
Kerosene	86.0	29.4	9.1
LPG	13.2	64.7	48.5
Electricity	58.1	97.1	100.0

Notes: Low = less than 4,000 KSh (1981), medium = 4,000–10,000 KSh, high = greater than 10,000 KSh. *Data source*: Milukas (1993).

Table 13.2 Estimated annual household energy consumption in Nakuru by monthly income category

	Income group		
	Low	*Medium*	*High*
Fuel use (GJ/household)			
Wood	5.2	0	0
Charcoal	14.5	10.7	11.8
Kerosene	2.7	2.3	0.55
LPG	0.16	1.7	6.6
Electricity	1.3	6.2	23
Petrol	0	7.3	35
Total	23.9	28.1	76.9

Notes: Low = 2,500 to 50,000 KSh (2010), medium = 30,000–70,000 KSh, high = 70,000–150,000 KSh. *Data source*: Chancel (2010).

must overcome before they can afford access to these fuel sources, a finding supported by other literature (e.g. Davis 1998, AED 2008, World Bank 2011).

13.4.2 Environmental impacts of energy use

Secondary cities, particularly in developing countries, often have markedly different energy systems structures when compared with primary cities, and these differences manifest themselves not only in the technologies and fuels used within a given city but also the impact of that system configuration on the local environment. As Milukas (1993: 544) notes:

> ... [T]his difference is manifested in a much greater reliance on indigenous biomass fuels of secondary cities. For a combination of reasons – price, habit, resource availability and infrastructure development – similar income groups in primate cities and in secondary cities make choices on fuel use that are different. The argument is made further that the danger of deforestation looms over the regions surrounding these cities due to this present pattern of energy demand.

At the time of the original study, it was noted that Kenya's forests – particularly those in the vicinity of Nakuru – were managed for timber production and were not intended for charcoal production or direct wood extraction. Yet, such fuel production was occurring and the Nakuru district saw woodland cover decline from 50 per cent to 18 per cent of the region's 729,000 hectares between 1974 and 1984. Milukas (1993: 555) notes several important drivers of this trade including the economics of competing land uses (driving charcoal production onto marginal lands) and the need to be near urban markets or a suitable means of transport. Estimating locally available land and energy needs, he hypothesizes that Nakuru's charcoal demands could just about be sustainably satisfied – if one ignores the fact that the rural residents of the Nakuru district also need energy supplies. The conclusion, in 1993, was that the city of Nakuru was effectively mining the surrounding areas for wood and the author warned that '[w]ithout changes in the means of supplying Nakuru with energy and the nature of energy demand in the city, eventual deforestation of the surrounding region is inevitable'.

Fast-forward to 2010 and what did Chancel see? An estimated four-fold increase in total biomass use, resulting in the predicted deforestation. Using aerial photography, it was estimated that forest cover in the Nakuru municipality fell from 192,300 hectares in 1990, to 81,976 hectares in 2009 and that the rate of deforestation has in fact increased in recent years. By estimating the local forest yield and local wood consumption, and comparing with the area observed to be devoted to charcoal production, Chancel (2010: 46) concludes that 'the biomass energy needs of Nakuru are met on an unsustainable basis (i.e. trees are cut down without replanting) and that energy needs for Nakuru account for all deforestation in the area'.

It should be noted that these environmental impacts are primarily local. As shown in Figure 13.5, one can conceive of a theoretical model in which low income cities primarily effect their local environments, whereas wealthier cities can afford to displace their polluting activities to the wider world. This links back to the ideas of a city as a thermodynamic system; in other words, a city exporting its wastes to maintain internal order. As discussed in Chapter 3, London deforested its hinterland back in the sixteenth century but we now think of its environmental impact as primarily being one of global climate change. These distinctions can be seen by comparing per capita greenhouse gas emissions in both cities: 5.9 tCO_{2e} per capita for London (GLA 2008); 3.7 tCO_{2e} per capita in Nakuru (assuming that trees for fuel production are not replanted as observed, Chancel (2010)).

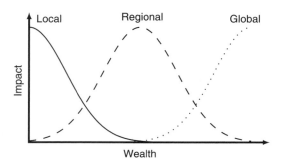

Figure 13.5 Schematic showing the varying scales of environmental impacts from cities. Based on McGranahan *et al.* (2005).

13.4.3 Technology and policy solutions

The Nakuru case study presents at least two notable policy challenges. The first is one of access: how can lower income households afford commercial fuels and benefit from their reduced health impacts and improved convenience? Second, how can the environmental impact of Nakuru's energy use be mitigated? Specifically, can the deforestation observed around Nakuru be brought under control and the forest allowed to grow back?

The solutions are of course connected. Revisiting the tables above, Chancel estimates that total annual household consumption of biomass (including the wood needed to make the charcoal) decreased from 7.5 tonnes in 1980 to 3.9 tonnes in 2010. This improvement is significant and was the result of the adoption of a more efficient stove technology, known as a *jiko*. Introduced to the Kenyan urban market in the early 1980s, the ceramic *jiko* has an efficiency of 20 to 40 per cent (in converting charcoal into useful heat) while the traditional metallic stove has an efficiency of 10–20 per cent. The shift from a traditional cookstove to an efficient one thus leads to the saving of 30–60 per cent the amount of fuel previously required (Ezzati *et al.* 2000). Before 1987, only 13 per cent of Kenyan households had access to this technology but in 2010, 91 per cent of low income Nakuru households had one. With a low cost of $2–3, repaid by fuel savings in 2–3 months, the main policy here has been to increase household awareness and the availability of such stoves. However, as noted above, there was a four-fold increase in total biomass consumption between the two study periods and this is the result of urban growth, with the number of households in Nakuru rising from 16,650 in 1980 to 122,875 in 2010. The challenge is therefore to provide this growing population with improved energy services, when technological improvements such as the *jiko* are helpful but insufficient to the scale of the challenge.

Chancel (2010) focused on the difficulty of addressing the high up-front costs associated with connection charges or appliances for modern energy fuels, like LPG or electricity. This can be clearly illustrated with the case of lighting. The running costs of an electricity-based system are significantly lower than current choices, e.g. a compact fluorescent lightbulb costs 0.1 KSh per lumen-hour whereas paraffin lighting is 93.9 KSh per lumen-hour. But the up-front costs are substantial with an electricity connection costing an estimated 35,000 KSh ($434).

Solutions must therefore combine both appropriate technologies and business models. One option assessed by Chancel is household-scale biodigesters. These devices use anaerobic digestion to convert household organic wastes into biogas, which can then be burned for cooking, and it is estimated that they could provide 70 per cent of an average low income household's cooking energy requirements, while at the same time providing a sanitary disposal of kitchen and human wastes. To make the economics work however, a biodigester solution must compete with the estimated 840 KSh low income households currently spend on charcoal. One way to arrange this is to have a non-governmental organization (e.g. a development charity) or utility pay the upfront installation costs of approximately 15,000 KSh, which would be recouped through monthly payments of 600 KSh per household. Assuming that these organizations could access credit at an interest rate of 10 per cent, the commercial loan could be repaid in roughly 30 months. Such a solution would also yield significant environmental benefits, decreasing the amount of charcoal required by an estimated 90 per cent and saving 2.7 tCO_{2e} per household annually. Similar structures could be designed to address other energy services like lighting and small electrical appliances (e.g. radios or mobile phones).

13.4.4 Summary

Energy transitions in the cities of developing countries are distinct from those discussed in more developed cities as the primary concerns are providing the health and economic benefits of high-quality energy sources while protecting the local environment, as opposed to energy security and global environmental concerns. The Nakuru case study showed that, in these situations, the 'energy ladder' model of urban energy transitions is an oversimplification. High income households continue to use lower quality fuels, often for specific purposes like cooking traditional foods. However, these households are able to afford the connection charges and up-front equipment costs associated with electricity and LPG.

For low income households, their energy needs are met largely by biomass which has led to significant deforestation around Nakuru over the past three decades. Although technology innovations like the *jiko* stove have improved the efficiency of charcoal use, the growth of the city has led to continued pressure on local resources. The challenge is therefore to develop innovative technologies and business models that allow low income households to afford the transition to higher quality energy systems. These issues are not limited to Nakuru, as population forecasts suggest that it will be these secondary cities in developing countries that grow most quickly in the next 20 years and thus determine the future of millions of new urban inhabitants.[2]

13.5 Conclusion

Urban energy systems face a number of significant challenges in the coming decades. For cities in developed countries, existing energy infrastructures will need to be retrofitted and upgraded to mitigate climate change and ensure robust affordable performance. In developing countries however, the priority is simply to expand access to modern energy services in order to support economic and social development goals, such as alleviating the health impacts of indoor air pollution and the local environmental impacts of deforestation.

The concept of technological transitions helps us to understand these processes. A new energy system is rarely the result of simply swapping in a new more efficient technology for an older equivalent. Entire relationships between consumer and producer may need to be redefined and modes of service provision adapted to meet new challenges. The notions of the system *landscape*, dominant *regime* and emergent *niches* help us to identify the maturity of individual technologies and their potential future evolution.

The two case studies illustrated these concepts. In Copenhagen's district heating system, we saw how decades of consistent policy were needed to establish the dominant position of the CHP and district heating. These were active choices made by decision-makers at both national and urban scales to promote collective solutions that delivered overall system efficiency. The counter-example of London shows how different geographic circumstances, cultural and political traditions have led to a more fragmented system in which the technical advantages of CHP and district heating can only be realized in small niche applications.

The Nakuru case study, although set in a very different context, highlighted similar issues. It illustrated how simple technological improvements like the *jiko* stove can make a difference, but a limited one. Widespread access to modern energy services will depend on new business models that can explicitly address the high up-front connection charges and equipment costs that prevent the low income households from accessing improved energy services. It also showed the distinction between secondary cities in developing countries, where the energy systems are integrally linked with local land resources, and larger cities like London and Copenhagen which are driven by national and international forces.

Transitions in urban energy systems are therefore rooted in local circumstances and the choice of appropriate technology, policy and system configuration will need to be sensitive to this context. Although no single solution presents itself, these case studies have hopefully illustrated that there is significant potential to improve the efficiency and environmental impact of urban energy systems even in such widely diverse settings.

Notes

1 In the central City of Copenhagen, 98 per cent of the heat demand is met by district heating (Copenhagen Energy 2009).
2 Throughout this discussion, it has been assumed that population control is not a viable policy option. Certainly, this can be debated, and indeed has been since the time of Malthus and earlier, but implementing such a policy is likely to be very unpopular and difficult to enforce. If the problem is analysed using the IPAT framework (i.e. that environmental impact (I) = population (P) × affluence (A) × technology (T)) (Commoner 1972, Ehrlich and Holdren 1972), then the most feasible solutions are those that improve technological efficiency rather than decreasing wealth or population. While such an analysis is simplified, it does give insight on the present challenges facing Nakuru and other cities.

14 Cities of the future

David Fisk

Imperial College London

14.1 Introduction

Urban energy futures are just one part of the processes by which cities manage their infrastructure investments. The technologies in earlier chapters provide powerful tools for inserting a more integrated approach to energy use into a city's future. But cities are complex systems in a formal sense. They have sometimes been thought of as 'space machines' in the sense that they occupy or 'consume' the land they cover on the map, and in doing so, produce the goods and services the rest of the economy requires. For those inside the city providing those goods and services or supporting that production, it is a place to live, work and find recreation. This chapter briefly reviews the way in which cities go about planning infrastructure, principally the master planning process.

14.2 Being formal about futures

The key components of an energy infrastructure have long lives, sometimes a century or more. So, a vision of the future has been a natural component of policy announcements and investment decisions. Indeed 'policy' is often conjured up to remove investment uncertainty about the future. That confidence can be transitory. California was, as declared by its Governor in 2004, to be the first hydrogen economy – at least for a while! Framing some part of this as a vision as to what a city might become – 'Cities of the Future' – is relatively recent. The term peaks in the 1970s in Google books, and has since fallen back to its level in the 1920s (see Figure 14.1). The string 'future of cities' starts to rise in the 1960s, peaks in the 2000s but has since seen a decline. It is not that commentators had never discussed futures before the 1960s but until then, what constituted a city was usually seen as a consequence of external changes not itself a directive of change. Cities evolved. Larger cities met problems from their size that they either solved and so became larger, or failed to solve at least for a while, in which case smaller cities caught up. As their external terms of trade changed, so did the city. Urban settlements created demand for tranches of fuel with different qualities. Cities were not seen as sources of power generation and energy storage for the economy at large, so national energy policy could largely ignore detailed changes in cities taking place

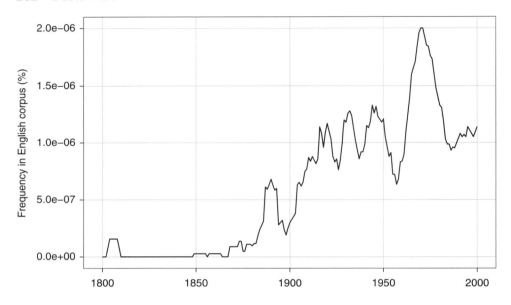

Figure 14.1 Frequency of the phrase 'cities of the future' in the English corpus, 1800–2000. *Data source*: Google (2010).

now and in the future. Even now, national energy forecasts tend to be 'delocalized'. That assumption may no longer be appropriate, as the optimization processes in earlier chapters reveal.

The slight decline in 'city futures' in discourse does not mean that the importance of investment in future city structures has changed. It is just that the context may have become more complex or uncertain since the 1990s. One or two of the most highly publicized future city projects have not yet fully materialized (e.g. the Dongtan eco-city near Shanghai and Masdar in Abu Dhabi). The value of a single clear future vision then seems more transitory. It is great to have an inspiring vision but it has to bring forward investment. This often seems hard to deliver. The public treasuries of cities in even the richest countries seem short of cash and lenders. But how then does the investor in energy systems integrate with the internal changes and developments in modern cities in an uncertain world?

14.3 Urban energy projections

The techniques used in national energy projections are not of great use even for a large metropolitan area. Not only are national energy statistics devoid of any sense of place, but the projection models similarly are geography free. These models usually use a 'neoclassical' economic growth model as an organizing framework. Now any growth modelling necessarily involves approximations. The real economy is a complex process and so not comfortably described by a row in a spreadsheet. Unfortunately, some of these model approximations are at their weakest in a spatial context, because space is a natural monopoly and models draw heavily on theories about idealized competitive markets.

A national model might posit an 'exogenous' GDP growth rate, grasp an energy intensity for GDP and hence produce a primary energy forecast. But there is little clue how this GDP is created across the network of the urban economy. Perhaps all cities are contributing to productivity, but perhaps some are growing and others declining. The national model gives no help. National models can guide large capital investment feeding into the top level of national energy grids but leave a gap for those about to make a more local investment. Such local investment could be a substantial part of the whole energy network in the future.

14.4 The 'concept city' future

Whereas national energy models can be colourless, the discourse about the future of cities is if anything too colourful, and so equally confusing, for investors. Consider the future city as presented in film. As with the 'concept city', the representation of the city is seldom the purpose of watching the movie. The film has a more immediate narrative to grip the audience. The 'city of the future' comes from the imagination of the story-boarder and set designer hung on the narrative line. Here, 'futures' are often a formal myth deployed not to conceptualize an investment plan but to safely discuss an issue in the present. To illustrate the point, consider three major films of the last century where the city remains a dominant visual image.

Fritz Lang's *Metropolis* could have been a vision of the year 2000 as seen from 1927, though why that would be of interest to an audience most of whom would be dead by then, is hard to argue. But it is easier to see why a scenario for the year 2000 could provide a platform to explore the direction of post-First World War Germany, with its widening gulf between the cultural elite and the manual labourers dehumanized by the machine. As it happens, there is a good deal of urban energy futures in the scenario. The set piece industrial explosion at the beginning of the movie takes place in its enormous centralized energy centre (see Figure 14.2). The most viewed clip for city planners is Lang's cityscape, a gigantic extrapolation of 1920s New York throbbing and swirling with people on the move in cohorts, cars, mass transit and the odd bi-plane – a lot of energy without much obvious purpose.

Almost as iconic, at least to architects, is Korda's *Things to Come* (Menzies 1936). Made in 1936, the movie traces a timeline of global events ending in 2036. Its depiction of an imminent industrialized war destroying urban infrastructure and the fabric of society from the air was probably more potent to the audience than the final Modernist vision of 2036 when Peace and Progress prevailed. But below rolling green hills, the subterranean future cityscape, clinically white and glass, tier upon tier, became the exemplar of the Modern Ideal. Form followed function. Tradition (and any sense of humour) was discarded. In a famous scene, an elder explains to his granddaughter, using something unerringly like a large iPad, the history of cities. Cities, she is told, were once 'above ground, all made of glass. They did not realize we could make a Sun of our own'. Urban energy sneaks its way into the futurology! The tension that brings the final scenes is between this clinical Progress and the Artists, and that soundly brought the discourse back to the intellectual debates of the 1930s. Not that it was a tension to bother Swiss architect Le Corbusier, who conceptualized large Modern urban

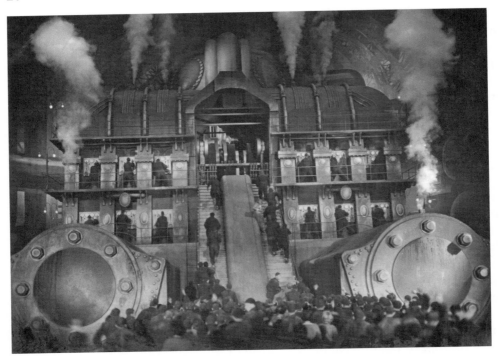

Figure 14.2 The 'energy centre' from Fritz Lang's 1927 film, *Metropolis*.
Still supplied courtesy of Eureka Entertainment Ltd. Reprinted with permission.

landscapes without much detailed thought to the individuality of the occupants. That prescription rapidly exhausted itself.

Ridley Scott's *Blade Runner* (Scott 1982) opens with a cityscape of Los Angeles in 2030, that is dark and heavy, punctuated by the sporadic flares of oil refineries. Urban energy begins the movie! This is hardly a 'post-industrial society'. The camera zooms in towards an enormous pyramid housing one of the city's mega-corporations. At the top, executives occupy vast gothic office spaces. The buildings tower over a sprawling chaotic warren of crowded streets and partly derelict hovels. Neon advertisements and loudhailer voice-overs are everywhere. With an echo of *Metropolis* the fate of the poorest looks not much better than today. As with Lang's film the central tension is not about housing conditions. It is elsewhere with the threat of machines so human-like that human identity becomes itself an issue. What is clear is that this cityscape does nothing to resolve the conflict.

This filmography is not complete. It would seem that, given a future storyline, the story-boarders have found it easy to create a vision of a future city as a backdrop. Cities then are seen as consequences of a social condition, reinforcing it and locking it in position. The point of exhibiting these three films spread over more than 80 years is to show that in some cases, 'Cities of the Future' are often not a true discussion of futures at all. Rather, they are a means of conducting a debate in a safe place about the present. Sometimes, when an issue is not acute, it is difficult to find space to debate it. When it does become acute, it may then be too inflammatory to even mention it. Projecting a chronic issue to the point at which it is acute, might be the way forward. Orwell's *1984* (1949) was not intended to be a prediction; he was debating issues of 1948. Albeit that current drab urban spaces scanned by CCTV and saturated by PR thought and Newspeak words like 'Zero Carbon' might suggest he was better at it than he imagined! Positing how we *might* want to live is a way of talking about who we are now. Asserting that a particular built form will evolve may be as much an assertion as to the inevitability of its implied social structure. This is a confusing factor in seeking the substance behind any discourse on cities of the future.

Competitions for 'Cities of the Future' attract the attention of large architectural practices, much as 'concept cars' attract product designers to show off their skills. They give concrete (*sic*) structure to the social constructs of the day and so in their own way, add to current debates. But that is not to say that they are fit for purpose to predict a city's future energy infrastructure needs. That was not their purpose. Styles of cities are not just an engineering issue. A form of space planning that produces low-rise compact urban spaces even featured in a US presidential campaign. Those who felt comfortable with that style of living and the personal values it implies to work would no doubt have voted for that candidate. Those more inclined to leave all this to the market and private covenants on land would have voted for someone else. These are important debates characteristically fought with bare knuckles. So the analyst needs to be alert to the risk that the assertions about possible external energy futures are not being used to lever other more immediate outcomes.

14.5 The single company town

The only part of a decision which is truly a decision is that part that is irreversible. So, whether the vision is for a zero carbon leafy garden city or a gas guzzling cluster of super-casinos, if the first step is to build an energy centre, the immediate resolution of the vision seems immaterial. The importance of the vision is more the degree to which it is shared, because then actors operating independently produce a coordinated outcome. If the town is to be a 'steel town', house builders will seek land up-wind but near enough for the workers to live. Rail to carry ore into town and roads to take away finished products will form major arteries used by other industrial enterprises. So, while a vision is potentially ephemeral, different actors tend to lock it in place. This seems to be what Adam Smith had in mind when he referred to the 'invisible hand'. The vision has to come from somewhere. It may be that of the dominant landowner. Steel towns were often built by the national steel company. Several of the world's capital cities were built on land designated by Supreme Leaders to be the central seat of Government. Some of the new cities of the future will be built to serve one role in their national economy.

The question that then defines the land use pattern in this particular case, is whether this external rationale exhibits economies of scale. If so, as in a steel mill, the activity clusters. If not, as in many service industries, the activity disperses. The point is that for the city to remain viable, it has to be in a form of equilibrium inside the economy in which it is embedded. If not, its products will be produced elsewhere and its population will migrate. So, even the simple company town cannot define itself entirely at its own discretion. There are also more widely held social norms to meet. The managers need to reaffirm the status they exercise when at work in where they live at home, and so on. That said, the single external purpose town is the easiest to comprehend. Its future is intimately aligned with its economic purpose. Whatever the changes in the means of production, they will feed through to the fate of the city (Kargon and Molella 2008).

14.6 Land use in large complexes

In contrast, larger urban settlements are more likely to accommodate several functions that interface with the wider community and labour market. Prosaically the owners of a large shopping mall will not simply let space randomly as in a strip mall, but assemble tenants in adjacent sites in a way that makes the whole more effective commercially than the sum of the parts. So similarly, land is not just valuable for what its owner can do on it but the access it provides to other adjacent land uses. That access is not always welcomed by other land owners. For example, that 'access' might be the dispersive properties of air to dilute the smell of an animal rendering plant. For this reason, in some countries, land owners enter into treaties or 'covenants' with their neighbours, that limit activities. Industrial complexes mindful of the stricter air quality requirements that apply where children might be exposed could equally well seek covenants that limited adjacent use to industrial applications.

In this version of land use, the hydrogen economy or any other specific energy network is perfectly feasible. An energy company might simply buy all the land it intends to serve, insert a requirement that any land use must connect itself to the hydrogen network and then sell the land back to the market. If hydrogen was indeed the ideal energy vector, then having secured the system the land would sell if anything, at a premium. This scenario is somewhat fanciful but it is not far from the requirement in some Scandinavian cities that any new housing development must connect to the district heating system. The problem of course is that since 'place' is a natural monopoly, the smart speculator buys just the plots that are critical to completing the system and holds out for the highest price. So this approach is feasible, but, unless the land is purchased surreptitiously, it is not very fast. Nor are such covenants likely to be future-proof. They signal to the investor what cannot happen, not what might.

14.7 Master planning

Markets are not effective at creating efficient system infrastructure, even if they are good at using it. Indeed, even the IT systems that are the backbone of the great financial trading centres of the world are all mutualized in one way or another. Adam Smith, writing at a time when the competence of government depended on the limited capabilities of simple monarchs, nevertheless concluded that roads had to be provided by the Crown, no doubt on a plan with the rapid movement of artillery in mind. The south London rail network – the Los Angeles freeway of the nineteenth century – is a confusing tangle of competing private systems when compared with the underground system to the North. When there is not a dominating land-use interest and there are numerous land owning interests, owners are usually persuaded that some kind of overarching land use plan is better for their own land use than the chaos without any.

So-called 'master planning' is the tool used by most local and regional governments for this purpose. The investor in energy infrastructure will need to integrate with this process. Master plans frequently use a 30-year vision but essentially prescribe decisions for the next decade. Decadal scales are common 'payback times' for fixed investments and reflect the time necessary to negotiate conflicting interests. It would normally draw on detailed data of current land use and resource consequences such as water and power demand. For example, Baltimore's master plan in 2010 was built on an earlier information gathering phase 'Live Earn Play Learn' that took place in 2008 (City of Baltimore 2008). A master plan would seldom specify detailed built form, but overarching figures such as areas for permitted use types (industrial, light industrial, etc.), or perhaps maximum heights or sometimes occupant densities. These would be used as drivers for estimates of services, such as transport, and checked for consistency. Thus, the transport network ought to be able to deliver the volume of travellers consistent with the occupation density of the commercial area. Noise and pollution levels in an area designated for housing ought to be consistent with a normal domestic life.

Planners are aided in this task by many modelling tools. These are usually treated as aids rather than prescriptive. It is possible to 'backcast' if there is consensus as to where the city expects to be. But making firm predictions of the future of a society and then telling the society this is where it will be in the future by the force of destiny, is somewhat tautological. It would have involved using the outcome of past choices to apparently provide no choice in futures. There are some invariants that enable the planners' models to have substance. After all, citizens have only 24 hours in a day. The purpose of the city is not just that people can work but that they can work together and enjoy each others' company. So for example, 'rush hours' and travel diaries provide useful constraints on separation and mean speed of travel between social spaces. Energy demand did not always appear as a constraint or determinant in twentieth-century master plans. The reliability of power does sometimes have an influence on building height. No-one in Lagos wants to be trapped in a lift while the National Electric Power Authority tries to recover from a brown out! For an oil rich nation, Nigeria's capital is determinedly low-rise.

The master plan, as well as codifying land use to inform more detailed investment, will normally designate changes of use and areas for major development. For new energy infrastructure, these areas are clearly of most significance. For example, in Barcelona, a large industrial area which had become increasingly derelict was 'designated' as an area for 'industrial innovation' and has involved clearing an enormous area of some 200 hectares for new state-of-the-art building and residences.

A city can be the only final judge of the quality of its plan, but some criteria can be suggested. From an external perspective, it should have a degree of internal dynamic consistency. This is a strong added value to the energy infrastructure investor who, focused on the costs of pipes and wires, might miss wider social dynamics that could undermine the investment. The external world is uncertain enough without the plan itself creating uncertainties about the future through unresolved conflicts. For example, suppose the draft plan was to decide that the less well-off should benefit from cheap heat from a heat from a waste plant, by placing social housing on the low-cost land adjacent to it. Would transport to work be so expensive that only the unemployed could afford to live there? Would the waste plant provide unneeded social stigma that its use in an industrial site would not entail? What would be the consequences for the housing if the plant's feedstock became expensive to burn because of competing disposal methods or new pollution standards? The questions are not entirely hypothetical. In the 1970s, one of the largest planned social re-housing projects in Europe at that time (Hulme in Manchester) fell into rapid social decay when the oil crisis meant that its district heating system based on the availability of cheap low-grade heating oil made life on the estate impossibly expensive and tenants with good records were relocated elsewhere.

14.8 Future scenarios as an urban planning tool

The fate of cities, and areas within them, is necessarily less easy to forecast than broad-brush national economic forecasts because local economies can be more volatile. Sometimes an area designated for regeneration simply fails to ignite. Sometimes it is so successful that it brings another nearby area into economic decline. Because the local world is so much more uncertain, cities often use scenario planning as a way of conducting debate prior to drawing up the master plan. In business plan usage, scenarios are self-consistent narratives against which a strategy can be tested. They are not strictly forecasts, as few things in the future would be consistent. The organizing narratives can embrace the wide differences of view in the populace.

Thus, scenarios might be used to test the robustness of a local housing strategy. As in a business application, it would be as much to detect the ability of the strategy to move forward when events were more favourable than anticipated as to retreat when events turned averse. This is a slightly different use of 'scenario' from that common in national energy policy. There the scenario set might be different rates of GDP growth and various side constraints with optimizing energy models run under each assumption to see how different the next immediate step might be. That is actually a little unrealistic since it is not obvious why the system is so locked into its scenario that investors can plan with such confidence in each case. At the city level, this academic quibble is hardly relevant as the politics are usually just too volatile to use optimizing economic models in this sense. Infrastructure demand, especially energy, is probably more robust than other indicators but the central narratives tend to be social and the energy analyst needs to become used to working within this material. Box 14.1 gives one set of scenario narratives developed for London Boroughs in 2008. It is quite possible that in a city the size of London, fragments of these scenarios could appear in different parts. That is not their point.

Box 14.1 Using scenarios to think about London's future

The following scenarios were prepared by the London Collaborative, on behalf of Capital Ambition, as part of 'an ambitious programme designed to improve the capacity of London's public sector to work across boundaries of place, profession and organisation in order to meet future challenges facing the capital'. (The London Collaborative 2008: 3). A total of six scenarios were devised as part of the programme, three of which are highlighted below. Note that these scenarios are not predictions of what will happen in London, but consistent narratives about the ways in which current trends could develop over time.

Scenario 1: Full Speed Ahead In this scenario, London is characterized by high population growth, a prevailing social attitude of 'tolerant co-existence' and a very strong economy (i.e. one with growth in output and employment exceeding the most optimistic of current predictions). Significant and continuing investment in London's transport infrastructure enables continuing population and workforce mobility. There is a mature, well-established market for public service provision, in which private and third-sector providers are major players.

Scenario 2: Hitting the Buffers This scenario explores the impact on London of increasing outward migration but steady population growth, due to a higher than expected birth rate and a prolonged recession. These trends combine with others to create a prevailing social climate of fear, suspicion and competition between communities. The capital has long been overtaken by international competitors and is no longer the destination of choice for economic migrants, let alone tourists. The poorest people live increasingly beyond the margins of formal society in health and housing conditions that seem to belong in a past century.

Scenario 3: Steady Ahead In this scenario, London is characterized by population growth in the middle band of current predictions, a prevailing social attitude of 'tolerant co-existence' and medium economic growth (i.e. with growth in output and employment in-line with current long-term predictions). The last 15 years have seen a shift of people, money and power from the centre to the suburbs, and from there onwards down to local wards and communities. While this has had many positive benefits, London as a whole is something of a patchwork, and standards vary considerably. Sustainability and the environment have continued their steady rise up the political and personal agendas.

Given the somewhat ambivalent use of 'scenario' in most national energy policies, it is an entertaining work-out for the energy analyst to use the three films discussed earlier as planning scenarios. The films provide totally self-consistent visions which differ widely in their organizing themes. As a consequence, it is easy to elaborate each vision to provide any further detail the analyst might want. Thus, suppose we were to use them to test the robustness of an energy service company (Esco) business model. Neither three scenarios are at all likely but they tease out how the model needs to be flexible. There are no right answers in this type of exercise. So, for example, *Metropolis* clearly has an energy centre but while energy services have meaning to its elite, there is a gross basic needs market for the workers that is quite a different business. How far would an Esco model be able to divide its functions in this way? Would its shareholders prefer to act as an energy supplier to the discerning and outsource the running of the system to a facilities company? How would the need for such flexibility be reflected in how it raised capital? How would it respond to a competitor seeking to attack the least attractive of the two markets?

14.9 Special factors in urban scenarios

As argued above, urban settlements are embedded in a wider economy and many of the elements that appear in broader scenario planning reappear in urban scenarios. Urban settlements taken as a group face many of the challenges they faced in the past. Demographic change, coupled with increased labour productivity of land means that most of the world's future population growth will be in urban settlements. Urban settlement size tends to follow roughly 'Zipf's law' – the second largest city has half the population of the first and so on (Semboloni 2008). This probably reflects the largely random way employment opportunities are created for new arrivals by those already there. Larger cities create proportionately more employment and so creep ahead of others. However, as a group, medium-sized cities create just as many opportunities. City creation only occurs plausibly at the few tens of thousands. So the implication is that the characteristic city of the twenty-first century is likely to be one of the many more modest sized than the few international mega-cities. There remains a question as to whether housing and population change much. For some countries, the inward migration of rural dwellers to towns and cities to undertake menial or marginal employment will outstrip the rate housing stock formation. The 2001 Delhi master plan in part served to legitimize a significant number of informal settlements that had clustered around the city. At the opposite extreme, some European and US cities have demographic projections of shrinking populations.

More recent scenarios embed external concerns about greenhouse gas emissions and plans to 'decarbonize' energy consumption. For example, London has a strategy for decentralizing power generation with 25 per cent of its power from such sources by 2025 (GLA 2010). This kind of target 'external' to the main master planning nevertheless has significant land-use implications and needs to be melded into the mainstream thinking. Demand-side energy security almost entirely resides at the city level. It would therefore be valuable to add stress tests to scenarios to understand the resilience of the urban networks. Of especial note is the ability of the city to work with varying degrees of success with limited availability of transport fuels. Technologies also play an important part.

14.10 Conclusion

The need to rationalize land use has always dictated some planning or coordination between adjacent plots in urban settlements. The fundamental sticky nature of the market for land has meant that such coordination has always required some long-term view. The most common mechanism for undertaking this process is through a master plan, characteristically looking 10 years ahead in the context of a longer-term vision.

Master plans contain much of what would be needed to devise an urban energy use strategy. Urban areas are often open to much more volatile conditions than higher level forecasts, and scenario analysis is often deployed. It would normally use social organizing themes with additions from the wider world. These provide an opportunity to test the robustness of infrastructure development plans. At the moment, urban energy provision is driven by the master plan's other assumptions. What may be the next stage is that the optimal urban energy system itself becomes part of the deterministic framework.

15 Conclusion

James Keirstead and Nilay Shah
Imperial College London

In the preface, we noted that this book represents something of a milestone. When we first began working on urban energy systems in 2006, a good part of the first year was spent scouring the literature for inspiration and advice. Although we eventually did uncover many valuable studies, it was difficult at first to find much of this material. There's no *Journal of Urban Energy Systems* and review articles tended to focus on specific issues, like district energy systems or 'energy ladder' development transitions, without an overarching framework. Although this makes sense when one thinks about the number of disciplines that have contributions to make in this field, the lack of a 'one-stop-shop' does make it difficult for new researchers to begin exploring the topic.

Fast-forward five years however, and there have been a number of high profile publications on urban energy systems. The International Energy Agency's 2008 *World Energy Outlook*, for example, presented some of the first global empirical data on urban energy use and helped to highlight the importance of this scale for global energy and climate concerns (IEA 2008b). Similarly, a major assessment report on climate change and cities flagged up both the governance issues surrounding urban energy systems and the threats they face in the coming decades (Rosenzweig *et al.* 2011). Methodologically, innovations like activity-based models of urban energy use are very recent and optimization approaches have continued to evolve, providing researchers with a rich choice of methods. We hope that we have provided readers with a sense of this progression in the literature reviews throughout this book, and a sense of excitement about where the field is heading.

This final chapter gives us a chance to review the major themes presented in the book and highlight our overall conclusions. We focus on four key areas, corresponding to each of the four parts of the book: the importance of urban energy systems, the evolving technologies for urban energy systems, analytical methods and the wider issues that need to be considered when implementing solutions. As the subtitle suggests, these themes are not independent and should be considered together as part of an integrated approach to urban energy systems.

15.1 The continuing importance of urban energy systems

With over half of the world's population now living in cities, the urban environment is increasingly recognized as part of the solution to global challenges such as energy and climate change. At this scale, problems which can seem intractable for a nation as a whole can be broken down into more digestible pieces, concentrating on the resources, skills, needs and desires of specific communities. We discussed how the economic and social ambitions of citizens and firms drive the patterns of energy consumption observed in cities, and how local decision-makers engage with energy and climate change issues as specific challenges to their communities. The twenty-first century is set to be an urban century: the challenge addressed by this book is how to ensure that these cities have clean, secure, and affordable energy services.

We defined an urban energy system as *the combined processes of acquiring and using energy to satisfy the energy service demands of a given urban area*. This captures key attributes such as the extensive supply chains of urban energy services, the delicate balance of energy supply and demand, the patterns of activities that lead to energy demands and the importance of local context in judging what solutions are appropriate. The brief history of London's urban energy system illustrated many of these concepts. For example, when local wood resources were exhausted, coal was imported by sea and the abundant supply of this new fuel supported a range of new energy services, including faster transportation, gas networks and street lighting. Inventors, consumers, government policies and local geography all helped to shape the precise network configurations that emerged and led to the modern city.

Several theoretical concepts were also introduced to understand urban energy systems. From a physical perspective, cities can be viewed as thermodynamic, metabolic and above all, complex systems. This means that we need to study the flows of materials and wastes across open and ill-defined boundaries, to appreciate the inevitable inefficiencies that result from energy conversion processes, and to consider the unexpected ways in which individual components might interact within a larger system. Urban energy systems were also shown to be social systems as well as technological systems, both at the scale of basic household technologies and in the construction and planning of large infrastructures.

These trends and concepts have led to the emerging field of urban energy systems analysis. Cities will be at the forefront of efforts to improve the efficiency of energy use, to reduce greenhouse gas emissions, and to respond to the demands of a changing climate and societies for some time to come. Fortunately, many opportunities for dramatic performance improvements exist, drawing on innovations in process systems engineering, information and communications technology, energy technologies and other fields which simply did not exist when the current infrastructure of many of the world's largest cities was built over the past 200 years. Our goal in this book is therefore to illustrate this potential and to provide readers with the tools they need to understand these changes.

15.2 Urban energy use and technologies

The second part of the book focused on urban energy use and technologies. Broadly speaking, these technologies can be thought of as intermediaries between the energy services we desire and the demand for energy resources and fuels. With more efficient technologies, the same service can be delivered for less input energy, thereby reducing the operating costs and environmental impacts of urban energy systems.

In the buildings sector, the primary energy service demands are for heat, cooling and electric power (e.g. for lighting and appliances). The basic science behind the technologies that provide these services is well known and it is technically possible to achieve significant improvements in energy efficiency through building retrofits. However, buildings also have important social dimensions. In the domestic sector, buildings can reflect our status within society, thus shaping our decisions about whether or not a particular retrofit measure is desirable. Similarly, in the commercial sector, there may be multiple agents involved in the retrofit process and it can be difficult to align all of their interests in pursuit of energy efficiency.

Two major energy supply technologies were examined. The first was distributed multi-generation and district energy systems, a series of technologies for providing heat, cooling and power to buildings or urban districts. By using more of the useful energy within the source fuel, and by taking advantage of the increased demand density of urban areas, these technologies can deliver primary energy efficiencies of 80 per cent or higher. The second related area was urban renewables, and urban biomass in particular. Although the density of urban energy demand is often orders of magnitude greater than the available renewable energy resource, cities can still reduce their carbon emissions by sourcing renewable energy from within their urban boundaries as well as in the near hinterland. Urban biomass is a particularly interesting example, with close links to distributed multi-generation systems, as the diversity of the resource and conversion pathways means that the analyst must be aware of the local constraints that will guide the best solution for a particular case.

While many of the technologies for improved stationary energy use are well known, the transport sector offers the promise of truly futuristic technologies. These range from high-speed maglev personal transport pods, to ICT-enabled smart vehicles and infrastructures, through to innovative business models that enable urban residents and firms to meet their mobility needs with lower energy demands, even when using existing vehicles and infrastructures. Again, an understanding of consumer behaviour is vital here, particularly when trying to assess how the electronic provision of services (e.g. online shopping) might affect travel patterns and the resultant energy demands.

These chapters also hinted at a range of analytical techniques and methodological considerations. Perhaps the most important one is the choice of an appropriate reference case when evaluating the benefits of a new technology. For example, the 'separate production' method is an integral part of assessing distributed multi-generation technologies and for transport technologies, in particular plug-in hybrid vehicles, one must consider the entire fuel supply chain. This clearly extends the scope of the analysis beyond the city where the vehicles are actually fuelled and used, to the upstream power stations and fuel refineries wherever they may be. In terms of specific techniques, there was a clear role for optimization modelling in the planning of building retrofits and bioenergy systems, computationally-efficient black-box modelling and parametric studies for assessing the benefits of distributed multi-generation, and stated preference modelling when trying to understand whether or not consumers will purchase a technology for which markets may not yet even exist.

15.3 Analysing urban energy systems

Given the complexity of urban energy systems, computational models are an essential tool for assessing new system concepts, testing policy interventions and conducting other experiments which may be impossible using a real city. A review of current practice suggests that there are two main modelling approaches relevant to urban energy systems modelling: optimization models for systems integration and planning (largely on the supply side), and activity-based micro-simulation models for capturing a rich representation of behaviour and energy demands. This part of the book then demonstrated several components from Imperial College's SynCity modelling framework, which attempts to link these two approaches.

From the optimization perspective, the TURN model was presented as a generic framework within which multiple resource conversion pathways can be assessed and optimal systems designed. For example, this approach can be used to examine the various routes for bioenergy within a new eco-town development or to help plan the retrofitting of an existing urban area. These problems are normally configured with the goal to minimize life-cycle costs, subject to environmental performance constraints, but optimization can also be used to examine environmental performance explicitly, as measured, for example, by material flow, environmental footprint or life-cycle analysis techniques. A comparative analysis of different environmental impacts from a single energy system demonstrated how there is rarely a single best solution, and choices must be made about which goals to prioritize.

Most of these models require temporally and spatially distributed patterns of energy demands as input. While these can be observed from real cities in some cases, the use of activity-based micro-simulation was highlighted as a highly promising technique for generating these data. These techniques have a long history within the land use-transport literature and have only recently been applied to the question of urban energy systems. However, they are likely to be a mainstay of serious analyses for years to come, as they uniquely offer a powerful framework that captures both the direct and indirect aspects of changes in policy, technology and consumer and firm behaviour.

A common challenge with these models is their complexity, making it difficult for both experts and non-experts to understand the results and use them for decision-making. Uncertainty and sensitivity analysis were therefore introduced as important 'meta' techniques – that is, analytical techniques that every modeller should use regardless of the exact modelling technique employed. A distinction was drawn between uncertainty analysis (describing the variability of a model's output) and sensitivity analysis (attributing that variability back to the model inputs), and a case study from Newcastle-upon-Tyne demonstrated how these methods can be employed to inform robust energy systems planning. Ultimately, these tools complete the loop described by the theoretical process of modelling: creating a formal representation of a natural system, experimenting with this simplified model, and then carefully interpreting the meaning of the model results for the natural system.

15.4 Implementing solutions

In the introduction section, it was emphasized that urban energy systems are about more than specific technologies and the modelling techniques used to design system configurations. This final part of the book has therefore examined the wider elements of an integrated approach to urban energy systems, focusing on the processes by which changes in urban energy systems occur.

Technological transitions theory was first referenced as a vital means of understanding how new innovations get adopted within large complex infrastructures. The case of district heating in Copenhagen demonstrated this clearly, and provided specific lessons such as the need for a consistent long-term policy environment if large energy-efficient infrastructure systems are to develop. In Nakuru, Kenya, the situation was slightly different as the main interest was providing basic energy services to a rapidly growing city. However, the common theme with both cases is that local circumstances need to be considered in any solution. All urban energy systems will represent the links between people, technology, infrastructures and the environment, but these relationships will vary from city-to-city and the analyst needs to be aware of this context when proposing specific solutions.

When planning the future of urban energy systems, it was shown how formal methods for thinking about urban futures – be they artistic representations, master plans or narrative scenarios – are traditionally focused on other elements of the city, in particular social and economic processes. Yet, these visions have clear implications for energy systems that are often visible just below the surface. The hope is that some of the modelling techniques discussed earlier will enable these connections to be made visible, thus enabling citizens, utilities and governments to consider how their decisions about energy systems today will affect their ability to achieve wider social and economic goals.

15.5 Putting it all together

These four themes can be summarized graphically in Figure 15.1. In our view, this is not a prescriptive framework for the analysis of urban energy systems – given their complexity and diversity, that is surely something of a losing proposition – but hopefully it is a helpful guide, both for navigating this book and performing one's own studies. The various techniques and theories suggested in these pages are really only a starting point and there is a wide literature to be explored, even though it is rarely labelled as 'urban energy systems' *per se*.

The coming decades will be a time of remarkable change for cities and their energy systems. With growing urban populations, continuing technological innovation, a changing climate and an ever-dynamic global energy market, the only thing that seems certain is that the historic template followed by many cities with now mature energy infrastructures (e.g. London, New York, Tokyo) is unlikely to be appropriate in future. Well-established cities will have to adapt their infrastructures to new conditions and

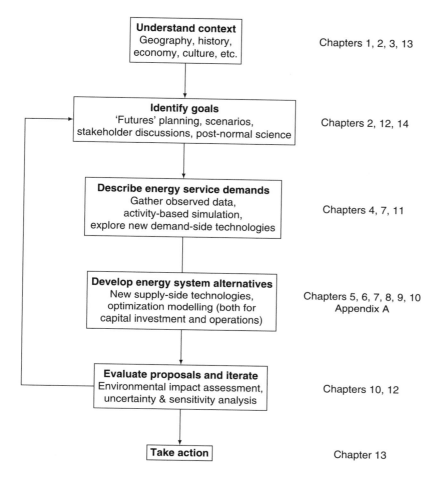

Figure 15.1 An integrated approach to urban energy systems, with reference to specific book chapters.

developing cities will have to meet their energy needs with innovative new designs. Fortunately, there exists a range of technologies and analysis techniques that hold the promise for highly efficient urban energy systems that will continue to power our commercial centres and industry, warm our homes and get us where we need to go in future.

Appendix A Optimization techniques

Nouri Samsatli, Mark G. Jennings and James Keirstead

Imperial College London

Optimization (also known as mathematical programming) has grown into a powerful method for aiding decision-making in many domains. It involves the formulation of an abstract model of the system under study (as opposed to a physical or analogue model). Mathematical symbols may be used to represent known (scalars and parameters) and unknown (variables) quantities of the abstracted system, while equality and inequality constraints represent relationships and boundaries within the system. Typically, an objective function including decision variables is minimized (or maximized) such that an optimal solution is provided. A history of the theory, development and application of optimization is not given here. Readers are directed to the references below. Instead, a short overview of optimization is given such that the previous analysis chapters may be placed in their proper context.

A.1 Background references and tools

Automation is meant to relieve man of particular mental tasks and optimization is typically applied to help make decisions on quantifiable systems where a large number of alternative choices exist (Dantzig 1963). Initially only small sized programming problems could be considered (10 equations in 1950, 10,000 equations by the mid-1960s (Wolfe 1963, 1968)), but progress in solution algorithms, commercial software and digital computing has led to the current situation where hundreds of thousands of equations may be solved relatively quickly. Optimization is currently used by various practitioners for different applications and research objectives, in disciplines such as operational research, quantitative finance, chemical engineering, electrical engineering and computer science. The benefits of optimization are thought to be (von Neumann and Morgenstern 1944, Dantzig 1963, Moore 1968, Williams 1999, 2009):

- Clear formulation of the concepts and problems
- The production of specified objectives
- Quantitative analysis of the structure of the system under study
- New insight on system operations
- Assessment of alternative decisions, strategies and/or controls in light of uncertainty and risk
- Potential improvements to operational efficiency.

The drawbacks of optimization are thought to be as follows:

- Only certain classes of problem have optimal solutions (e.g. LP, MILP, well-bounded NLP)
- Logical constraints have no unique formulation, and no analytical solution
- Many classes of problem are difficult to solve in efficient periods of time (polynomial time)
- RAM or disk memory of central processing units impact storage capacity
- No feedback loops in linear deductive models.

Given the growth of interest in optimization over the past 70 years, it was thought most useful to provide a categorized table of reference books rather than a detailed narrative, which is outside the scope of this book. Selected optimization references can be found in Table A.1 and while non-exhaustive, it should be useful to readers seeking to discover more about a particular topic in optimization.

In practice, optimization modelling is usually done with the aid of specialized computing packages. The current popularity of optimization is such that there are over 200 commercial linear optimization packages available (Bertsimas and Tsitsiklis 1997). These packages allow for easy formulation of mathematical programs, feature the latest solution algorithms and offer advanced features for report generation and analysis. Some notable examples include GAMS (http://www.gams.com), AIMMS (http://www.aimms.com), AMPL (http://www.ampl.com) and many more.

Table A.1 References for optimization

Optimization topic	Reference book
Theory	Overview of mathematical origins (Dantzig 1963)
	Theory of games (von Neumann and Morgenstern 1944)
	Linear programming and extensions (Dantzig 1963)
	Convex analysis (Rockafellar 1970)
	Logic and integer programming (Williams 2009)
Computing devices	Overview of mathematical computations (Dantzig 1963)
	History of electronic digital computers (Rojas and Hashagen 2002)
Algorithms	Simplex algorithm (Dantzig 1963)
	Details of other pertinent algorithms (Minoux 1986, Saunders 2012)
	Evolutionary optimization methods (Goldberg 1989)
Introductory text	Introduction to linear optimization (Bertsimas and Tsitsiklis 1997)
	Integer and combinatorial optimization (Nemhauser and Wosley 1989)
	Nonlinear optimization (Floudas 1995)
	Operations research (Moore 1968, Taha 1982)
Model formulation	Linear and integer programs (Williams 1999)
	Modeling languages in mathematical optimization (Kallrath 2004)
Applications	Practical applications (Bixby 2002, Reklaitis *et al.* 2006)
Other	Online source for optimization (see University of Wisconsin Madison 2012)

A.2 Model classifications

Optimization is a very diverse field, and means different things to different disciplines. In general, there are accepted classifications under which programs may be considered. This section provides a brief introduction to the basic types of optimization modelling so that readers might appreciate which formulation is most appropriate for particular problem types.

A.2.1 Variable and linearity classification

Two features of a model's formulation are typically used in determining its classification: the type of decision variables to be optimized and the linearity of the constraints and objective function. Regarding the decision variables, a mathematical program may contain only continuous variables, only discrete variables (e.g. binary $0, 1$ or integer variables), or a combination of continuous and discrete variables. Problems containing only integer variables (including binary variables) are known as integer programs (IPs).

Problems may be further classified according to whether or not there are non-linear terms in the constraints and objective function. A problem comprising only linear constraints and objective function is known as a linear programming (LP) problem. For example, if a model is seeking to minimize the value of $y = x_1 + x_2$ where all three terms are decision variables, then this would be a linear model. If however, any constraint or the objective function contains a non-linear combination of variables, then this is a non-linear program (NLP); for example, $y = x_1 x_2$.

NLPs are much harder to solve than LPs. In particular, if the NLP is non-convex then there may be multiple *local* optima and there is no guarantee that the *global* optimum will be found. Conversely, LPs can always be solved to global optimality because they are always convex. Strictly speaking, integer programming is a form of non-linear optimization.

Model classifications are typically given as abbreviations summarizing these attributes. For example, a model with both continuous and integer variables would be known as an MILP (mixed integer linear program) if the constraints and objective were all linear or an MINLP (mixed integer non-linear program) if not. MILPs will be discussed in more detail in the next section due to their usefulness in modelling urban energy systems.

A.2.2 Risk and uncertainty

Models can also be classified based on their treatment of risk and uncertainty. In simple applications, optimization models are *deterministic* because they are given one set of inputs and produce a single output. In other words, there is no uncertainty within the relationships in the model or indeed within the input parameters.

However, in systems where there is some uncertainty about future outcomes, *stochastic* programming techniques may be used to model random data. When outcomes are not certain, there is embedded risk in any decision taken. Risk may be considered on a short-term or a long-term basis, reflecting the uncertainty surrounding particular choices (Birge and Louveaux 2011). Typically, probability distributions are used for representing short-term randomness and scenarios employed for representing long-term uncertainty. Risk may be modelled as the likely loss or downside if an expected target is not met: for example, see Rockafellar (2000) for details of conditional value at risk, and Eppen *et al.* (1989) for details of downside risk. The attitude of the decision-maker to risk can be modelled, for instance, by use of minimax functions (von Neumann and Morgenstern 1944, Luce and Raiffa 1957).

Assuming neutral decision-makers, stochastic programming techniques quantify the uncertainty due to random data (although robust stochastic programming seeks to include the risk tolerance of the decision maker (Mulvey *et al.* 1995)). Stochastic programming uses the same theory as deterministic programming, but allows for parameter uncertainty. Stochastic programming can be single-stage or multi-stage in structure. Single-stage stochastics include the uncertainty and the decisions in the same stage, while multi-stage problems may allow for decisions to be taken after uncertainty in the first stage becomes known (Beale 1955, Dantzig 1955). The general techniques used to perform optimization under uncertainty are as follows (Sahinidis 2004, Birge and Louveaux 2011):

- Stochastic programming

    - Probability distributions of data and/or scenario-based approaches
    - Programming with recourse
    - Probabilistic programming

- Fuzzy-based programming

    - Flexible programming
    - Possibilistic programming

- Stochastic dynamic programming.

A.3 Mixed integer linear programming (MILP)

A large number of UES applications can be effectively modelled with an MILP formulation. MILPs can be expressed formally as:

$$\min f(\mathbf{x}, \mathbf{y})$$
$$\text{subject to} \quad g(\mathbf{x}, \mathbf{y}) \leq \mathbf{b}$$
$$x_{min} \leq \mathbf{x} \leq x_{max}$$
$$\mathbf{y} \in \{0, 1\}^q$$

where $\mathbf{x}$ is a vector of continuous variables, bounded by the limits x_{min} and x_{max}; $\mathbf{y}$ is a vector of zero-one variables of length q; $f(.)$ is the linear objective function; $g(.)$ and $\mathbf{b}$ represent the linear constraints.

The key feature of this formulation is that the integer-valued variables **y** can be used to represent a number of useful real-world conditions found in urban energy systems including: the presence or absence of a pipe connection, the decision to invest at a particular time, or the number of discrete technologies to purchase. The continuous variables **x** represent quantities such as operating rates which can vary smoothly between two limits.

A.3.1 Benefits and drawbacks

Much of the power of MILP lies in its use to formulate logical constraints. Integer variables enable the modelling of indivisibilities (e.g. one combined heat and power plant or none) and non-convexities, e.g. by use of separable programming and special ordered sets (Miller 1963, Williams 1999). There is considerable flexibility available from the use of logical constraints in particular. A simple example for urban energy systems is: if no combined heat and power plant is built in this area, then it will not be possible to supply heat via district heating pipelines from the plant. Boolean algebra may be used to represent such logical conditions. Examples of logical conditions abound based on Boolean operations: or, and, not, implies, and if and only if. For further details, see Williams (1999, 2009).

The major drawback to using MILP for modelling systems is that there is no standard formulation or analytical solution (Nemhauser and Wosley 1989, Williams 2009). In general, MILPs are much harder to solve than LPs. In LP models, there is a dual solution for the primal solution such that marginal valuations (i.e. shadow prices) may be computed (Dantzig 1963). In MILP, there is no obvious symmetric dual, so marginal valuations may not be computed. Some of the economic analytics available in LPs may not therefore be applied to MILPs. With these two points in mind, the next section offers strategies for strong formulations of MILPs.

A.3.2 Formulation strategies for MILP

Formulating an MILP involves accurate descriptions of the interrelations between components of a system. The purpose of such a program is to provide (Dantzig 1963: 2):

> a statement of the actions to be performed, their timing, and their quantity (called a "program" or "schedule"), which will permit the system to move from a given status toward the defined objective.

Including the appropriate components of a system, and in sufficient detail, to meet this purpose is a skill developed by the modeller with practice and experience. Most importantly, the model developer must manage a trade-off between tractable formulations and representational fidelity. The goal of the modeller is therefore to provide a strong formulation, i.e. the smallest convex set of linear equality and inequality constraints that contains all the feasible integer points.

Key model parameters that affect the running time of LP models include the number of constraints and the sparsity of the constraint matrix (Bertsimas and Tsitsiklis 1997).

Solution times of LPs may be reduced by making the model compact but in MILPs it is often a better strategy to introduce extra binary variables. Additional binary variables can be used to reduce the number of branches to be searched in a tree search strategy. With respect to the number of constraints, strategies for strong formulations LPs and MILP also diverge. In LPs, reducing the number of constraints often reduces the solution times. However, in MILPs, additional constraints can be used to reformulate the relaxed MILP (see the next section for an explanation) such that the convex hull of feasible continuous variables may be closer to the convex hull of feasible integer variables. The optimality gap between a relaxed MILP and the optimal MILP (found in the output of a model run) is an indicator of how strong a formulation is: a 50 per cent optimality gap may be reduced closer to an acceptable 5 per cent gap with the use of additional constraints. Details of other formulation strategies can be found in Williams (1999).

A.3.3 *Solution strategies*

Linear programs can be easily solved by means of the simplex algorithm (Dantzig 1963). However, for MILPs, things are more difficult. Formally, there exists no systematic methods in polynomial time for generating the linear equality and inequality constraints that define the convex polyhedron which contains the convex hull of integer points. As such, existing integer programming solution algorithms either can take up to an exponential time to solve in the worst case, or they provide an approximate solution.

Many methods have been developed for solving integer programming problems. The following list provides a brief overview (Bertsimas and Tsitsiklis 1997).

EXACT ALGORITHMS

Optimal solution guaranteed, but in the worst case will take an exponential number of iterations.

- *Rounding method*: rounding up or down continuous solutions to the nearest integer solution. No obvious way of moving from the LP solution to the IP solution.
- *Tree search*: intelligent enumerative technique using branches of integer solutions, e.g. branch and bound technique (Land and Doig 1960).
- *Cutting-plane method*: progressively excluding non-optimal integer solutions of the initial polyhedron until the convex hull of integer solutions is found (Gomory 1958).
- *Dynamic programming*: sequentially solves integer problems (Howard 1966).
- *Interior point methods*: (Karmarker 1984).

APPROXIMATION ALGORITHMS

Sub-optimal solution by means of a bound on the sub-optimal solution, provided in polynomial time.

- Epsilon approximation (Ibarra and Chul 1975).

HEURISTIC ALGORITHMS

Sub-optimal solution but without guarantees on its quality. Running time is not guaranteed to be polynomial, but good solutions may be found quickly.

- Simulated annealing (Kirkpatrick *et al.* 1983).
- Genetic algorithms (Mitchell 1998).

The branch-and-bound method is perhaps the most popular approach for solving MILPs. As the name suggests, this method presents the problem as a binary tree. Each integer variable represents a node on the tree, where each subsequent branch represents an integer value for the variable (e.g. one branch for $y_i = 0$, another for $y_i = 1$). To find the optimal solution, an LP is first solved by converting the integer variables to continuous variables (a so-called 'relaxed' problem); this solution provides a lower bound on the MILP model. Various heuristics can then be used to traverse the tree efficiently and locate the optimal solution. The efficiency of the algorithm will depend on the quality of the bounds available at each node of the solution tree.

   This algorithm and many others are typically implemented within commercial MILP solvers such as GAMS/CPLEX.

Bibliography

Acha, S., Green, T.C. and Shah, N. (2010) 'Techno-economical tradeoffs from embedded technologies with storage capabilities on electric and gas distribution networks'. In *IEEE PES 2010*. Minneapolis, MN: IEEE.

—————— (2011a) 'Optimal charging strategies of electric vehicles in the UK power market'. In *Proceedings of ISGT 2011*. Anaheim, CA: IEEE PES.

Acha, S., van Dam, K.H., Keirstead, J. and Shah, N. (2011b) 'Integrated modelling of agent-based electric vehicles into optimal power flow studies'. In *21st International Conference and Exhibition on Electricity Distribution (CIRED 2011)*. Frankfurt, DE.

Acha, S. (2010) *Impacts of Embedded Technologies on Optimal Operation of Energy Service Networks*. Ph.D. thesis, Imperial College London.

Ackroyd, P. (2000) *London: the Biography*. London: Chatto and Windus.

Adamo, L., Cammarata, G., Fichera, A. and Marletta, L. (1997) 'Improvement of a district heating network through thermoeconomic approach'. *Renewable Energy*, 10(2–3): 213–216.

AED (2008) 'Powering and empowering development: Increasing access to electricity in Angola'. Online. Accessed: 26 July 2012. http://test.aed.org/Publications/upload/Powering_and_Empowering_Dev.pdf

Aiel (2011) *The potentials of biofuels from urban forestry (La produzione di combustibili legnosi dalla selvicoltura urbana)*. AIEL Associazione. http://www.aiel.cia.it

Alarcon-Rodriguez, A., Ault, G. and Galloway, S. (2010) 'Multi-objective planning of distributed energy resources: A review of the state-of-the-art'. *Renewable and Sustainable Energy Reviews*, 14(5): 1353–1366.

Alberti, M. and Waddell, P. (2000) 'An integrated urban development and ecological simulation model'. *Integrated Assessment*, 1: 215–227.

Allen, R. (2010) *The British Industrial Revolution in Global Perspective*. Cambridge: Cambridge University Press.

Almeida, A.M., Ben-Akiva, M., Pereira, F.C., Ghauche, A., Guevara, C., Niza, S. and Zegras, C. (2009) 'A framework for integrated modelling of urban systems'. In *Proceedings of the 45th ISOCARP Congress*. Porto, PT: ISOCARP. http://www.isocarp.net/Data/case_studies/1623.pdf

Alonso, W. (1964) *Location and Land Use*. Cambridge: Harvard University Press.

Amano, Y., Ito, K., Yoshida, S., Matsuo, K., Hashizume, T., Favrat, D. and Maréchal, F. (2010) 'Impact analysis of carbon tax on the renewal planning of energy supply system for an office building'. *Energy*, 35(2): 1040–1046.

Amsa Gruppo (2009) *Valutazioni preliminari per la raccolta di olio vegetale esausto da utilizzare in un impianto di cogenerazione di energia elettrica e calore*. Amsa Gruppo. http://www.fast.mi.it/pubblicazioni/biocarburanti/amsa.pdf

Anderstig, C. and Mattsson, L.G. (1991) 'An integrated model of residential and employment location in a metropolitan region'. *Papers in Regional Science*, 70(2): 167–184.

Arentze, T. and Timmermans, H. (2000) 'The ALBATROSS system'. In *ALBATROSS: A Learning Based Transportation Oriented Simulation System*, edited by T. Arentze and H. Timmermans, pp. 81–107. Eindhoven, NL: European Institute of Retailing and Services Studies.

Arsenault, R. (1984) 'The end of the long hot summer: the air conditioner and Southern culture'. *The Journal of Southern History*, 50(4): 597–628.

Arup (2011) *Climate Action in Megacities: C40 Cities Baseline and Opportunities*. Technical report, Arup, C40 Cities Climate Leadership Group. http://www.arup.com/~/media/Files/PDF/Publications/Research_and_whitepapers/ArupC40ClimateActionInMegacities.ashx

ASHRAE (2006) *ASHRAE Standard 100-2006 Energy Conservation in Existing Buildings*. American Society of Heating, Refrigerating, and Air-Conditioning Engineers.

Ausubel, J.H., Marchetti, C. and Meyer, P.S. (1998) 'Toward green mobility: the evolution of transport'. *European Review*, 6(2): 137.

Ayers, R. (1991) 'Evolutionary economics and environmental imperatives'. *Structural Change and Economic Dynamics*, 2(2): 255–275.

Ayres, R. and Ayres, L. (2002) *A Handbook of Industrial Ecology*. London: Edward Elgar Publishing Ltd.

Azapagic, A. and Clift, R. (1995) 'Life cycle assessment and linear programming – environmental optimisation of a product system'. *Computers and Chemical Engineering*, 19(S): S229–S234.

Baccini, P. (1997) 'A city's metabolism: towards the sustainable development of urban energy systems'. *Journal of Urban Technology*, 4(2): 27–39.

Bahaj, A., Myers, L. and James, P. (2007) 'Urban energy generation: influence of micro-wind turbine output on electricity consumption in buildings'. *Energy and Buildings*, 39(2): 154–165.

Bai, X. (2007) 'China's solar-powered city'. Accessed: 19 June 2012. http://www.renewableenergyworld.com/rea/news/article/2007/05/chinas-solar-powered-city-48605

Banham, R. (1984) *The Architecture of the Well-Tempered Environment*, 2nd edn. Chicago: University of Chicago Press.

Banister, D., Stead, D., Steen, P., Akerman, J., Dreborg, K., Nijkamp, P. and Schleicher-Tappeser, R. (2000) *European Transport Policy and Sustainable Mobility*. London: Spon Press.

Banister, D. (2008) 'The sustainable mobility paradigm'. *Transport Policy*, 15(2): 73–80.

Barnes, D.F., Krutilla, K. and Hyde, W.F. (2005) *The urban household energy transition: social and environmental impacts in the developing world*. Washington, DC: Resources for the Future.

Barredo, J.I., Kasanko, M., McCormick, N. and Lavalle, C. (2003) 'Modelling dynamic spatial processes: simulation of urban future scenarios through cellular automata'. *Landscape and Urban Planning*, 64(3): 145–160.

Barrett, J., Vallack, H., Jones, A. and Haq, G. (2002) *A Material Flow Analysis and Ecological Footprint of York*. York: Stockholm Environment Institute.

Barty-King, H. (1984) *New Flame*. London: Graphmire Ltd.

Batty, M. (1971) 'Modelling cities as dynamic systems'. *Nature*, 231(5303): 425–428.

———— (1976) *Urban Modelling: Algorithms, Calibrations, Predictions*. Cambridge: Cambridge University Press.

———— (2005) 'Agents, cells, and cities: new representational models for simulating multiscale urban dynamics'. *Environment and Planning A*, 37(8): 1373–1394.

———— (2006) 'Rank clocks'. *Nature*, 444(7119): 592–596.

———— (2007) *Cities and Complexity: Understanding cities with cellular automata, agent-based models, and fractals*. Cambridge, MA: MIT Press.

———— (2008) 'Cities as complex systems: Scaling, interactions, networks, dynamics and urban morphologies'. *UCL Working Paper Series*, 131: 0–62.

Batty, M., Xie, Y. and Sun, Z. (1999) 'Modeling urban dynamics through GIS-based cellular automata'. *Computers, Environment and Urban Systems*, 23(3): 205–233.

Beale, E.M.L. (1955) 'On minimizing a convex function subject to linear inequalities'. *Journal of the Royal Statistical Society. Series B (Methodological)*, 17(2): 173–184.

BEE (2008) 'Status of biomass resource assessments'. Accessed: 18 June 2012. http://www.eu-bee.com/default.asp?sivuID=24352

Beggs, S., Cardell, S. and Hausman, J. (1981) 'Assessing the potential demand for electric cars'. *Journal of Econometrics*, 17(1): 1–19.

Beherc, D., Jindal, S. and Malhotrd, H. (1994) 'Ventilatory function in nonsmoking rural Indian women using different cooking fuels'. *Respiration*, 61(2): 89–92.

Beller, M. (1976) 'Reference energy system methodology'. In *81st National Meeting of the American Institute of Chemical Engineers*. Uphaven, NY: Brookhaven National Laboratory. http://www.osti.gov/bridge/servlets/purl/7191575-SyhUpT/7191575.pdf

Benonysson, A., Bøhm, B. and Ravn, H.F. (1995) 'Operational optimization in a district heating system'. *Energy Conversion and Management*, 36(5): 297–314.

Ben-Akiva, M. and Lerman, S. (1985) *Discrete Choice Analysis*. Cambridge, MA: MIT Press.

Berg, N. (2011) 'Defining cities in a metropolitan world'. Accessed: 27 July 2012. http://www.theatlanticcities.com/neighborhoods/2011/09/defining-cities-metropolitan-world/102/

Berndes, G., Hansson, J., Egeskog, A. and Johnssons, F. (2010) 'Strategies for 2nd generation biofuels in EU – co-firing to stimulate feedstock supply development and process integration to improve energy efficiency and economic competitiveness'. *Biomass and Bioenergy*, 34(2): 227–236.

Berndes, G., Hoogwijk, M. and Richard, V.D.B. (2003) 'The contribution of biomass in the future global energy supply: a review of 17 studies'. *Biomass and Bioenergy*, 25(1): 1–28.

Bertalanffy, L.V. (1950) 'An outline of general system theory'. *The British Journal for the Philosophy of Science*, I(2): 134–165.

Bertsimas, D. and Tsitsiklis, J.N. (1997) *Introduction to Linear Optimization*. Nashua, NH: Athena Scientific.

Bettencourt, L.M.A., Lobo, J., Helbing, D., Kühnert, C. and West, G.B. (2007) 'Growth, innovation, scaling, and the pace of life in cities'. *Proceedings of the National Academy of Sciences of the United States of America*, 104(17): 7301–7306.

Better Buildings Partnership (2010) *Low Carbon Retrofit Toolkit: A Roadmap to Success*. Technical report, Accenture. http://www.betterbuildingspartnership.co.uk/download/bbp-low-carbon-retrofit-toolkit.pdf

Bhat, C., Guo, J., Srinivasan, S. and Sivakumar, A. (2004) 'A comprehensive econometric microsimulator for daily activity-travel patterns'. *Transportation Research Record*, 1894: 57–66.

Bhat, C. and Waller, T. (2008) 'CEMDAP: A second generation activity-based travel modeling system for metropolitan areas'. *Presented at the 2008 TRB Innovations in Travel Modeling conference*. Portland, OR.

Biezma, M. and Cristobal, J. (2006) 'Investment criteria for the selection of cogeneration plants – a state of the art review'. *Applied Thermal Engineering*, 26(5–6): 583–588.

Birge, J.R. and Louveaux, F. (2011) *Introduction to Stochastic Programming*, 2nd edn. New York: Springer.

Bixby, R.E. (2002) 'Solving Real-World Linear Programs: A Decade and More of Progress'. *INFORMS*, 50(1): 3–15.

Bojic, M., Trifunovic, N. and Gustafsson, S.I. (2000) 'Mixed 0–1 sequential linear programming optimization of heat distribution in a district-heating system'. *Energy and Buildings*, 32(3): 309–317.

Borbely, A. and Kreider, J. (2001) 'Distributed generation: an introduction'. In *Distributed Generation: The Power Paradigm of the New Millennium*, edited by A. Borbely and J. Kreider, chapter 1, pp. 1–52. Boca Raton, FL: CRC Press.

Bose, R.K. and Anandalingam, G. (1996) 'Sustainable urban energy-environment management with multiple objectives'. *Energy*, 21(4): 305–318.

BP (2011) *Statistical Review of World Energy 2011*. London: BP.

BRE (2002) *Carbon dioxide emissions from non-domestic buildings: 2000 and beyond (BR 442)*. Technical report, Building Research Establishment.

——— (2006) *Conventions for U value calculations*. Technical report, Building Research Establishment. http://www.bre.co.uk/filelibrary/pdf/rpts/BR_443_%282006_Edition%29.pdf

——— (2008) *GB domestic energy fact file 2008*. Technical report, Building Research Establishment. http://www.bre.co.uk/filelibrary/pdf/rpts/Fact_File_2008.pdf

Bridgwater, A.V., Toft, A.J. and Brammer, J.G. (2002) 'A techno-economic comparison of power production by biomass fast pyrolysis with gasification and combustion'. *Renewable and Sustainable Energy Reviews*, 6(3): 181–246.

Bridgwater, A.V. (2011) 'Review of fast pyrolysis of biomass and product upgrading'. *Biomass and Bioenergy*, 38(March): 1–27.

Bringezu, S. (2000) 'Industrial ecology and material flow analysis'. Online. Accessed: 26 July 2012. http://www.greenleaf-publishing.com/content/pdfs/iebring.pdf

Brooks, A.N. (2002) *Vehicle-to-grid demonstration project: Grid Regulation Ancillary Service with a Battery Electric Vehicle*. Technical report, California Air Resources Board, San Dimas, CA. http://www.smartgridnews.com/artman/uploads/1/sgnr_2007_12031.pdf

Brownstone, D., Bunch, D.S., Golob, T.F. and Ren, W. (1996) 'A transactions choice model for forecasting demand for alternative-fuel vehicles'. *Research in Transportation Economics*, 4: 87–129.

Brownsword, R.A., Fleming, P.D., Powell, J.C. and Pearsall, N. (2005) 'Sustainable cities: modelling urban energy supply and demand'. *Applied Energy*, 82(2): 167–180.

Bruce, N., Perez-Padilla, R. and Albalak, R. (2000) 'Indoor air pollution in developing countries: a major environmental and public health challenge'. *Bulletin of the World Health Organization*, 78(9): 1078–1092.

Bruckner, T., Groscurth, H. and Kümmel, R. (1997) 'Competition and synergy between energy technologies in municipal energy systems'. *Energy*, 22(10): 1005–1014.

Bruckner, T., Morrison, R., Handley, C. and Patterson, M. (2003) 'High-resolution modeling of energy-services supply systems using decco: Overview and application to policy development'. *Annals of Operations Research*, 121: 151–180.

Buettner, T. (2007) 'An overview of world urbanization prospects'. In *Urbanization, Development Pathways and Carbon Implications workshop*. Tsukuba, JP. http://www.gcp-urcm.org/files/A20070328/buettner.pdf

Bulkeley, H. and Betsill, M. (2003) *Cities and Climate Change: Urban sustainability and global environmental governance*. New York: Routledge.

Bull, M. (2005) 'No dead air! the iPod and the culture of mobile listening'. *Leisure Studies*, 24(4): 343–355.

C40 Cities (2011) 'C40 cities: Climate leadership group'. Accessed: 12 June 2012. http://www.c40cities.org

Cairns, S., Sloman, L., Newson, C., Anable, J., Kirkbride, A. and Goodwin, P. (2008) 'Smarter choices: Assessing the potential to achieve traffic reduction using "soft measures"'. *Transport Reviews*, 28(5): 593–618.

Calfee, J.E. (1985) 'Estimating the demand for electric automobiles using fully disaggregated probabilistic choice analysis'. *Transportation Research Part B: Methodological*, 19(4): 287–301.

Caller, B. (2012) 'Email correspondence'. Personal communication.

Cantor, J. (2011) *Air Source Heat Pumps – Friend or Foe: A review of current technology and its viability*. Technical report, AECB. http://www.aecb.net/PDFs/HeatPumpsArticleJuly2011.pdf

Carnell, R. (2009) *triangle: Provides the standard distribution functions for the triangle distribution*. R package version 0.5. http://cran.r-project.org/web/packages/triangle/

Carpaneto, E., Chicco, G., Mancarella, P. and Russo, A. (2011) 'Cogeneration planning under uncertainty. Part II: Decision theory-based assessment of planning alternatives'. *Applied Energy*, 88(4): 1075–1083.

Caulfield, B., Farrell, S. and McMahon, B. (2010) 'Examining individuals preferences for hybrid electric and alternatively fuelled vehicles'. *Transport Policy*, 17(6): 381–387.

CCC (2008) 'Power'. Online. Accessed: 27 July 2012. http://www.theccc.org.uk/sectors/power

Chancel, L. (2010) *Urban energy transition in the developing world: The case of Nakuru, Kenya*. MSc thesis, Imperial College London.

Chappells, H. and Shove, E. (2005) 'Debating the future of comfort: environmental sustainability, energy consumption and the indoor environment'. *Building Research & Information*, 33(1): 32–40.

Chester, M. and Horvath, A. (2009) *Life-cycle Energy and Emissions Inventories for Motorcycles, Diesel Automobiles, School Buses, Electric Buses, Chicago Rail, and New York City Rail*. Technical report, University of California Berkeley, Centre for Future Urban Transport. http://escholarship.org/uc/item/6z37f2jr

Chevalier, C. and Meunier, F. (2005) 'Environmental assessment of biogas co- or tri-generation units by life cycle analysis methodology'. *Applied Thermal Engineering*, 25(17–18): 3025–3041.

Chicco, G. and Mancarella, P. (2006) 'From cogeneration to trigeneration: Profitable alternatives in a competitive market'. *IEEE Transactions on Energy Conversion*, 21(1): 265–272.

———— (2007) 'Trigeneration primary energy saving evaluation for energy planning and policy development'. *Energy Policy*, 35(12): 6132–6144.

———— (2008a) 'Assessment of the greenhouse gas emissions from cogeneration and trigeneration systems. Part I: Models and indicators'. *Energy*, 33(3): 410–417.

———— (2008b) 'A unified model for energy and environmental performance assessment of natural gas-fueled poly-generation systems'. *Energy Conversion and Management*, 49(8): 2069–2077.

———— (2009a) 'Distributed multi-generation: A comprehensive view'. *Renewable and Sustainable Energy Reviews*, 13(3): 535–551.

———— (2009b) 'Matrix modelling of small-scale trigeneration systems and application to operational optimization'. *Energy*, 34(3): 261–273.

Chinese, D., Meneghetti, A. and Nardin, G. (2004) 'Diffused introduction of organic rankine cycle for biomass-based power generation in an industrial district: a systems analysis'. *International Journal of Energy Research*, 28(11): 1003–1021.

Chua, K., Chou, S. and Yang, W. (2010) 'Advances in heat pump systems: A review'. *Applied Energy*, 87(12): 3611–3624.

CIBSE (2004) *CIBSE Guide F: Energy Efficiency in Buildings*. London: The Chartered Institution of Building Service Engineers.

City of Baltimore (2008) 'Planning/comprehensive master plan'. Accessed: 27 July 2012. http://www.baltimorecity.gov/Government/AgenciesDepartments/Planning/ComprehensiveMasterPlan.aspx

CIWM (2002) *A Resource Flow and Ecological Footprint Analysis of Greater London*. Technical report, Chartered Institute of Wastes Management, Best Foot Forward, Northampton.

Clarke, K.C. and Gaydos, L.J. (1998) 'Loose-coupling a cellular automaton model and GIS: long-term urban growth prediction for San Francisco and Washington/Baltimore'. *International Journal for Geographical Information Science*, 12(7): 699–714.

Climate Policy Initiative (2011) *Drivers of Thermal Retrofit Decisions – A Survey of German Single- and Two-Family Houses*. Technical report, Climate Policy Initiative. http://climatepolicyinitiative.org/wp-content/uploads/2011/12/Drivers-of-Thermal-Retrofit-Decisions-A-Survey.pdf

Cochrane, R. (1986) *Cradle of Power: the story of Deptford power stations*. Central Electricity Generating Board. http://swehs_archive.swelocker.co.uk/news25su.pdf

Commoner, B. (1972) 'A bulletin dialogue on "the closing circle", response'. *Bulletin of the Atomic Scientists*, 28(5): 17, 42–56.

Consoli, F., Allen, D., Boustead, I., Fava, J., Franklin, W., Jensen, A., Oude, N.D., Parrish, R., Perriman, R. and Postelthwaite, D. (1994) 'Guidelines for life cycle assessment: A "code of practice"'. *Environmental Science and Pollution Research*, 1(1): 55.

Cooper, G. (2002) *Air-Conditioning America: Engineers and the Controlled Environment, 1900–1960*. Baltimore, MD: John Hopkins University Press.

Copenhagen Energy Summit (2009) 'Global district energy climate awards: Waste to heat in Vienna'. Accessed: 26 July 2012. http://www.copenhagenenergysummit.org/index.php?action=award

Copenhagen Energy (2009) 'Copenhagen district heating system'. Online. Accessed: 26 July 2012. http://www.copenhagenenergysummit.org/applications/Copenhagen,Denmark-DistrictEnergyClimate Award.pdf

CTR (2004) *The Main District Heating Network in Copenhagen*. Copenhagen, DK: Centralkommunernes Transmissionsselskab I/S. http://www.ctr.dk/Images/Publikationer/Themaindistrictheatingnetworkin cph-UK.pdf

———— (2009) *CTR Annual Report and Financial Statements 2009*. Copenhagen, DK: Centralkom-munernes Transmissionsselskab I/S. http://www.ctr.dk/Images/%C3%85rsberetninger/Aarsberetning_2009_Engelsk.pdf

Czachorski, M., Silvis, J., Barkalow, G., Spiegel, L. and Coldwell, M. (2008) *Development of an energy module for I-PLACE3S*. Technical Report CEC-500-2008-024, California Energy Commission.

Dagsvik, J.K., Wennemo, T., Wetterwald, D.G. and Aaberge, R. (2002) 'Potential demand for alternative fuel vehicles'. *Transportation Research Part B: Methodological*, 36(4): 361–384.

Danish Energy Saving Trust (2011) 'Installing district heating'. Accessed: 27 July 2012. http://www.savingtrust.dk/consumer/products/indoor-climate/district-heating/conversion-to-district-heating

Dantzig, G. (1963) *Linear Programming and Extensions*. Princeton: Princeton University Press.

Dantzig, G.B. (1955) 'Linear programming under uncertainty'. *Management Science*, 1: 197–206.

Daou, K., Wang, R. and Xia, Z. (2006) 'Desiccant cooling air conditioning: a review'. *Renewable and Sustainable Energy Reviews*, 10(2): 55–77.

Darvill, T. (1987) *Prehistoric Britain*. Abingdon: Routledge.

Davis, M. (1998) 'Rural household energy consumption'. *Energy Policy*, 26(3): 207–217.

DCLG (2011) 'Live tables on dwelling stock (including vacants)'. Online. Accessed: 27 July 2012. http://www.communities.gov.uk/housing/housingresearch/housingstatistics/housingstatisticsby/stockincludingvacants/livetables/

DEA (2011) 'Heat supply: Goals and means over the years'. Accessed: 27 July 2012. http://www.ens.dk/EN-US/SUPPLY/HEAT/GOALS_AND_MEANS/Sider/Forside.aspx

DECC (2008) 'Total final energy consumption at sub-national level'. Accessed: 27 July 2012. http://www.decc.gov.uk/en/content/cms/statistics/regional/total_final/total_final.aspx

———— (2009) *Biomethane into the Gas Network: A Guide for Producers*. Technical report, Dept of Energy and Climate Change, London. http://www.decc.gov.uk/assets/decc/whatwedo/ukenergysupply/energymarkets/gas_markets/nonconventional/1_20091229125543_e_@@_biomethaneguidance.pdf

———— (2011a) *Digest of UK Energy Statistics*. Technical report, Dept of Energy and Climate Change, London. http://www.decc.gov.uk/en/content/cms/statistics/publications/dukes/dukes.aspx

———— (2011b) *Great Britain's housing energy fact file*. London: U.K. Department of Energy and Climate Change.

———— (2012) *The Future of Heating: A strategic framework for low carbon heat in the UK*. London: Dept of Energy and Climate Change. http://www.decc.gov.uk/en/content/cms/consultations/cons_smip/cons_smip.aspx

Defra (2011) *Environmental reporting: Defra's greenhouse gas (GHG) conversion factors for company reporting*. Technical report, Dept for Environment, Food and Rural Affairs. http://www.defra.gov.uk/environment/business/reporting/conversion-factors.htm

Denmark, S. (2009) 'FAM55N: Households 1. January by region, type of household and size'. Accessed: 27 July 2012. http://www.statistikbanken.dk/statbank5a/SelectVarVal/Define.asp?MainTable=FAM55N&PLanguage=1&PXSId=0

Der Meijden, C.M.V., Veringa, H.J. and Rabou, L.P.L.M. (2010) 'The production of synthetic natural gas (SNG): A comparison of three wood gasification systems for energy balance and overall efficiency'. *Biomass and Bioenergy*, 34(3): 302–311.

de la Barra, T. (1989) *Integrated Land Use and Transport Modelling*. Cambridge: Cambridge University Press.

de Rocquigny, E., Devictor, N. and Tarantola, S. (eds) (2008) *Uncertainty in Industrial Practice: A guide to quantitative uncertainty management*. Chichester: John Wiley & Sons, Ltd.

Dhakal, S. and Hanaki, K. (2002) 'Improvement of urban thermal environment by managing heat discharge sources and surface modification in Tokyo'. *Energy and Buildings*, 34(1): 13–23.

Diamond, J. (2005) *Collapse: How Societies Choose to Fail or Survive*. London: Penguin Books.

Diwekar, U.M. (2008) *Introduction to Applied Optimization*. Dordrecht, NL: Springer.

Dixon, T. (2009) 'Urban land and property ownership patterns in the UK: trends and forces for change'. *Land Use Policy*, 26S: S43–S53.

DOE (2010) *Buildings Energy Data Book*. Richland, WA: U.S. Department of Energy. http://buildingsdatabook.eren.doe.gov/

DTI (2006) *United Kingdom Generic Distribution System (UKGDS) Phase One Profile Data*. London: DTI Centre for Distributed Generation and Sustainable Electrical Energy.

———— (2007) *Meeting the Energy Challenge: A White Paper on Energy*. London: Dept of Trade and Industry. http://www.dti.gov.uk/energy/whitepaper/page39534.html

Dunnett, A.J., Adjiman, C.S. and Shah, N. (2008) 'A spatially explicit whole-system model of the lignocellulosic bioethanol supply chain: an assessment of decentralised processing potential'. *Biotechnology for Biofuels*, 1: 13.

Duvigneaud, P. and Denaeyer-De Smet, S. (1977) *L'ecosysteme urbain bruxellois: Traveaux de la Section Belge du Programme Biologique International*. Brussells: Edition Duculot.

E4Tech (2009) *Biomass supply curves for the UK: a report for DECC*. Technical report, E4Tech Ltd.

Ebadian, M., Sowlati, T., Sokhansanj, S., Stumborg, M. and Townley-Smith, L. (2011) 'A new simulation model for multi-agricultural biomass logistics system in bioenergy production'. *Biosystems Engineering*, 110(3): 280–290.

Eberhard, M. and Tarpenning, M. (2006) 'The 21st century electric car'. Accessed: 27 July 2012. http://www.evworld.com/library/Tesla_21centuryEV.pdf

Echenique, M.H., Hargreaves, A.J., Mitchell, G. and Namdeo, A. (2012) 'Growing cities sustainably: Does urban form really matter?' *Journal of the American Planning Association*, 78(2): 121–137.

EcoRec (2012) 'Ecorec'. Accessed: 27 July 2012. http://www.ecorec.it/

EEA (1999) *Environment in the European Union at the Turn of the Century*. Copenhagen, DK: European Environment Agency.

———— (2000) *Household and Municipal Waste: Comparability of Data in EEA member countries*. 3/2000. Copenhagen, DK: European Environment Agency. http://www.eea.europa.eu/publications/Topic_report_No_32000

———— (2001) *Biodegradable municipal waste management in Europe*. 15/2001. Copenhagen, DK: European Environment Agency. http://www.eea.europa.eu/publications/topic_report_2001_15

———— (2007) *Estimating the environmentally compatible bioenergy potential from agriculture*. 12/2007. Copenhagen, DK: European Environment Agency.

Egger, S. (2006) 'Determining a sustainable city model'. *Environmental Modelling and Software*, 21(9): 1235–1246.

Ehrenfield, J. and Gertler, N. (1997) 'Industrial ecology in practice: The evolution of interdependence at Kalundborg'. *Journal of Industrial Ecology*, 1(1): 67–79.

Ehrgott, M. (2005) *Multicriteria Optimization*, 2nd edn. Berlin: Springer.

Ehrlich, P.R. and Holdren, J.P. (1972) 'A bulletin dialogue on "the closing circle", critique'. *Bulletin of the Atomic Scientists*, 28(5): 16,18–27.

EIA (2011) *Annual Energy Review*. Washington, DC: U.S. Department of Energy.

Eighmy, T.T. and Kosson, D.S. (1996) 'USA national overview on waste management'. *Waste Management*, 16(5–6): 361–366.

Eijck, J.V. and Romijn, H. (2008) 'Prospects for Jatropha biofuels in Tanzania: An analysis with strategic niche management'. *Energy Policy*, 36(1): 311–325.

Ekechukwu, O., Madu, A., Nwanya, S. and Agunwamba, J. (2011) 'Optimization of energy and manpower requirements in Nigerian bakeries'. *Energy Conversion and Management*, 52(1): 564–568.

Element Energy (2010) *Electric Vehicles in the UK and Republic of Ireland: Greenhouse gas emissions reductions and infrastructure needs*. Element Energy. http://www.element-energy.co.uk/wordpress/wp-content/uploads/2012/05/EVs-in-the-UK-and-ROI_final-report_10.12.10.pdf

Entec (2010) *Royal Borough of Kensington and Chelsea: Heat Mapping Study*. Mayor of London. http://www.londonheatmap.org.uk/Content/borough_heat_map.aspx

EPD Hong Kong (2010) 'Green Hong Kong: Carbon audit'. Online. Accessed: 14 June 2012. http://www.epd.gov.hk/epd/english/climate_change/ca_intro.html

Eppen, G.D., Marting, R.P. and Schrage, L. (1989) 'A scenario approach to capacity planning'. *Operations Research*, 37(4): 517–527.

EST (2010) *Getting warmer: a field trial of heat pumps*. Technical report, Energy Savings Trust. http://www.energysavingtrust.org.uk/Media/Generate-your-own-energy/PDFs/Full-heat-pump-field-trial-report

———— (2011) 'Home Energy Efficiency Database (HEED)'. Online. Accessed: 27 July 2012. http://heed.est.org.uk/

Eurostat (2012) 'Environmental data centre on waste'. Online. Accessed: 27 July 2012. http://epp.eurostat.ec.europa.eu/portal/page/portal/waste/introduction/

EU (2007) *State of European Cities Report: Adding value to the European Urban Audit*. May. European Union. http://ec.europa.eu/regional_policy/sources/docgener/studies/pdf/urban/stateofcities_2007.pdf

———— (2009) 'Directive 2009/28/EC of the European Parliament and of the Council on the promotion of the use of energy from renewable sources'. Online. Accessed: 27 July 2012. http://eur-lex.europa.eu/LexUriServ/LexUriServ.do?uri=Oj:L:2009:140:0016:0062:en:PDF

Evans, R., Guy, S. and Marvin (1999) 'Making a difference: sociology of scientific knowledge and urban energy policies'. *Science, Technology and Human Values*, 24(1): 105–131.

Ezzati, M., Mbinda, B.M. and Kammen, D.M. (2000) 'Comparison of emissions and residential exposure from traditional and improved cookstoves in Kenya'. *Environmental Science and Technology*, 34(4): 578–583.

Falk, B.B. (1994) 'Opportunities for the woodwaste resource'. *Forest Products Journal*, 47(6): 17–22.

Fatih Demirbas, M. (2009) 'Biorefineries for biofuel upgrading: A critical review'. *Applied Energy*, 86: S151–S161.

Fermi, E. (1956) *Thermodynamics*. New York: Dover Publications Inc.

Filchakova, N., Robinson, D. and Scartezzini, J.L. (2008) 'Quo vadis thermodynamics and the city: a critical review of applications of thermodynamic methods to urban systems'. *International Journal of Ecodynamics*, 2(4): 222–230.

FIPER (2011) 'L'incentivazione delle fer nel settore del riscaldamento-raffreddamento'. Online. Accessed: 27 July 2012. http://www.fiper.it/it/biblioteca.html

Fischer, G. and Schrattenholzer, L. (2001) 'Global bioenergy potentials through 2050'. *Biomass and Bioenergy*, 20(3): 151–159.

Fisk, D. (1981) *Thermal Control of Buildings*. London: Applied Science Publishers.

—— (2003) 'Energy demand 30 years from now'. *Journal for the Foundation of Science and Technology*, 17(8): 10–12.

—— (2005) 'Energy demand – rethinking from basics'. In *Bowman Lecture*. De Montford University.

—— (2008) 'What are the risk-related barriers to, and opportunities for, innovation from a business perspective in the UK, in the context of energy management in the built environment?' *Energy Policy*, 36(12): 4615–4617.

Floudas, C.A. (1995) *Nonlinear and Mixed-Integer Optimization: Fundamentals and applications*. Oxford: Oxford University Press.

Foresight (2008) *Powering our Lives: Sustainable Energy Management and the Built Environment*. London: Government Office for Science.

Forrester, J. (1969) *Urban Dynamics*. Cambridge, MA: Productivity Press.

Fouquet, R. and Pearson, P.J.G. (2006) 'Seven centuries of energy services: The price and use of light in the United Kingdom (1300–2000)'. *The Energy Journal*, 27(1): 139–177.

Fouquet, R. (2008) *Heat, Power and Light: Revolutions in Energy Services*. Cheltenham, UK: Edward Elgar Publishing Ltd.

Freese, B. (2006) *Coal: A Human History*. London: Arrow Books.

Freppaz, D., Minciardi, R., Robba, M., Rovatti, M., Sacile, R. and Taramasso, A. (2004) 'Optimizing forest biomass exploitation for energy supply at a regional level'. *Biomass and Bioenergy*, 26(1): 15–25.

Friotherm (2005) 'Värtan Ropsten: The largest sea water heat pump facility worldwide, with 6 Unitop 50FY and 180 MW total capacity'. Accessed: 27 July 2012. http://www.friotherm.com/downloads/vaertan_e008_uk.pdf

Frombo, F., Minciardi, R., Robba, M. and Sacile, R. (2009) 'A decision support system for planning biomass-based energy production'. *Energy*, 34(3): 362–369.

Fujita, M., Krugman, P.R. and Venables, A. (1999) *The Spatial Economy: Cities, regions and international trade*. Cambridge, MA: MIT Press.

FURORE (2003) *R & D Technology Roadmap – A contribution to the identification of key technologies for a sustainable development of European road transport*. Future Road Vehicle Research. http://www.furore-network.com/documents/furore_rod_map_final.pdf

Galloway, J.A., Keene, D. and Murphy, M. (1996) 'Fuelling the city: production and distribution of firewood and fuel in London's region, 1290–1400'. *The Economic History Review*, 49(3): 447–472.

Gann, D.M. (2000) *Building Innovation: complex constructs in a changing world*. London: Thomas Telford Publishing.

García, A.J., Esteban, M.B., Márquez, M.C. and Ramos, P. (2005) 'Biodegradable municipal solid waste: characterization and potential use as animal feedstuffs'. *Waste Management*, 25(8): 780–787.

García, C.A., Manzini, F. and Islas, J. (2010) 'Air emissions scenarios from ethanol as a gasoline oxygenate in Mexico City Metropolitan Area'. *Renewable and Sustainable Energy Reviews*, 14(9): 3032–3040.

Geels, F. (2002) 'Technological transitions as evolutionary reconfiguration processes: a multi-level perspective and a case-study'. *Research Policy*, 31(8–9): 1257–1274.

——— (2004) 'From sectoral systems of innovation to socio-technical systems: Insights about dynamics and change from sociology and institutional theory'. *Research Policy*, 33(6–7): 897–920.

Geidl, M. and Andersson, G. (2007) 'Optimal power flow of multiple energy carriers'. *IEEE Transactions on Power Systems*, 22(1): 145–155.

Geidl, M. (2007) *Integrated modeling and optimization of multi-carrier energy systems.* Ph.D. thesis, Eidgenössische Technische Hochschule, Zürich.

Ghauche, A. (2010) *Integrated Transportation and Energy Activity-Based Model.* MSc thesis, MIT.

Giddens, A. (2009) *Sociology*, 6th edn. Cambridge: Polity Press.

Giddings, B., Hopwood, B. and O'Brien, G. (2002) 'Environment, economy and society: fitting them together into sustainable development'. *Sustainable Development*, 10(4): 187–196.

Girardet, H. (1995) *The Gaia Atlas of Cities.* London: Gaia Books Limited.

——— (2009) *A Renewable World: Energy, Ecology, Equality.* London: Green Books. http://www.worldfuturecouncil.org/a_renewable_world.html

Girardin, L., Marechal, F., Dubuis, M., Calame-Darbellay, N. and Favrat, D. (2010) 'EnerGis: A geographical information based system for the evaluation of integrated energy conversion systems in urban areas'. *Energy*, 35(2): 830–840.

GLA (2001) *Development of a renewable energy assessment and targets for London.* London: Greater London Authority.

——— (2003a) *City Solutions: new and emerging technologies for sustainable waste management.* London: Greater London Authority.

——— (2003b) *London's Ecological Footprint: a review.* Technical report, Greater London Authority, London. http://www.london.gov.uk/mayor/economic_unit/docs/ecological_footprint.pdf

——— (2008) 'London energy and greenhouse gas emissions inventory 2008'. Online. Accessed: 26 July 2012. http://www.london.gov.uk/priorities/environment/climate-change/leggi

——— (2010) *Delivering London's energy future: The Mayor's draft Climate Change Mitigation and Energy Strategy for consultation with the London Assembly and functional bodies.* London: Greater London Authority.

Global Footprint Network (2007) 'Humanity's footprint 1961–2003'. Accessed: 27 July 2012. http://www.footprintnetwork.org/gfn_sub.php?content=global_footprint

Goldberg, D.E. (1989) *Genetic Algorithms in Search, Optimization, and Machine Learning.* Boston: Addison-Wesley Pub. Co.

Goldman, T. and Gorham, R. (2006) 'Sustainable urban transport: Four innovative directions'. *Technology in Society*, 28(1–2): 261–273.

Gomory, R. (1958) 'Essentials of an algorithm for integer solutions to linear programs'. *Bulletin of the American Mathematical Society*, 64(5): 275–278.

Google (2010) 'Google Books Ngram Viewer'. Online. Accessed: 27 July 2012. http://books.google.com/ngrams/

Graham, S. and Marvin, S. (1994) 'More than ducts and wires: Post-Fordism, cities and utility networks'. In *Managing Cities: The New Urban Context*, edited by P. Healy, S. Cameron, S. Davoudi, S. Graham and A. Madani-Pour, pp. 169–190. Chichester: John Wiley & Sons, Ltd.

Graham, S. (2000) 'Introduction: Cities and infrastructure'. *International Journal of Urban and Regional Research*, 24(1): 114–119.

Greening, L.A., Greene, D.L. and Difiglio, C. (2000) 'Energy efficiency and consumption – the rebound effect – a survey'. *Energy Policy*, 28(6–7): 389–401.

Griskevicius, V., Tybur, J.M. and den Bergh, B.V. (2010) 'Going green to be seen: Status, reputation, and conspicuous conservation'. *Journal of Personality and Social Psychology*, 98(3): 292–304.

Grohnheit, P.E. and Mortensen, B.O.G. (2003) 'Competition in the market for space heating. district heating as the infrastructure for competition among fuels and technologies'. *Energy Policy*, 31(9): 817–826.

Groscurth, H., Bruckner, T. and Kümmel, R. (1995) 'Modeling of energy-services supply systems'. *Energy*, 20(9): 941–958.

Grübler, A. (1998) *Technology and Global Change*. Laxenburg, AT: International Institute of Applied Systems Analysis.

Gustafsson, S.I., Karlsson, B.G. and Sjöholm, B.H. (1987) 'Differential rates for district heating and the influence on the optimal retrofit strategy for multi-family buildings'. *Heat Recovery Systems and CHP*, 7(4): 337–341.

Gustafsson, S. (1992) 'Optimization of building retrofits in a combined heat and power network'. *Energy*, 17(2): 161–171.

Guy, S. and Marvin, S. (2001) 'Sustainable urban future: constructing sustainable urban futures: from models to competing pathways'. *Impact Assessment and Project Appraisal*, 19(2): 131–139.

Hägerstrand, T. (1970) 'What about people in regional science?' *Papers of the Regional Science Association*, 14(7–21): 7–21.

Halliday, S. (2003) *The Great Stink of London and the Cleansing of the Victorian Metropolis*. Stroud: Sutton Publishing.

Hall, P. (2002) *Cities of Tomorrow*, 3rd edn. Oxford: Blackwell.

Hamelinck, C.N. and Faaij, A.P. (2002) 'Future prospects for production of methanol and hydrogen from biomass'. *Journal of Power Sources*, 111(1): 1–22.

Hamilton, I., Davies, M., Steadman, P., Stone, A., Ridley, I. and Evans, S. (2009) 'The significance of the anthropogenic heat emissions of London's buildings: a comparison against captured shortwave solar radiation'. *Building and Environment*, 44(4): 807–817.

Hammer, S.A., Keirstead, J., Dhakal, S., Mitchell, J., Colley, M., Connell, R., Gonzalez, R., Herve-Mignucci, M., Parshall, L., Schulz, N. and Hyams, M. (2011) 'Climate change and urban energy systems'. In *Climate Change and Cities: First Assessment Report of the Urban Climate Change Research Network*, edited by C. Rosenzweig, W.D. Solecki, S.A. Hammer and S. Mehrotra, pp. 85–112. Cambridge: Cambridge University Press.

Hanya, T. and Ambe, Y. (1976) 'A study on the metabolism of cities'. In *Science for a better environment*, pp. 228–233. Tokyo: HESC Science Council of Japan.

Hart, D. (2009) *Don't Worry About the Government? The LEED-NC "Green Building" Rating System and Energy Efficiency in U.S. Commercial Buildings*. Technical report, Massachusetts Institute of Technology. http://web.mit.edu/ipc/publications/pdf/09-001.pdf

Harvey, D. (2006) *A Handbook on Low-energy Buildings and District Energy Systems*. London: Earthscan.

Hatcher, J. (1993) *The History of the British Coal Industry: Volume I Before 1700*. Oxford: Clarendon Press.

Havelsky, V. (1999) 'Energetic efficiency of cogeneration systems for combined heat, cold and power production'. *International Journal of Refrigeration*, 22(2): 479–485.

Hawkes, A., Munuera, L. and Strbac, G. (2011) *Low Carbon Residential Heating: Grantham Briefing Paper No.6*. Technical report, Grantham Institute for Climate Change, London.

Hawkes, A. (2010) 'Estimating marginal CO_2 emissions rates for national electricity systems'. *Energy Policy*, 38(10): 5977–5987.

Hendriks, C., Obernosterer, R., Muller, D., Kytzia, S., Baccini, P. and Brunner, P. (2002) 'Material flow analysis: A tool to support environmental policy decision making. Case studies on the city of Vienna and the Swiss lowlands'. *Local Environment*, 5: 311–328.

Henning, D., Amiri, S. and Holmgren, K. (2006) 'Modelling and optimisation of electricity, steam and district heating production for a local Swedish utility'. *European Journal of Operational Research*, 175(2): 1224–1247.

Henning, D. (1997) 'MODEST: An energy-system optimisation model applicable to local utilities and countries'. *Energy*, 22(12): 1135–1150.

Hickman, R. and Banister, D. (2007) 'Looking over the horizon: Transport and reduced CO_2 emissions in the UK by 2030'. *Transport Policy*, 14(5): 377–387.

Hiremath, R., Shikha, S. and Ravindranath, N. (2007) 'Decentralized energy planning; modeling and application – a review'. *Renewable and Sustainable Energy Reviews*, 11(5): 729–752.

Hobsbawm, E. (2008) *The Age of Revolution*. London: Abacus.

Hodder, I. (2006) *Çatalhöyük: The Leopard's Tale*. London: Thames and Hudson.

Hoekman, S.K., Broch, A. and Robbins, C. (2011) 'Hydrothermal carbonization (HTC) of lignocellulosic biomass'. *Energy & Fuels*, 25(4): 1802–1810.

Hofstedder, P. (1988) *Perspectives in Life Cycle Impact Assessment: A Structured Approach to Combine Models of the Technosphere, Ecosphere and Valuesphere*. Boston, MA: Kluwer Academic Publishers.

Holman, J. (1997) *Heat Transfer*, 8th edn. London: McGraw-Hill.

Hong, S.H., Gilbertson, J., Oreszczyn, T., Green, G. and Ridley, I. (2009) 'A field study of thermal comfort in low-income dwellings in England before and after energy efficient refurbishment'. *Building and Environment*, 44(6): 1228–1236.

Hoogwijk, M. (2003) 'Exploration of the ranges of the global potential of biomass for energy'. *Biomass and Bioenergy*, 25(2): 119–133.

Horlock, J. (1997) *Cogeneration: Combined Heat and Power (CHP)*. Malabar, FL: Krieger.

Houwing, M., Negenborn, R.R. and Schutter, B.D. (2011) 'Demand response with micro-CHP systems'. *Proceedings of the IEEE*, 99(1): 200–213.

Howard, R.A. (1966) 'Dynamic programming'. *Management Science*, 12(5): 317–348.

Huangfu, Y., Wu, J., Wang, R. and Xia, Z. (2007) 'Experimental investigation of adsorption chiller for micro-scale BCHP system application'. *Energy and Buildings*, 39(2): 120–127.

Huang, S.L. and Chen, C.W. (2005) 'Theory of urban energetics and mechanisms of urban development'. *Ecological Modelling*, 189(1–2): 49–71.

Hughes, T. (1983) *Networks of Power: Electrification in Western Society, 1880–1930*. Baltimore, MD: Johns Hopkins University Press.

Hugo, A., Ciumei, C., Buxton, A. and Pistikopoulos, E. (2004) 'Environmental impact minimization through material substitution: a multi-objective optimization approach'. *Journal of Green Chemistry*, 6: 407–417.

Hunt, J.D., Kriger, D.S. and Miller, E.J. (2005) 'Current operational urban land-use-transport modelling frameworks: a review'. *Transport Reviews*, 25(3): 329–376.

Hunt, J.D. and Simmonds, D.C. (1993) 'Theory and application of an integrated land-use and transport modelling framework'. *Environment and Planning B: Planning and Design*, 20(2): 221–244.

Ibarra, O.H. and Chul, E.K. (1975) 'Fast approximation algorithms for the knapsack and sum of subset problems'. *Journal of the Association for Computing Machinery*, 22(4): 463–468.

ICLEI, IDRC and UNEP (1996) *The Local Agenda 21 Planning Guide: An introduction to sustainable development planning*. International Council for Local Environmental Initiatives. http://web.idrc.ca/openebooks/448-2

IEA (2003) *Municipal solid waste and its role in sustainability*. Paris: International Energy Agency. www.ieabioenergy.com/media/40_IEAPositionPaperMSW.pdf

——— (2005) *Strategies to manage heat losses, ANNEX VII, 2005, 8DHC-05.07*. Paris: International Energy Agency.

——— (2008a) *Improved Cogeneration and Heat Utilization in DH Networks*. Paris: International Energy Agency. http://www.iea-dhc.org/reports/pdf/DRAFT_IEA_DHC_CHP_AnnexVIII_8DHC_08_02_Improved_Cogeneration_and_Heat_Utilization_in_DH_Networks_updated_04Apr2008.pdf

——— (2008b) *World Energy Outlook*. Paris: International Energy Agency.

——— (2009) *World Energy Outlook*. Paris: International Energy Agency.

——— (2010a) *Energy Technology Perspectives 2010*. Technical report, International Energy Agency, Paris.

——— (2010b) *Technology Roadmap – Electric and plug-in hybrid electric vehicles brochure*. Paris: International Energy Agency. http://www.iea.org/papers/2009/EV_PHEV_brochure.pdf

——— (2011a) *International Energy Agency: World Energy Balances*. Paris: ESDS International, University of Manchester.

——— (2011b) *Technology Roadmap – Biofuels for transport*. Paris: International Energy Agency. http://www.iea.org/papers/2011/biofuels_roadmap.pdf

——— (2011c) *Technology roadmap – biofuels for transport foldout*. Online. Accessed: 27 July 2012. http://www.iea.org/papers/2011/Biofuels_Foldout.pdf

——— (2011d) *Technology Roadmap – Electric and plug-in hybrid electric vehicles*. Paris: International Energy Agency. http://www.iea.org/papers/2011/EV_PHEV_Roadmap.pdf

———— (2012) 'Task 37: Energy from biogas and landfill gas'. Online. Accessed: 27 July 2012. http://www. iea-biogas.net/

Incropera, F. and DeWitt, D. (2002) *Fundamentals of Heat and Mass Transfer*, 5th edn. Chichester: John Wiley & Sons, Ltd.

ISQ (2009) 'OIL PRODIESEL – integrated waste management system for the reuse of used frying oils to produce biodiesel for municipality fleet of Oeiras'. Online. Accessed: 26 July 2012. http://ec.europa.eu/environment/life/project/Projects/index.cfm?fuseaction=search.dspPage&n_proj_id=2828

Jablonski, S., Strachan, N., Brand, C., Pantaleo, A. and Bauen, A. (2008) *A systematic assessment of bioenergy representation in the UK-MARKAL model: insights on the formulation of bioenergy scenarios*. Technical report, Transport Studies Unit, University of Oxford. Working paper No. 1040. http://www. tsu.ox.ac.uk/pubs/1040-jablonski-etal.pdf

Jaccard, M.K. (2005) *Sustainable Fossil Fuels: The Unusual Suspect in the Quest for Clean and Enduring Energy*. Cambridge: Cambridge University Press.

Jenkins, N., Ekanayake, J. and Strbac, G. (2010) *Distributed Generation*. London: The Institution of Engineering and Technology.

Jennings, M., Fisk, D. and Shah, N. (2011a) 'Optimal scheduling of low carbon investment decisions for a social housing refurbishment case study'. In *The 24th International Conference on Efficiency, Cost, Optimization, Simulation and Environmental Impact of Energy Systems*, edited by M. Bojić, N. Lior, J. Stevanović, P. Gordana and S. Vladimir, pp. 2023–2036. Novi Sad, Serbia.

Jennings, M., Munuera, L., Fisk, D. and Shah, N. (2011b) 'Optimal investment planning for low carbon domestic heating options in London: 2010–2050'. In *World Engineers' Convention 2011*. Geneva, CH.

Joerges, B. (1988) 'Large technical systems: Concepts and issues'. In *The Development of Large Technical Systems*, edited by R. Mayntz and T.P. Hughes, chapter 1, pp. 9–36. Frankfurt am Main: Campus Verlag, Schriften des Max-Planck-Instituts für Gesellschaftsforschung.

Jollands, N., Dowling, P., Alber, G., Dhakal, S., Hammer, S., Schulz, N.B., Ruth, M., Wescott, W. and Kerr, T. (2010) 'Hung out to dry? Cities and their energy use and CO_2 emissions'. *Energy Policy*, Submitted.

Jones, P., Dix, M., Clarke, M. and Heggie, I. (1983) *Understanding Travel Behaviour*. Aldershot: Gower.

Jonsson, A. and Hillring, B. (2006) 'Planning for increased bioenergy use: evaluating the impact on local air quality'. *Biomass and Bioenergy*, 30(6): 543–554.

Kaikko, J. and Backman, J. (2007) 'Technical and economic performance analysis for a microturbine in combined heat and power generation'. *Energy*, 32(4): 378–387.

Kallrath, J. (2004) *Modeling Languages in Mathematical Optimization*. Norwell, MA: Kluwer Academic Publishers.

Kargon, R.H. and Molella, A.P. (2008) *Invented Edens: Techno-Cities of the Twentieth Century*. Cambridge, MA: MIT Press.

Karmarker, N. (1984) 'A new polynomial-time algorithm for linear programming'. *Combinatorica*, 4(4): 373–395.

Kazuaki, M. (2006) 'Sustainable urban planning based on integrated land-use and transportation models'. In *Presented at Asian Development Bank*. http://www.adb.org/urbandev/documents/SustainableUrbanPlanning-K.Miyamoto.ppt

Keeley, L. (1997) 'Frontier warfare in the early neolithic'. In *Troubled Times: Violence and Warfare in the Past*, edited by D. Martin and D. Frayer, pp. 303–320. Amsterdam, NL: OPA.

Keirstead, J. and Calderon, C. (2012) 'Capturing spatial effects, technology interactions, and uncertainty in urban energy and carbon models: retrofitting newcastle as a case-study'. *Energy Policy*, 46: 253–267.

Keirstead, J., Jennings, M. and Sivakumar, A. (2012a) 'A review of urban energy system models: Approaches, challenges and opportunities'. *Renewable and Sustainable Energy Reviews*, 16(6): 3847–3866.

Keirstead, J., Samsatli, N., Pantaleo, A.M. and Shah, N. (2012b) 'Evaluating biomass energy strategies for a UK eco-town with an MILP optimization model'. *Biomass and Bioenergy*, 39: 306–316.

Keirstead, J., Samsatli, N., Shah, N. and Weber, C. (2012c) 'The impact of CHP (combined heat and power) planning restrictions on the efficiency of urban energy systems'. *Energy*, 41(1): 93–103.

Keirstead, J. and Schulz, N.B. (2010) 'London and beyond: taking a closer look at urban energy policy'. *Energy Policy*, 38(9): 4870–4879.

Keirstead, J. and Shah, N. (2011) 'Calculating minimum energy urban layouts with mathematical programming and monte carlo analysis techniques'. *Computers, Environment and Urban Systems*, 35(5): 368–377.

Keirstead, J. and Sivakumar, A. (2012) 'Using activity-based modeling to simulate urban resource demands at high spatial and temporal resolutions'. *Journal of Industrial Ecology*, In press.

Kemin (2009) 'The Danish example'. Online. Accessed: 26 July 2012. http://www.kemin.dk/en-US/facts/danishexample/Sider/TheDanishExample.aspx

Kemp, R., Schot, J. and Hoogma, R. (1998) 'Regime shifts to sustainability through processes of niche formation: the approach of strategic niche management'. *Technology Analysis and Strategic Management*, 10(2): 175.

Kennedy, B. and Fletcher, R. (1991) 'Conservation voltage reduction (CVR) at Snohomish County PUD'. *Transactions on Power Systems*, 6(3): 986–998.

Kennedy, C., Cuddihy, J. and Engel-Yan, J. (2008) 'The changing metabolism of cities'. *Journal of Industrial Ecology*, 11(2): 43–59.

Kennedy, C., Steinberger, J., Gasson, B., Hansen, Y., Hillman, T., Havránek, M., Pataki, D., Phdungsilp, A., Ramaswami, A. and Mendez, G.V. (2009) 'Methodology for inventorying greenhouse gas emissions from global cities'. *Energy Policy*, 38(9): 4828–4837.

Kesicki, F. (2010) 'Marginal abatement cost curves for policy making – expert-based vs. model-derived curves'. In *33rd IAEE International Conference*, pp. 1–19.

Kirkpatrick, S., Gelatt, C.D. and Vecchi, M.P. (1983) 'Optimization by simulated annealing'. *Science*, 220(4598): 671–680.

KNBS (2005) 'Kenya integrated household survey'. Online. Accessed: http://www.knbs.or.ke/pdf/Basic%20Report%20%28Revised%20Edition%29.pdf

Knowledge of London (2011) 'The London omnibus'. Online. Accessed: 27 July 2012. http://www.knowledgeoflondon.com/buses.html

Kölbl, R. and Helbing, D. (2003) 'Energy laws in human travel behaviour'. *New Journal of Physics*, 5(03): 48–48.

Kondili, E., Pantelides, C.C. and Sargent, R.W.H. (1993) 'A general algorithm for short-term scheduling of batch operations—I. MILP formulation'. *Computers & Chemical Engineering*, 17: 211–227.

Koomey, J., Webber, C.A., Atkinson, C.S. and Nicholls, A. (2001) 'Addressing energy-related challenges for the US buildings sector: results from the clean energy futures study'. *Energy Policy*, 29(14): 1209–1221.

Korytarova, K. (2006) *Evaluation of KfW soft loans for building modernisation: Withing the framework of the Aid-EE project*. Technical report, AID-EE. http://www.aid-ee.org/documents/000003KfWbuildingprogramme-Germany.pdf

Kostantinidis, M., Samsatli, N.J., Keirstead, J.E. and Shah, N. (2010) 'Modelling of integrated municipal solid waste to energy technologies in the urban environment'. In *Proceedings of the 3rd International Conference on Engineering for Waste and Biomass Valorisation*, pp. 1–6. Beijing.

Kotkin, J. (2005) *The City: a Global History*. New York: Weidenfeld & Nicolson.

Kühnert, C., Helbing, D. and West, G.B. (2006) 'Scaling laws in urban supply networks'. *Physica A: Statistical Mechanics and its Applications*, 363(1): 96–103.

Kumar, A., Cameron, J.B. and Flynn, P.C. (2004) 'Pipeline transport of biomass'. *Applied Biochemistry and Biotechnology*, 113–116(1): 27–39.

Kumar, A., Sokhansanj, S. and Flynn, P.C. (2006) 'Development of a multicriteria assessment model for ranking biomass feedstock collection and transportation systems'. *Applied Biochemistry and Biotechnology*, 129–132(1): 71–87.

Labay, D.G. and Kinnear, T.C. (1981) 'Exploring in the the consumer of decision energy process adoption solar systems'. *Journal of Consumer Research*, 8(3): 271–278.

Land, A.H. and Doig, A. (1960) 'An automatic method of solving discrete programming problems'. *Econometrica*, 28: 497–520.

Lang, F. (1927) *Metropolis*. Universum Film AG.

Lauer, M. (2009) 'Current terrestrial methods and activities for biomass potential assessment in Europe'. In *17th European Biomass Conference and Exhibition*. Hamburg, DE.

Lautso, K., Spiekermann, K., Wegener, M., Sheppard, I., Steadman, P., Martino, A., Domingo, R. and Gayda, S. (2004) *Planning and research of policies for land use and transport for increasing urban sustainability*. Technical report, PROPOLIS Final Report, European Commission, Energy, Environment and Sustainable Development Thematic Programme of the Fifth RTD Framework Programme.

Lazzarin, R. and Noro, M. (2006a) 'District heating and gas engine heat pump: economic analysis based on a case study'. *Applied Thermal Engineering*, 26(2–3): 193–199.

—— (2006b) 'Local or district heating by natural gas: Which is better from energetic, environmental and economic point of views?' *Applied Thermal Engineering*, 26(2–3): 244–250.

LBNL (2007) *Energy Use in China: Sectoral Trends and Future Outlook*. Berkeley, CA: Lawrence Berkeley National Laboratory.

—— (2009) *Building Commissioning: A Golden Opportunity for Reducing Energy Costs and Greenhouse Gas Emissions*. Berkeley, CA: California Energy Commission, Public Interest Energy Research.

LDA (2010) 'Pimlico and Whitehall decentralised energy'. Online. Accessed: 1 December 2011. http://www.lda.gov.uk/projects/pimlico-whitehall-decentralised-energy/index.aspx

Leach, G. (1992) 'The energy transition'. *Energy Policy*, 20(2): 116–123.

Leca, A. (2008) 'District heating efficiency and affordability: The case of Romania'. Online. Accessed: 15 February 2012. http://www.ghgprotocol.org/

Lee, J. (2003) 'Feeding the colleges: Cambridge's food and fuel supplies, 1450–1560'. *Economic History Review*, LV1(2): 243–264.

Lei, K. and Wang, Z. (2003) 'The analysis of ecological footprints of Macao in 2001'. *Journal of Natural Resources*, 18(2): 197–203.

Lenntorp, B. (1976) *A time-geographic study of movement possibilities of individuals*. Lund Studies In Geography, Human Geography, 44.

Lenzen, M. and Murray, S. (2001) 'A modified ecological footprint method and its application to Australia'. *Ecological Economics*, 37(2): 229–255.

Levine, M., Urge-Vorsatz, D., Blok, K., Geng, L., Harvey, D., Lang, S., Levermore, G., Mongameli Mehlwana, A., Mirasgedis, S., Novikova, A., Rilling, J. and Yoshino, H. (2007) 'Residential and commercial buildings'. In *Climate Change 2007: Mitigation. Contribution of Working Group III to the Fourth Assessment Report of the Intergovernmental Panel on Climate Change*, edited by B. Metz, O. Davidson, P. Bosch, R. Dave and L. Meyer. Cambridge: Cambridge University Press.

Liang, H., Long, W.D., Keirstead, J., Samsatli, N. and Shah, N. (2012) 'Urban energy system planning and Chinese low-carbon eco-city case study'. *Advanced Materials Research*, 433–440: 1338–1345.

Lindenberger, D., Bruckner, T., Groscurth, H.M. and Kümmel, R. (2000) 'Optimization of solar district heating systems: seasonal storage, heat pumps, and cogeneration'. *Energy*, 25(7): 591–608.

Lin, C., Wu, S., Lee, K., Lin, P. and Chang, J. (2007) 'Integration of fermentative hydrogen process and fuel cell for on-line electricity generation'. *International Journal of Hydrogen Energy*, 32(7): 802–808.

Li, X. and Yeh, A.G.O. (2000) 'Modelling sustainable urban development by the integration of constrained cellular automata and GIS'. *International Journal of Geographical Information Science*, 14(2): 131–152.

Lloyd's (2006) 'Climate change: Adapt or bust'. Online. Accessed: 26 July 2012. http://www.lloyds.com/~/media/3be75eab0df24a5184d0814c32161c2d.ashx

Lowry, I.S. (1964) *A Model of Metropolis*. Technical report, RAND Corporation.

Lozano, M.A., Ramos, J.C., Carvalho, M. and Serra, L.M. (2009) 'Structure optimization of energy supply systems in tertiary sector buildings'. *Energy and Buildings*, 41(10): 1063–1075.

LTOA (1991) *London Bioenergy Report*. London: London Tree Officers' Association.

Luce, R.D. and Raiffa, H. (1957) *Games and Decisions: Introduction and Critical Survey*. New York: John Wiley & Sons, Ltd.

Lucon, O. and Santos, E.d. (2005) 'The HORUS model – inventory of atmospheric pollutant emissions from industrial combustion in São Paulo, Brazil'. *Environmental Impact Assessment Review*, 25(2): 197–214.

Lutsey, N. and Sperling, D. (2008) 'America's bottom-up climate change mitigation policy'. *Energy Policy*, 36(2): 673–685.

MacKay, D.J. (2009) *Sustainable Energy: Without the Hot Air*. Cambridge: UIT.

Mackett, R. (1985) 'Integrated land use-transport models'. *Transport Reviews*, 5(4): 325–343.

Mackett, R.L. (1983) *The Leeds Integrated Land-use Transport Model (LILT)*. Technical report, Transport and Road Research Laboratory.

Madsen, I.T., Elleriis, J., Gullev, L., Andersen, F., Birnbaum, A. and Foged, M. (2009) 'Varmeplan Hovedstaden'. Online. Accessed: 27 July 2012. http://www.varmeplanhovedstaden.dk/english

Maidment, G. and Prosser, G. (2000) 'The use of CHP and absorption cooling in cold storage'. *Applied Thermal Engineering*, 20(12): 1059–1073.

Mairie de Paris (2012) 'Energie & plan climat'. Accessed: 14 June 2012. http://www.paris.fr/pratique/environnement/energie-plan-climat/p8411

Majer, E.L., Baria, R., Stark, M., Oates, S., Bommer, J., Smith, B. and Asanuma, H. (2007) 'Induced seismicity associated with enhanced geothermal systems'. *Geothermics*, 36: 185–222.

Malanima, P. (2006) 'Energy crisis and growth 1650–1850: the European deviation in a comparative perspective'. *Journal of Global History*, 1(1): 101–121.

Mancarella, P. and Chicco, G. (2008) 'Assessment of the greenhouse gas emissions from cogeneration and trigeneration systems. Part II: Analysis techniques and application cases'. *Energy*, 33(3): 418–430.

——— (2009a) *Distributed Multi-generation Systems: Energy models and analyses*. Hauppauge, NY: Nova Science Publishers.

——— (2009b) 'Global and local emission impact assessment of distributed cogeneration systems with partial-load models'. *Applied Energy*, 86(10): 2096–2106.

——— (2010) 'Distributed cogeneration: modelling of environmental benefits and impact'. Online. Accessed: 26 July 2012. http://sciyo.com/articles/show/title/distributed-cogeneration-modelling-of-environmental-benefits-and-impact

Mancarella, P., Gan, C. and Strbac, G. (2011) 'Fractal models for electro-thermal network studies'. In *Proceedings of the 17th Power Systems Computation Conference (PSCC)*. Stockholm, SE.

Mancarella, P. (2006) *From cogeneration to trigeneration: energy planning and evaluation in a competitive market framework*. Ph.D. thesis, Politecnico di Torino.

——— (2009) 'Cogeneration systems with electric heat pumps: Energy-shifting properties and equivalent plant modelling'. *Energy Conversion and Management*, 50(8): 1991–1999.

Manfren, M., Caputo, P. and Costa, G. (2011) 'Paradigm shift in urban energy systems through distributed generation: Methods and models'. *Applied Energy*, 88(4): 1032–1048.

Martinez, F.J. (1996) 'Mussa: a land use model for santiago city'. *Transportation Research Record*, 1552: 126–134.

Mayntz, R. and Hughes, T.P. (1988) 'Foreword'. In *The Development of Large Technical Systems*, edited by R. Mayntz and T.P. Hughes, pp. 5–8. Frankfurt am Main: Campus Verlag, Schriften des Max-Planck-Instituts für Gesellschaftsforschung.

Mayor of London (2004) *Green light to clean power: The Mayor's energy strategy*. London: Greater London Authority.

——— (2011) 'Delivering London's Energy Future: the Mayor's climate change mitigation and energy strategy'. Online. Accessed: 14 June 2012. http://www.london.gov.uk/who-runs-london/mayor/publication/climate-change-mitigation-energy-strategy

——— (2012) 'Climate change adaptation strategy'. Online. Accessed: 14 June 2012. http://www.london.gov.uk/climatechange/

McDonald, G. and Patterson, M. (2004) 'Ecological footprints and interdependencies of New Zealand regions'. *Ecological Economics*, 50(1–2): 49–67.

McFadden, D. (1978) 'Modelling the choice of residential location'. In *Spatial Interaction Theory and Planning Models*, edited by A. Karlquist, pp. 75–96. Amsterdam: Elsevier Science Ltd.

McGranahan, G., Marcotullio, P., Bai, X., Balk, D., Braga, T., Douglas, I., Elmqvist, T., Rees, W., Satterthwaite, D., Songsore, J., Zlotnik, H., Hassan, R., Scholes, R. and Ash, N. (2005) 'Urban systems'. In *Ecosystems and Human Well-being: Current State and Trends*, volume 1, chapter 27, pp. 795–825. Washington, DC: Island Press.

Meadows, D.H., Meadows, D.L., Randers, J. and Behrens, W.W. (1972) *The Limits to Growth*. New York: Universe Books.

Meadows, D.H., Meadows, D.L. and Randers, J. (1992) *Beyond the Limits: Global Collapse or a Sustainable Future*. London: Earthscan.

Menzies, W.C. (1936) *Things to Come*. United Artists.

Miller, C.E. (1963) 'The simplex method for local separable programming'. In *Recent Advances in Mathematical Programming*, edited by R.L. Graves and P. Wolfe, pp. 89–100. London: McGraw-Hill.

Miller, E. and Roorda, M. (2003) 'A prototype model of household activity-travel scheduling'. *Transportation Research Record*, 1831(1): 114–121.

Miller, E., Hunt, J.D., Abraham, J. and Salvini, P. (2004) 'Microsimulating urban systems'. *Computers, Environment and Urban Systems*, 28(1–2): 9–44.

Milne, G. and Boardman, B. (2000) 'Making cold homes warmer: the effect of energy efficiency improvements in low-income homes'. *Energy Policy*, 28(6–7): 411–424.

Milukas, M.V. (1993) 'Energy for secondary cities: the case of Nakuru, Kenya'. *Energy Policy*, 21(5): 543–558.

Minoux, M. (1986) *Mathematical programming: theory and algorithms*. New York: John Wiley & Sons, Ltd.

Mirzaei, P.A. and Haghighat, F. (2010) 'Approaches to study urban heat island: abilities and limitations'. *Building and Environment*, 45(10): 2192–2201.

Mitchell, M. (1998) *An Introduction to Genetic Algorithms*. Cambridge, MA: MIT Press.

Mithin, S. (2003) *After the Ice: A Global Human History, 20000–5000 BC*. London: Weidenfeld & Nicolson.

Modes, J. (2010) 'Energy efficiency policies and initiatives for existing buildings in Germany'. In *Conference on the Sustainable Buildings Network*. Paris, France. http://www.iea.org/work/2010/sbn/modes.pdf

Mohan, N. (1980) 'Improvement in energy efficiency of induction motors by means of voltage control'. *IEEE Transactions on Power Apparatus and Systems*, PAS-99(4): 1466–1471.

Monson, K.D., Esteves, S.R., Guwy, A.J. and Dinsdale, R.M. (2007) *Anaerobic Digestion of Biodegradable Municipal Wastes: A Review*. Technical report, University of Glamorgan. http://serc.research.glam.ac.uk/projects/adofbmw/

Moore, M. and Foster, P. (2011) 'China to create largest mega city in the world with 42 million people'. Online. Accessed: 27 July 2012. http://www.telegraph.co.uk/news/worldnews/asia/china/8278315/China-to-create-largest-mega-city-in-the-world-with-42-million-people.html

Moore, P.G. (1968) *Basic Operational Research*. London: Pitman Publishing Ltd.

Mori, Y., Kikegawa, Y. and Uchida, H. (2007) 'A model for detailed evaluation of fossil-energy saving by utilizing unused but possible energy-sources on a city scale'. *Applied Energy*, 84(9): 921–935.

Morris, A. (1994) *A History of Urban Form: Before the Industrial Revolutions*. Harlow: Longman.

Mor, S., Ravindra, K., Visscher, A.D., Dahiya, R.P. and Chandra, A. (2006) 'Municipal solid waste characterization and its assessment for potential methane generation: A case study'. *Science of the Total Environment*, 371(1–3): 1–10.

Moss, T., Guy, S. and Marvin, S. (2000) *Urban Infrastructure in Transition: Networks, Buildings and Plans*. London: Earthscan.

MSE (2010) 'Bilancio energetico nazionale 2010'. Online. Accessed: 27 July 2012. http://dgerm.sviluppoeconomico.gov.it/dgerm/ben/ben_2010.pdf

Muller, K. and Axhausen, K. (2011) 'Population synthesis for microsimulation: State of the art'. In *Transportation Research Board 90th Annual Meeting*. Washington, DC.

Mulvey, J.M., Vanderbei, R.J. and Zenios, S.A. (1995) 'Robust Optimization of Large-Scale Systems'. *INFORMS*, 43(2): 264–281.

NASA (2010) 'Technology readiness levels demystified'. Online. Accessed: 27 July 2012. http://www.nasa.gov/topics/aeronautics/features/trl_demystified.html

National Grid (2009) *The Potential for Renewable Gas in the UK*. Technical report, National Grid. http://www.nationalgrid.com/uk/Media+Centre/Documents/biogas.htm

Nemhauser, G.L. and Wosley, L.A. (1989) *Integer and Combinatorial Optimization*. New York: Wiley.

Newcombe, K., Kalina, J. and Aston, A. (1978) 'The metabolism of a city: the case of Hong Kong'. *Ambio*, 7: 3–15.

Newman, P. (1999) 'Sustainability and cities: extending the metabolism model'. *Landscape and Urban Planning*, 44: 219–226.

Niches (2010) *Guidelines for implementers of electric cars in car share clubs.* Technical report, NICHES+. http://www.niches-transport.org/fileadmin/NICHESplus/G4Is/21582_policynotesWG4_3.indd_low.pdf

Nicolis, G. and Prigogine, I. (1977) *Self-organization in nonequilibrium systems: from dissipative structures to order through fluctuations.* Chichester: Wiley.

Nordhaus, W.D. (1992) 'Lethal model 2: The limits to growth revisited'. *Brookings Papers on Economic Activity*, 2(1971): 1–59. http://www.brookings.edu/~/media/Files/Programs/ES/BPEA/1992_2_bpea_papers/1992b_bpea_nordhaus_stavins_weitzman.pdf

Nordqvist, J. (2006) *Evaluation of Japan's Top Runner Programme: Within the Framework of the AID-EE project.* Technical report, AID-EE. http://www.aid-ee.org/documents/018TopRunner-Japan.PDF

Nye, D.E. (1990) *Electrifying America: social meanings of a new technology, 1880–1940.* Cambridge, MA: MIT Press.

Odum, H.T. (1988) 'Self-organization, transformity, and information'. *Science*, 242(4882): 1132–1139.

OECD (1998) *Towards sustainable development: Environmental Indicators.* Paris: Organisation for Economic Cooperation and Development.

Ofgem (2008) 'Carbon Emissions Reduction Target (CERT) 2008–2011 Technical Guidance Manual'. Online. Accessed: 27 July 2012. http://www.ofgem.gov.uk/Sustainability/Environment/EnergyEff/InfProjMngrs/Documents1/TM%20Guidance.pdf

——— (2010) *Renewables Obligation: Annual Report 2008–2009.* Technical report, Office for Gas and Electricity Markets.

Olgyay, V. and Seruto, C. (2010) 'Whole building retrofits: A gateway to climate stablization'. *ASHRAE Transactions*, 116(2): 244–251.

ONS (2011) 'UK business: Activity, size and location, 2011'. Online. Accessed: 27 July 2012. http://www.ons.gov.uk/ons/publications/re-reference-tables.html?edition=tcm:77-227577

Ooka, R. and Komamura, K. (2009) 'Optimal design method for building energy systems using genetic algorithms'. *Building and Environment*, 44(7): 1538–1544.

Ormrod, R.K. (1990) 'Local context and innovation diffusion in a well-connected world'. *Economic Geography*, 66(2): 109–122.

Ornetzeder, M. (2001) 'Old technology and social innovations. inside the Austrian success story on solar water heaters'. *Technology Analysis and Strategic Management*, 13(1): 105–115.

Orwell, G. (1949) *Nineteen Eighty-Four.* London: Secker and Warburg.

Osbourn, D. and Greenho, R. (2007) *Mitchell's Introduction to Building*, 4th edn. Harlow: Pearson Education Limited.

Overgaard, J., Woods, P. and Riley, O. (2005) *Large or small-scale CHP/DH – a comparison.* Technical report, Danish Board of District Heating. http://www.dbdh.dk/artikel.asp?id=485&mid=22

Pantelides, C.C. (1994) 'Unified frameworks for optimal process planning and scheduling'. In *Proceedings of the 2nd Conference on Foundations of Computer-Aided Operations*, edited by D.W.T. Rippin and J. Hale, pp. 253–274. Crested Butte, CO.

Pendyala, R., Kitamura, R. and Reddy, D. (1995) 'A rule-based activity-travel scheduling algorithm integrating neural networks of behavioral adaptation'. In *EIRASS Conference on Activity-Based Approaches.* Eindhoven, NL.

Persson, U. and Werner, S. (2011) 'Heat distribution and the future competitiveness of district heating'. *Applied Energy*, 88(3): 568–576.

Philippen, C. (2011) *Greenhouse Gas Mitigation in Cities: The Relevance of Energy Supply Networks and Policy Implications.* IDEA League Studienarbeit, RWTH Aachen University and Imperial College London.

Pike Research (2010) *Energy Efficiency Retrofits for Commercial and Public Buildings.* Technical report, Pike Research. http://www.pikeresearch.com/research/energy-efficiency-retrofits-for-commercial-and-public-buildings

Pimlott, B. and Rao, N. (2002) *Governing London.* Oxford: Oxford University Press.

Pootakham, T. and Kumar, A. (2010) 'A comparison of pipeline versus truck transport of bio-oil'. *Bioresource Technology*, 101(1): 414–421.

Postgate, J. (1994) *Early Mesopotamia: society and economy at the dawn of history*. Abingdon: Routledge.

PRé Consultants (2001) *The Eco-Indicator 99, A damage oriented method for Life Cycle Impact Assessment, Methodology Report*. Technical report, PRé Consultants.

Pudjianto, D., Ramsay, C. and Strbac, G. (2007) 'Virtual power plant and system integration of distributed energy resources'. *IET Renewable Power Generation*, 1(1): 10.

Pujol, G. (2008) *sensitivity: Sensitivity Analysis*. R package version 1.4-0. http://cran.r-project.org/web/packages/sensitivity

Putman, S. (1995) 'EMPAL and DRAM location and land-use models: An overview'. In *Land use Modelling Conference*. Dallas, TX.

Rackham, O. (2010) *Woodlands*. London: Collins.

RAE (2010) *Electric Vehicles: Charged with Potential*. Technical report, Royal Academy of Engineering. http://www.raeng.org.uk/news/publications/list/reports/Electric_Vehicles.pdf

Ramaswami, A., Chavez, A., Ewing-Thiel, J. and Reeve, K.E. (2011) 'Two approaches to greenhouse gas emissions foot-printing at the city scale'. *Environmental Science and Technology*, 45(10): 4205–6.

Randolph, J. and Masters, G.M. (2008) *Energy for Sustainability: Technology, Planning, Policy*. Washington, DC: Island Press.

Ravetz, J. (1999) 'What is post-normal science?' *Futures*, 31(7): 647–653.

Reader, J. (2005) *Cities*. London: Vintage.

Reche-López, P., Ruiz-Reyes, N., García Galán, S. and Jurado, F. (2009) 'Comparison of metaheuristic techniques to determine optimal placement of biomass power plants'. *Energy Conversion and Management*, 50(8): 2020–2028.

Rees, W. and Wackernagel, M. (1996) 'Urban ecological footprints: Why cities cannot be sustainable – and why they are a key to sustainability'. *Environmental Impact Assessment Review*, 16(4–6): 223–248.

Reklaitis, G.V., Ravindran, A. and Ragsdell, K.M. (2006) *Engineering Optimization – Methods and Applications*. New York: John Wiley & Sons, Ltd.

Renfrew, C. (2008) *Prehistory: the Making of the Human Mind*. London: Phoenix.

Rentizelas, A., Karellas, S., Kakaras, E. and Tatsiopoulos, I. (2009) 'Comparative techno-economic analysis of ORC and gasification for bioenergy applications'. *Energy Conversion and Management*, 50(3): 674–681.

Rentizelas, A., Tolis, A. and Tatsiopoulos, I. (2008) 'Biomass district energy trigeneration systems: Emissions reduction and financial impact'. *Water Air Soil Pollution Focus*, 9(1–2): 139–150.

Richardson, I., Thomson, M., Infield, D. and Clifford, C. (2010) 'Domestic electricity use: a high-resolution energy demand model'. *Energy and Buildings*, 42(10): 1878–1887.

Richards, D., Brade, J., Allenby, R. and Frosch, R. (1994) 'The greening of industrial ecosystems: Overview and perspective'. In *The Greening of Industrial Ecosystems*, pp. 1–22. Washington, DC: National Academy Press.

Richter, S. and Hamacher, T. (2003) 'Langfristige Auswirkungen sich verändernder Stromkosten auf eine dezentrale Energieversorgung in urbanen Energiesystemen'. In *Power-Gen Europe 2003*. Max Planck Institut für Plasmaphysik. https://www.dpg-physik.de/dpg/gliederung/ak/ake/tagungen/vortragssammlung/04/13-Richter.pdf

Robinson, D. (2009) 'Case studies of interdisciplinary research projects on urban futures : process and lessons learned'. In *Understanding the Dimensions of Urban Futures Research: Tackling Complex Reality*, pp. 12–14. Zürich, CH. http://www.ags.ethz.ch/research/AGSatETH_urbanfutures_seminar.pdf

Robinson, D., Campbell, N., Gaiser, W., Kabel, K., Le-Mouel, A., Morel, N., Page, J., Stankovic, S. and Stone, A. (2007) 'SUNtool – a new modelling paradigm for simulating and optimising urban sustainability'. *Solar Energy*, 81(9): 1196–1211.

Rockafellar, R.T. (1970) *Convex Analysis*. Princeton, NJ: Princeton University Press.

——— (2000) 'Optimization of conditional value-at-risk'. *The Journal of Risk*, 2: 21–41.

Rodrigue, J.P., Comtois, C. and Slack, B. (2009) *The Geography of Transport Systems*, 2nd edn. Abingdon: Routledge.

Rogers, E.M. (2003) *Diffusion of Innovations*, 5th edn. London: Free Press.

Rohracher, H. and Ornetzeder, M. (2002) 'Green buildings in context: improving social learning processes between users and producers'. *Built Environment*, 28(1): 73–84.

Rojas, R. and Hashagen, U. (2002) *The First Computers: History and Architectures*. Cambridge, MA: MIT Press.

Rolfsman, B. (2004) 'Optimal supply and demand investments in municipal energy systems'. *Energy Conversion and Management*, 45(4): 595–611.

Roorda, M., Miller, E. and Habib, K. (2008) 'Validation of TASHA: A 24-h activity scheduling microsimulation model'. *Transportation Research Part A: Policy and Practice*, 42(2): 360–375.

Rosenthal, R.E. (2012) *GAMS: A User's Guide*. Washington, DC: GAMS Development Corporation. http://www.gams.com/dd/docs/bigdocs/GAMSUsersGuide.pdf

Rosenzweig, C., Solecki, W.D., Hammer, S.A. and Mehrota, S. (eds) (2011) *Climate Change and Cities: First Assessment Report of the Urban Climate Change Research Network*. Cambridge: Cambridge University Press.

Rosen, R. (1991) *Life Itself*. Columbia University Press.

Ryckebosch, E., Drouillon, M. and Vervaeren, H. (2011) 'Techniques for transformation of biogas to biomethane'. *Biomass and Bioenergy*, 35(5): 1633–1645.

R Development Core Team (2011) *R: A Language and Environment for Statistical Computing*. R Foundation for Statistical Computing, Vienna, Austria. http://www.R-project.org/

Sahely, H.R., Dudding, S. and Kennedy, C. (2003) 'Estimating the urban metabolism of Canadian cities: GTA case study'. *Canadian Journal for Civil Engineering*, 30: 468–483.

Sahinidis, N.V. (2004) 'Optimization under uncertainty: state of the art and opportunities'. *Computers and Chemical Engineering*, 28: 971–983.

Sakai, S., Sawell, S.E., Changler, A.J., Eighmy, T.T., Kosson, D.S., Vehlow, J., Sloot, H.A.v.d., Hartldn, J. and Hjelmar, O. (1996) 'World trends in municipal solid waste management'. *Waste Management*, 16(5/6): 341–350.

Saltelli, A., Ratto, M., Andres, T., Campolongo, F., Cariboni, J., Gatelli, D., Saisana, M. and Tarantola, S. (2008) *Global Sensitivity Analysis: The Primer*. London: Wiley Blackwell.

Sands, R. (2005) *Forestry in a Global Context*. Wallingford: CABI Publishing.

Santos, G., Behrendt, H. and Teytelboym, A. (2010) 'Part II: Policy instruments for sustainable road transport'. *Research in Transportation Economics*, 28(1): 46–91.

Saunders, M. (2012) 'Large-Scale Numerical Optimization: Class notes'. Online. Accessed: 27 July 2012. http://www.stanford.edu/class/msande318/

Saxena, R.C., Adhikari, D.K. and Goyal, H.B. (2009) 'Biomass-based energy fuel through biochemical routes: a review'. *Renewable and Sustainable Energy Reviews*, 13(1): 167–178.

SBGF, SGC and Swedish Gas Association (2008) *Biogas from manure and waste products, Swedish case studies*. Technical report, International Energy Agency. http://www.iea-biogas.net/_download/publi-task37/publi-member/Swedish_report_08.pdf

Schenk, N.J. (2006) *Modelling Energy Systems: a Methodological Exploration of Integrated Resource Management*. Ph.D. thesis, University of Groningen.

Schulz, N. (2010) 'Lessons from the London Climate Change Strategy: Focusing on combined heat and power and distributed generation'. *Journal of Urban Technology*, 17(3): 3–23.

Schwaegerl, C., Tao, L., Mancarella, P. and Strbac, G. (2011) 'A multi-objective optimization approach for assessment of technical, commercial and environmental performance of microgrids'. *European Transactions on Electrical Power*, 21(2): 1269–1288.

Sciubba, E. and Ulgiati, S. (2005) 'Emergy and exergy analyses: complementary methods or irreducible ideological options?' *Energy*, 30(10): 1953–1988.

Scott, R. (1982) *Blade Runner*. Warner Brothers Pictures.

SEDAC-CIESIN (2012) 'Global Rural-Urban Mapping Project (GRUMP)'. Online. Accessed: 26 July 2012. http://sedac.ciesin.org/gpw/index.jsp

Semboloni, F. (2008) 'Hierarchy, cities size distribution and Zipf's law'. *The European Physical Journal B*, 63(3): 295–301.

Severn Wye Energy Agency (2011) *Target 2050 future proofing homes in Stroud District and beyond*. Stroud: Stroud District Council.

Shiman, R. (1993) 'Explaining the collapse of the British electrical supply industry in the 1880s: Gas versus electric lighting prices'. *Business*, 22(1): 318–327.

Shove, E. (1998) 'Gaps, barriers and conceptual chasms: theories of technology transfer and energy in buildings'. *Energy Policy*, 26(15): 1105–1112.

Sieferle, R. (2001) *The Subterranean Forest: Energy Systems and the Industrial Revolution*. Cambridge: White Horse Press.

Siemens (2009) *Sustainable urban infrastructure: London Edition – a view to 2025*. Technical report, Siemens. http://www.siemens.com/entry/cc/features/urbanization_development/all/en/pdf/study_london_en.pdf

Simmonds, D.C. (1999) 'The design of the DELTA land-use modelling package'. *Environment and Planning B: Planning and Design*, 26(5): 665–684.

Simon, H. (1959) 'Theories of decision-making in economics and behavioral science'. *The American Economic Review*, 49(3): 253–283.

Simon, H.A. (1962) 'The architecture of complexity'. *Proceedings of the American Philosophical Society*, 106(6): 467–482.

Singh, T. (2009) 'What is the future of public transport?' Accessed: 26 July 2012. http://www.euinfrastructure.com/news/future-of-public-transport/

Sivakumar, A., Daina, N., Polak, J., Skippon, S., Stannard, J. and Vine, S.L. (2012) 'Impacts of experience on stated choice behaviour: An EV adoption study'. In *13th International Conference on Travel Behaviour Research*. Toronto, CA.

Sivakumar, A., Keirstead, J. and Polak, J.W. (2010a) 'Integrated modelling of urban energy systems: Results from a London case study'. In *European Transport Conference*. Glasgow.

Sivakumar, A., Keirstead, J. and Polak, J. (2010b) 'Integrated modelling of the demand and supply vectors in urban energy systems: Conceptual and modelling frameworks for the development of a new toolkit'. In *European Transport Conference*. Glasgow.

Sivakumar, A. and Polak, J.W. (2009) 'Modelling the endogeneity in activity participation and technology holdings: An exploration of data pooling techniques'. In *International Choice Modelling Conference*. Harrogate.

Skinner, I., Ferguson, M., Kroger, K., Kelly, C. and Bristow, A. (2003) *Critical Issues in Decarbonising Transport: Final report of the Theme 2 project T2.22*. Technical report, Tyndall Centre for Climate Change Research.

Slade, R., Bauen, A. and Gross, R. (2010) *The UK bio-energy resource-base to 2050*. Technical Report UKERC/WP/TPA/2010/002, UK Energy Research Centre.

Slater, S. and Dolman, M. (2009) *Strategies for the Uptake of Electric Vehicles and Associated Infrastructure Implications*. Technical report, Committee on Climate Change. http://hmccc.s3.amazonaws.com/Element_Energy_-_EV_infrastructure_report_for_CCC_2009_final.pdf

Smeets, E.M.W., Faaij, A.P.C., Lewandowski, I.M. and Turkenburg, W.C. (2007) 'A bottom-up assessment and review of global bio-energy potentials to 2050'. *Progress in Energy and Combustion Science*, 33(1): 56–106.

Smil, V. (1994) *Energy in World History*. Boulder, CO: Westview Press Inc.

———— (2010) *Energy Transitions: History, Requirements, Prospects*. Santa Barbara, CA: Praeger.

Smith, A. (2002) *The Theory of Moral Sentiments (Cambridge Texts in the History of Philosophy)*. Cambridge: Cambridge University Press.

Smits, R. (2002) 'Innovation studies in the 21st century: Questions from a user's perspective'. *Technological Forecasting and Social Change*, 69(9): 861–883.

Sorrell, S. (2007) *The Rebound Effect: an assessment of the evidence for economy-wide energy savings from improved energy efficiency*. Technical report, UK Energy Research Centre. http://www.ukerc.ac.uk/Downloads/PDF/07/0710ReboundEffect/0710ReboundEffectReport.pdf

Southampton CC (2010) 'Southampton CHP and geothermal scheme'. Online. Accessed: 27 July 2012. http://www.southampton.gov.uk/s-environment/energy/Geothermal/

Southworth, F. (1995) *A Technical Review of Urban Land Use-Transportation Models as Tools for Evaluating Vehicle Travel Reduction Strategies*. Technical Report ORNL-6881, US Department of Energy.

Stanhill, G. (1977) 'A urban agro-ecosystem: the example of nineteenth-century Paris'. *Agro-Ecosystems*, 3: 269–284.

Stefanis, S.K., Livingston, A. and Pistikopoulos, E. (1995) 'Minimizing the environmental impact of process plants: a process systems methodology'. *Computers and Chemical Engineering*, 19(S): S39–S44.

Strauch, D., Moeckel, R., Wegener, M., Gräfe, J., Muhlhans, H., Rindsfuser, G. and Beckmann, K.J. (2005) 'Linking transport and land use planning: the microscopic dynamic simulation model ILUMASS'. In *Geodynamics*, edited by P.M. Atkinson, G.M. Foody, S.E. Darby and F. Wu, pp. 295–311. Boca Raton, FL: CRC Press.

Strbac, G. (2008) 'Demand side management: Benefits and challenges'. *Energy Policy*, 36(12): 4419–4426.

Stumpf, M.P.H. and Porter, M.A. (2012) 'Critical Truths About Power Laws'. *Science*, 335(6069): 665–666.

Sundberg, G. and Karlsson, B. (2000) 'Interaction effects in optimising a municipal energy system'. *Energy*, 25(9): 877–891.

Taha, H. (1982) *Operations Research: An introduction*, 3rd edn. New York: MacMillan Publishing Co.

Taylor, S. (2002) *The Moving Metropolis: A history of London's transport since 1800*. London: Laurence King Publishing.

The City of New York (2011) 'PlaNYC'. Accessed: 14 June 2012. http://www.nyc.gov/html/planyc2030/html/home/home.shtml

The Council of the European Union (1999) 'Council Directive 1999/31/EC of 26 April 1999 on the landfill of waste'. *Official Journal of the European Communities*, 182: 1–19. http://eur-lex.europa.eu/LexUriServ/LexUriServ.do?uri=CELEX:31999L0031:EN:NOT

The London Collaborative (2008) *State of Play June 2008*. The London Collaborative. http://www.youngfoundation.org/files/images/stateofplay_web.pdf

Thompson, M., Ellis, R. and Wildavsky, A. (1990) *Cultural Theory*. Boulder, CO: Westview Press Inc.

Tilley, D.R. (2004) 'Howard T. Odum's contribution to the laws of energy'. *Ecological Modelling*, 178(1–2): 121–125.

Tobler, W.R. (1970) 'A computer movie simulating urban growth in the Detroit region'. *Economic Geography*, 46 (Supplement: Proceedings. International Geographical Union. Commission on Quantitative Methods): 234–240.

Toke, D. and Fragaki, A. (2008) 'Do liberalised electricity markets help or hinder CHP and district heating? the case of the UK'. *Energy Policy*, 36(4): 1448–1456.

Torrens, P. (2000) *How cellular models of urban systems work (1. Theory)*. Technical Report 28, UCL Casa Working Paper Series. http://discovery.ucl.ac.uk/1371/

Train, K. (1980) 'The potential market for non-gasoline-powered automobiles'. *Transportation Research Part A: General*, 14(5–6): 405–414.

—— (2003) *Discrete Choice Methods with Simulation*. Cambridge: Cambridge University Press.

Turner, G.M. (2008) 'A comparison of The Limits to Growth with 30 years of reality'. *Global Environmental Change*, 18(3): 397–411.

Turvey, R. (2005) 'Horse traction in Victorian London'. *The Journal of Transport History*, 26(2): 38–59.

Tveit, T., Savola, T., Gebremedhin, A. and Fogelholm, C. (2009) 'Multi-period MINLP model for optimising operation and structural changes to CHP plants in district heating networks with long-term thermal storage'. *Energy Conversion and Management*, 50(3): 639–647.

Tyson, K.S., Bozell, J., Wallace, R., Petersen, E. and Moens, L. (2004) *Biomass Oil Analysis : Research Needs and Recommendations*. Technical Report NREL/TP-510-34796, National Renewable Energy Laboratory. http://www.nrel.gov/docs/fy04osti/34796.pdf

University of Wisconsin Madison (2012) 'NEOS Wiki'. Online. Accessed: 27 July 2012. http://www.neos-guide.org/NEOS/index.php/NEOS:About

Unruh, G.C. and Einstein, A. (2000) 'Understanding carbon lock-in'. *Energy Policy*, 28(12): 817–830.

UNSD (2011) 'Population density and urbanization'. Online. Accessed: 27 July 2012. http://unstats.un.org/unsd/demographic/sconcerns/densurb/densurbNotes.htm

UN Habitat (2007) 'Climate Change Statement by Brian Williams, UN Commission on Sustainable Development, 15th Session'. Online. Accessed: 27 July 2012. http://www.unhabitat.org/content.asp?cid=4756&catid=356&typeid=8&subMenuId=0

UN (2011) 'World urbanization prospects: The 2011 revision'. Online. Accessed: 27 July 2012. http://esa.un.org/unpd/wup/index.htm

Uslu, A., Faaij, A. and Bergman, P. (2008) 'Pre-treatment technologies, and their effect on international bioenergy supply chain logistics. techno-economic evaluation of torrefaction, fast pyrolysis and pelletisation'. *Energy*, 33(8): 1206–1223.

US-Canada Power System Outage Task Force (2004) *Final Report on the August 14, 2003 Blackout in the United States and Canada: Causes and Recommendations*. Technical report, US Department of Energy, Natural Resources Canada. https://reports.energy.gov/

US EPA (2011) 'Catalogue of CHP technologies'. Online. Accessed: 27 July 2012. http://www.epa.gov/chp/basic/catalog.html

US FHWA (2005) 'Traffic congestion and reliability: Trends and advanced strategies for congestion mitigation'. Online. Accessed: 27 July 2012. http://www.ops.fhwa.dot.gov/congestion_report/chapter4.htm

Vallios, I., Tsoutsos, T. and Papadakis, G. (2009) 'Design of biomass district heating systems'. *Biomass and Bioenergy*, 33(4): 659–678.

Van Der Stelt, M.J.C., Gerhauser, H., Kiel, J.H.A. and Ptasinski, K.J. (2011) 'Biomass upgrading by torrefaction for the production of biofuels: a review'. *Biomass and Bioenergy*, 35(9): 3748–3762.

van Dyken, S., Bakken, B.H. and Skjelbred, H.I. (2010) 'Linear mixed-integer models for biomass supply chains with transport, storage and processing'. *Energy*, 35(3): 1338–1350.

Veblen, T.B. (1899) *The Theory of the Leisure Class: An economic study in the evolution of institutions*. New York: MacMillan Publishing Co.

Verbong, G. and Geels, F. (2007) 'The ongoing energy transition: Lessons from a socio-technical, multi-level analysis of the Dutch electricity system (1960–2004)'. *Energy Policy*, 35(2): 1025–1037.

Verbruggen, A. (1982) 'A system model of combined heat and power generation in district heating'. *Resources and Energy*, 4(3): 231–263.

Vlahogianni, E.I., Golias, J.C. and Karlaftis, M.G. (2004) 'Short-term traffic forecasting: Overview of objectives and methods'. *Transport Reviews*, 24(5): 533–557.

Vleuten, E.v.d. and Raven, R. (2006) 'Lock-in and change: Distributed generation in Denmark in a long-term perspective'. *Energy Policy*, 34(18): 3739–3748.

Von Meier, A. (2006) *Electric Power Systems: A conceptual introduction*. Chichester: John Wiley & Sons, Ltd.

von Neumann, J. and Morgenstern, O. (1944) *Theory of Games and Economic Behaviour*. Princeton, NJ: Princeton University Press.

von Neumann, J. (2002) *The Computer and the Brain*, 2nd edn. New Haven, CT: Yale University Press.

von Thünen, J. (1826) *Der isolierte Staat*. Jena: G. Fischer.

Wackernagel, M., Lewan, L. and Borgstromm Hansson, C. (1999) 'Evaluating the use of natural capital with the ecological footprint, applications in Sweden and subregions'. *Ambio*, 28(7): 604.

Wackernagel, M. (1998) 'The ecological footprint of Santiago de Chile'. *Local Environment: The International Journal of Justice and Sustainability*, 3(1): 7–25.

Waddell, P. (2002) 'UrbanSim: Modeling urban development for land use, transportation, and environmental planning'. *Journal of the American Planning Association*, 68(3): 297–314.

——— (2005) 'Building an integrated model: Some guidance'. In *TRB Workshop 162 on Integrated Land Use-Transport Models*. Washington, DC.

Wagner, P. and Wegener, M. (2007) 'Urban land use, transport and environment models: experiences with an integrated microscopic approach'. *disP – The Planning Review*, 170(3/2000): 46–56.

Wang, R. and Oliveira, R. (2006) 'Adsorption refrigeration – an efficient way to make good use of waste heat and solar energy'. *Progress in Energy and Combustion Science*, 32(4): 424–458.

Warren-Rhodes, K. and Koenig, A. (2001) 'Escalating trends in the urban metabolism of Hong Kong: 1971–1997'. *Ambio*, 30(7): 429–438.

Warr, B., Ayres, R., Eisenmenger, N., Krausmann, F. and Schandl, H. (2010) 'Energy use and economic development: A comparative analysis of useful work supply in Austria, Japan, the United Kingdom and the US during 100 years of economic growth'. *Ecological Economics*, 69(10): 1904–1917.

WCED (1987) *Our common future*. Oxford: World Commission on Environment and Development, Oxford University Press.

Weber, C., Keirstead, J., Samsatli, N., Shah, N. and Fisk, D. (2010) 'Trade-offs between layout of cities and design of district energy systems'. In *23rd International Conference on Efficiency, Cost, Optimization, Simulation and Environmental Impact of Energy Systems*. Lausanne, CH.

Weber, C. and Shah, N. (2011) 'Optimisation based design of a district energy system for an eco-town in the United Kingdom'. *Energy*, 36: 1292–1308.

Weber, M., Roth, G. and Wittich, C. (1968) *Economy and Society: an outline of interpretive sociology*. New York: Bedminster Press.

Wegener, M. (1994) 'Operational urban models: State of the art'. *Journal of the American Planning Association*, 60(1): 17–29.

——— (2004) 'Overview of land use transport models'. In *Handbook of Transport Geography and Spatial Systems*, edited by D.A. Hensher and K. Button, chapter 9, pp. 127–146. Oxford: Pergamon, Elsevier Science Ltd.

Weisman, A. (2007) *The World Without Us*. London: Virgin Books.

Weiss, M., Junginger, M., Patel, M.K. and Blok, K. (2010) 'A review of experience curve analyses for energy demand technologies'. *Technological Forecasting and Social Change*, 77(3): 411–428.

Wemhoner, C. and Afjei, T. (2003) 'Country report Switzerland on task 1: System analysis and state of the art in standardisation'. In *IEA Heat Pump Programme*. Paris: Interational Energy Agency.

West, G.B., Brown, J.H. and Enquist, B.J. (1997) 'A general model for the origin of allometric scaling laws in biology'. *Science*, 276(5309): 122–126.

White, D.J. (1990) 'A bibliography on the applications of mathematical programming multiple-objective methods'. *Journal of the Operational Research Society*, 41(8): 669–691.

Widén, J. and Wäckelgård, E. (2010) 'A high-resolution stochastic model of domestic activity patterns and electricity demand'. *Applied Energy*, 87(6): 1880–1892.

Williams, H. (1999) *Model Building in Mathematical Programming*, 4th edn. Chichester: John Wiley & Sons, Ltd.

Williams, H.P. (2009) *Logic and Integer Programming*. London: Springer.

Williams, M. (2002) *Deforesting the Earth: From Prehistory to Global Crisis*. Chicago: University of Chicago Press.

Williams, T. (1981) *A History of the British Gas Industry*. Oxford: Oxford University Press.

Wilson, M. (2007) "Hot Tub' Nuclear Reactor Could Power Cities'. Online. Accessed: 27 July 2012. http://gizmodo.com/326125/hot-tub-nuclear-reactor-could-power-cities

Wilton, P.C. and Pessemier, E.A. (1981) 'Forecasting the ultimate acceptance of an innovation: the effects of information'. *Journal of Consumer Research*, 8(2): 162–171.

Wiltsee, G. (1999) *Urban Wood Waste Resource Assessment*. Technical Report NREL/SR-570-25918, National Renewable Energy Laboratory. http://www.osti.gov/energycitations/product.biblio.jsp?osti_id=6088357

Wit, M.D. and Faaij, A. (2010) 'European biomass resource potential and costs'. *Biomass and Bioenergy*, 34(2): 188–202.

Wolfe, P. (1963) 'Methods of nonlinear programming'. In *Recent Advances in Mathematical Programming*, edited by R. Graves and P. Wolfe. London: McGraw-Hill.

——— (1968) 'Introduction: Philip Wolfe's review of nonlinear programming'. In *Optimization: Symposium of the Institute of Mathematics and Its Applications*, edited by R. Fletcher. University of Keele: Academic Press.

Wolfram, S. (1984) 'Universality and complexity in cellular automata'. *Physica D: Nonlinear Phenomena*, 10(1–2): 1–35.

Wolman, A. (1965) 'The metabolism of cities'. *Scientific American*, 213(3): 178–190.

Wolmar, C. (2004) *The Subterranean Railway: How the London Underground Was Built and How it Changed the City Forever*. London: Atlantic Books.

Wooldridge, J. (2004) *Econometric Analysis of Cross Section and Panel Data*. Cambridge, MA: MIT Press.

World Bank (2005) 'Poverty statistics for Kenya'. Online. Accessed: 26 July 2012. http://povertydata.worldbank.org/

——— (2008) *World Development Report 2009: Reshaping Economic Geography, Part 1*. Washington, DC: World Bank. http://www.worldbank.org/wdr2009

———— (2011) *One Goal, Two Paths*. Washington, DC: World Bank. http://elibrary.worldbank.org/content/book/9780821388372

Wright, M.M., Brown, R.C. and Boateng, A.A. (2008) 'Distributed processing of biomass to bio-oil for subsequent production of Fischer-Tropsch liquids'. *Biofuels, Bioproducts and Biorefining*, 2(3): 229–238.

Wrigley, E. (2010) *Energy and the English Industrial Revolution*. Cambridge: Cambridge University Press.

WRI/WBCSD (2011) 'The Greenhouse Gas Protocol Initiative'. Online. Accessed: 26 July 2012. http://www.ghgprotocol.org/

WSA (2008) *Feasibility study of real time parking information at Metrorail parking facilities (Virginia Stations)*. Technical report, Wilbur Smith Associates. http://www.wmata.com/pdfs/planning/Real_Time_Parking_Study.pdf

Wulfinghoff, D. (1999) *Energy Efficiency Manual*. Wheaton, MD: Energy Institute Press.

Wüstenhagen, R., Markard, J. and Truffer, B. (2003) 'Diffusion of green power products in Switzerland'. *Energy Policy*, 31(7): 621–632.

Wu, D. and Wang, R. (2006) 'Combined cooling, heating and power: a review'. *Progress in Energy and Combustion Science*, 32(5–6): 459–495.

WWF (2006) *Living Planet Report*. Gland, CH: WWF International. http://assets.panda.org/downloads/living_planet_report.pdf

Yamamoto, H., Yamaji, K. and Fujino, J. (2000) 'Scenario analysis of bioenergy resources and CO_2 emissions with a global land use and energy model'. *Applied Energy*, 66(4): 325–337.

Zavadskas, E.K., Kaklauskas, A. and Raslanas, S. (2004) 'Evaluation of investments into housing renovation'. *International Journal of Strategic Property Management*, 8(3): 177–190.

Zhai, J., LeClaire, N. and Bendewald, M. (2011) 'Deep energy retrofit of commercial buildings: a key pathway toward low-carbon cities'. *Carbon Management*, 2(4): 425–430.

Zhang, H.F., Ge, X.S. and Ye, H. (2007) 'Modeling of a space heating and cooling system with seasonal energy storage'. *Energy*, 32(1): 51–58.

Zhang, Q., Li, J., Chen, Y.D. and Chen, X. (2011) 'Observed changes of temperature extremes during 1960–2005 in China: natural or human-induced variations?' *Theoretical and Applied Climatology*, 106(3–4): 417–431.

Zinko, H., Bøhm, B., Kristjansson, H., Ottosson, U., Rämä, M. and Siplilä, K. (2008) 'District heating distribution in areas with low heat demand density'. In *IEA RD Programme on District Heating and Cooling including the integration of CHP*, p. 134. Paris: International Energy Agency.

Zucchetto, J. (1975) 'Energy economic theory and mathematical models for combining the systems of man and nature. Case study, the urban region of Miami'. *Ecological Modelling*, 1: 241–268.

Index